*Science and
Technology in
World History*

Science and Technology in World History

AN INTRODUCTION

*James E. McClellan III
and Harold Dorn*

THE JOHNS HOPKINS UNIVERSITY PRESS
Baltimore and London

© 1999 The Johns Hopkins University Press
All rights reserved. Published 1999
Printed in the United States of America on acid-free paper
9 8 7 6 5

The Johns Hopkins University Press
2715 North Charles Street
Baltimore, Maryland 21218-4363
www.press.jhu.edu

Library of Congress Cataloging-in-Publication Data will be found
at the end of this book.

A catalog record for this book is available from the British Library.

ISBN 0-8018-5868-2
ISBN 0-8018-5869-0 (pbk.)

Contents

Preface and Acknowledgments

This book was written as an introduction for lay readers and under-graduate students to provide the "big picture" that an educated person might wish to have of the history of science and technology. It was not written for scholars or experts, and its character as a textbook is self-evident. The presentation grows out of our extensive experience of en-gaging undergraduates in these matters. The hard knocks of the class-room have suggested both the essential lessons and what materials and examples work well in conveying those lessons.

We owe thanks to valued friends on whom we imposed to read por-tions or all of the manuscript and whose critical comments substan-tially improved the final effort; in this connection we cannot omit mention of Dr. Philip R. Reilly, Prof. James E. McClellan Jr., Jackie McClellan, and Messrs. Michael Feldstein, Paul Kay, and Jeff Ruth. Similarly, we are indebted to a number of professional colleagues whose specialized expertise saved us from more than one gaffe, and here we must name Professors Mott T. Green and James Evans of the Univer-sity of Puget Sound, Dr. Murray C. McClellan of Boston University, Dr. Deepak Kumar of NISTADS in New Delhi, and two anonymous reviewers.

Dr. Robert J. Brugger of the Johns Hopkins University Press started us down this rocky road, and his substantive and stylistic suggestions along the way resulted in many improvements. Others at the Press dis-played outstanding professionalism in addition to a welcomed toler-ance toward a cantankerous authorial couple, and we have especially to thank Kimberly F. Johnson, Lee Campbell Sioles, Anita Walker Scott, and Therese D. Boyd for their several contributions.

The people and resources at the New York Public Library, the Amer-ican Museum of Natural History, and the libraries of Columbia Uni-versity were instrumental in helping us locate essential scholarly and iconographic materials, for which we express our sincere appreciation. Ms. Carol Perkins and the staff of Williams Library at Stevens Institute of Technology likewise proved invaluable as well as unusually efficient in negotiating interlibrary loans for us, and we have the pleasure of acknowledging their many kindnesses again in print. Our friends at Modernage Photographic Services on Sixth Avenue in Manhattan did their usual high-quality thing and made this a more appealing book than it otherwise would have been. Mr. Shyam Laxminarayan helped us in checking Web resources. Mr. William L. Nelson drew the maps; Dr. Andrew P. Rubenfeld prepared the index.

And finally, we thank our students at Stevens Institute of Technology who tested drafts of this book: they let us know when we succeeded in communicating the story clearly and when we did not. What follows is not error free, and we alone are responsible for its shortcomings.

*Science and
Technology in
World History*

The Guiding Themes

The twentieth century has witnessed a fateful change in the relationship between science and society. In World War I scientists were conscripted and died in the trenches; in World War II they were exempted as national treasures and committed to secrecy, and they rallied behind the war effort. The explanation of the change is not hard to find—governments came to believe that theoretical research can produce practical improvements in industry, agriculture, and medicine. That belief was firmly reinforced by developments such as the discovery of antibiotics and the application of nuclear physics to the production of atomic weapons. Science has become so identified with practical benefits that the dependence of technology on science is commonly assumed to be a timeless relationship and a single enterprise. Science and technology, research and development—these are assumed to be almost inseparable twins. They rank among the sacred phrases of our time. The belief in the coupling of science and technology is now petrified in the dictionary definition of technology as applied science, and journalistic reports under the rubric of "science news" are, in fact, often accounts of engineering rather than scientific achievements.

That belief, however, is an artifact of twentieth-century cultural attitudes superimposed without warrant on the historical record. Although the historical record shows that in the earliest civilizations under the patronage of pharaohs and kings, and in general whenever centralized states arose, knowledge of nature was exploited for useful purposes; even then it cannot be said that science and technology were systemically and closely related. By the same token, in ancient Greece (where theoretical science had its beginning), among the scholastics of the Middle Ages, in the time of Galileo and Newton, and even for Darwin and his contemporaries in the nineteenth century, science constituted a learned calling whose results were recorded in scientific publications, while technology was understood as the crafts practiced by unschooled artisans. Until the second half of the nineteenth century few artisans or engineers attended a university or, in many cases, received any formal schooling at all. Conversely, the science curriculum of the university centered largely on pure mathematics and what was often termed natural philosophy—the philosophy of nature—and was written in technical terms (and often language) foreign to artisans and engineers.

In some measure, the wish engenders the thought. Science has undoubtedly bestowed genuine benefits on humankind in this century, and it has fostered the hope that research can be channeled in the direction of social utility. But a more secure understanding of science, one less bound by the cultural biases of our time, can be gained by viewing it through the lens of history. Seen thus, with its splendid achievements but also with its blemishes and sometimes in an elitist posture inconsistent with our democratic preferences, science becomes a multidimensional reality rather than a culture-bound misconception. At the same time, a more accurate historical appreciation of technology will place proper emphasis on independent traditions of skilled artisans whose talents crafted everyday necessities and amenities throughout the millennia of human existence. Such a historical reappraisal will also show that in many instances technology directed the development of science, rather than the other way around.

In order to develop the argument that the relationship between science and technology has been a historical process and not an inherent identity, we trace the joint and separate histories of science and technology from the prehistoric era to the present. In this way we intend to review the common assumption that technology is applied science and show, instead, that in most historical situations prior to the twentieth century science and technology have progressed in either partial or full isolation from each other—both intellectually and sociologically. In the end, an understanding of the historical process will shed light on the circumstances under which science and technology have indeed merged over the past hundred years.

PART I

From Ape to Alexander

Technology in the form of stone tools originated literally hand in hand with humankind. Two million years ago a species of primate evolved which anthropologists have labeled Homo habilis, *or "handy man," in recognition of its ability, far beyond that of any other primate, to fashion tools. Over the next 2,000 millennia our ancestors continued to forage for food, using a tool kit that slowly became more elaborate and complex. Only toward the end of that long prehistoric era did they begin to observe the natural world systematically in ways that appear akin to science. Even when a few communities gave up the foraging way of life, around 12,000 years ago, in favor of farming or herding and developed radically new tools and techniques for earning a living, they established societies that show no evidence of patronizing scientists or fostering scientific research. Only when civilized—city-based— empires emerged in the ancient Near East did monarchs come to value higher learning for its applications in the management of complex societies and found institutions for those ends. The ancient Greeks then added natural philosophy, and abstract theoretical science took its place as a component of knowledge. An account of these developments forms the subject matter of part 1.*

CHAPTER 1

Humankind Emerges: Tools and Toolmakers

Scholars customarily draw a sharp distinction between *prehistory* and *history*. Prehistory is taken to be the long era from the biological beginnings of humankind over 2 million years ago to the origins of civilization about 5,000 years ago in the first urban centers of the Near East. The transition to civilization and the advent of written records traditionally mark the commencement of history proper.

Prehistory, because of the exclusively material nature of its artifacts, mainly in the form of stone, bone, or ceramic products, has inescapably become the province of the archaeologist, while the historical era, with its documentary records, is the domain of the historian. However, the single label "prehistory" obscures two distinctly different substages: the *Paleolithic,* or Old Stone Age, which held sway for around 2 million years, is marked by rudimentary stone tools designed for collecting and processing wild food sources, while the succeeding *Neolithic,* or New Stone Age, which first took hold in the Near East around 12,000 years ago, entailed substantially more complex stone implements adapted to the requirements of an economy of low-intensity food production in the form of gardening or herding.

The technologies of both the Paleolithic and Neolithic eras have left a rich legacy of material artifacts. In contrast, only a feeble record exists of any scientific interests in these preliterate societies, mainly in the form of astronomically oriented structures. Thus, at the very outset, the evidence indicates that science and technology followed separate trajectories during 2,000 millennia of prehistory. Technology—the crafts—formed an essential element of both the nomadic food-collecting economy of Paleolithic societies and the food-producing activities in Neolithic villages, while science, as an abstract interest in nature, was essentially nonexistent, or, at any rate, has left little trace.

The Arrival of Handyman

By most accounts human beings appeared on Earth only recently, as measured on the scales of cosmic, geologic, or evolutionary time. As scientists now believe, the cosmos itself originated with the "Big Bang" some 12 to 15 billion years ago. Around 4 billion years ago the earth took shape as the third in a string of companion planets to an ordinary star near the edge of an ordinary galaxy; soon the self-replicating chemistry of life began. Biological evolution then unfolded over the next millions and billions of years. In the popular imagination the age of the dinosaurs exemplifies the fantastic history of life in past ages, and the catastrophic event—probably a comet or an asteroid colliding with the earth—that ended the dinosaur age 65 million years ago illustrates the vicissitudes life suffered in its tortuous evolution. The period that followed is known as the age of mammals because these animals flourished and diversified in the niche vacated by the dinosaurian reptiles. By about 4 million years ago a line of "ape-men" arose in Africa—the australopithecines—our now-extinct ancestral stock.

Figure 1.1 depicts the several sorts of human and prehuman species that have arisen over the last 4 million years. Experts debate the precise evolutionary paths that join them, and each new fossil discovery readjusts the details of the story; yet its broad outlines are not in dispute.

The figure shows that anatomically modern humans, *Homo sapiens sapiens,* or the "wise" variety of "wise Man," evolved from a series of human and prehuman ancestors. Archaic versions of modern humans made their appearance after about 500,000 years ago, with the Neanderthals being an extinct race of humans that existed mainly in the cold of Europe between 135,000 and 35,000 years ago. Scholars differ over the modernity of Neanderthals and whether one would or would not stand out in a crowd or in a supermarket. Many scientists look upon them as so similar to ourselves as to form only an extinct variety or race of our own species, and so label them *Homo sapiens neanderthalensis.* Others think Neanderthals more "brutish" than anatomically modern humans and therefore regard them as a separate species, *Homo neanderthalensis.*

Preceding *Homo sapiens,* the highly successful species known as *Homo erectus* arose around 2 million years ago and spread throughout the Old World (the continents of Africa, Europe, and Asia). Before that, the first species of human being, *Homo habilis,* coexisted with at least two other species of upright hominids, the robust and the gracile forms of the species *Paranthropus.* At the beginning of the sequence stood the ancestral genus *Australopithecus* (or "Southern Ape") that includes *Australopithecus afarensis*—represented by the fossil "Lucy."

This sequence highlights several points of note. First is the fact of human evolution, that we arose from more primitive forebears. Among the more significant indicators of this evolution is a progression in brain

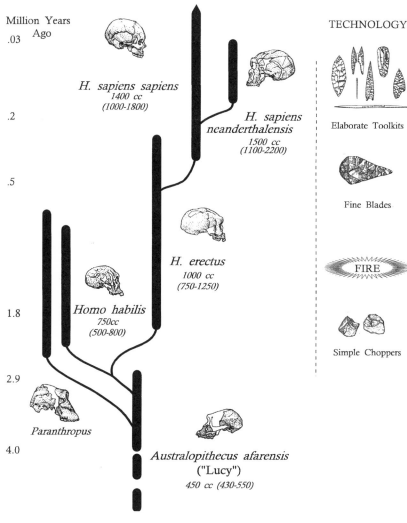

HUMAN EVOLUTION

Million Years Ago
.03

.2

.5

1.8

2.9

4.0

H. sapiens sapiens
1400 cc
(1000-1800)

H. sapiens neanderthalensis
1500 cc
(1100-2200)

H. erectus
1000 cc
(750-1250)

Homo habilis
750cc
(500-800)

Paranthropus

Australopithecus afarensis
("Lucy")
450 cc (430-550)

TECHNOLOGY

Elaborate Toolkits

Fine Blades

FIRE

Simple Choppers

Fig. 1.1. Human evolution. Modern humans (*Homo sapiens sapiens*) evolved from earlier, now extinct, human and prehuman ancestors. (Plants and animals are classified according to the binomial nomenclature of genus and species: genus being general groups of related species, and species being specific interbreeding populations of individuals. Thus, *Homo* is the genus, and *sapiens* the species; the third name indicates a subspecies.) In general, brain size and technological sophistication increased over time, but there is no strict correlation between species and technologies. For example, *Paranthropus* and *Homo habilis* may both have used simple choppers; *H. erectus* and archaic *H. sapiens* cannot be distinguished by their respective fine-blade tool kits. Aspects of this picture are matters of debate, notably the relationship of Neanderthals to modern humans. New findings regularly shed new light on the details of human biological and cultural evolution.

size, from around 450 cubic centimeters (cc) in the case of prehuman Lucy, only slightly larger than the brain of a modern chimpanzee, through an average of 750 cc for *Homo habilis*, 1000 cc for *Homo erectus*, to around 1400 cc for humanity today. An as-yet-unexplained irony of this "progression" is that Neanderthals had slightly larger brains than today's humans.

Bipedality—or walking upright on two feet—represents another defining feature of this evolutionary sequence. Experts debate whether Lucy and her kin were fully bipedal, but her successors certainly were. An upright stance allows the hand and arm to become a multipurpose utensil for grasping and carrying items. Lucy and her type had probably adopted male-female cooperation, at least temporary pair-bonding, and a "family" structure for raising offspring.

From the point of view of the history of technology, however, the most important lesson to be drawn from figure 1.1 concerns tool use among our ancestors. It used to be thought that tool use—technology—is an exclusively human characteristic; the oldest fossil of the human genus, *Homo habilis,* received its name ("handy man") both because of its "human" skeletal features and because it was discovered along with simple stone choppers. However, the older notion can no longer be maintained. Indeed, the origin of technology is rooted in biology. Some nonhuman animals create and use tools, and technology as a cultural process transmitted from generation to generation arises occasionally among monkey and ape communities. Chimpanzees in the wild sometimes "fish" for termites by carefully preparing a twig, inserting it into a termite nest, and licking off the insects that cling to it. Since the activity is not instinctive but is instead taught to juveniles by their mothers, it must be regarded as cultural, unlike, say, the instinct of bees to build hives. Reportedly, chimpanzees have also culturally transmitted knowledge of medicinal plants, so it may be possible to identify the origins of medical technology outside of the human genus, too. Perhaps the best documented feats of technical innovation and cultural transmission in the animal world concern a single female, Imo, the "monkey genius" of a colony of Japanese macaques. Incredibly, Imo made two separate technical discoveries. First she discovered that to remove sand from potatoes thrown on the beach she could wash them in the sea rather than pick off the sand with her fingers. Then, in an even more remarkable display of ingenuity, Imo found that to separate *rice* from sand she did not have to pick out the individual grains; the mixture can be dropped into water where the sand will sink, and the rice will float and can be easily recovered. Both techniques were adopted by younger members of the troop as well as by older females and passed on to the next generation.

Claims have been made that not only *Homo habilis* but also species of *Paranthropus* probably made stone implements and may have used fire. Furthermore, little correlation exists between species type and different types of tool kits. For example, Neanderthal tools varied little from the precedents set by *Homo erectus.* The record reveals only a weak correlation between biological species and the tool kit used.

That said, however, making and using tools and the cultural transmission of technology became essential to the human mode of existence and was practiced in *all* human societies. Moreover, humans seem to be the only creatures who fashion tools to make other tools. Without tools humans are a fairly frail species, and no human society has ever survived without technology. Humankind owes its evolutionary success in large measure to mastery and transmission of tool-making and -using, and thus human evolutionary history is grounded in the history of technology.

Control of fire represented a key new technology for humankind. Fire provided warmth. Fire made human migration into colder climes possible, opening up huge and otherwise inhospitable areas of the globe for human habitation. The technology of fire also supplied artificial light, thus extending human activity after dark and into dark places, such as caves. Fire offered protection against wild animals. Fire permitted foods to be cooked, which lessened the time and effort required to eat and digest meals. Fire-hardened wooden tools became possible. And fire no doubt served as a hearth and a hub for human social and cultural relations for a million years. Their practical knowledge of fire gave early humans a greater degree of control over nature. *Homo erectus* was an exceptionally successful animal, at least as measured by its spread across the Old World from Africa to Europe, Asia, Southeast Asia, and archipelagoes beyond. That success in large measure depended on mastering fire.

The grasping hand constitutes one human "tool" that evolved through natural selection; speech is another. Speech seems to be a relatively recent acquisition, although paleontologists have not yet reached agreement on how or when it first appeared. Speech may have evolved from animal songs or calls; novel brain wiring may have been involved. But, once acquired, the ability to convey information and communicate in words and sentences must have been an empowering technology that produced dramatic social and cultural consequences for humanity.

A turning point occurred around 40,000 years ago. Previously, Neanderthals and anatomically modern humans had coexisted for tens of thousands of years in the Middle East and in Europe. Around 35,000 years ago Neanderthals became extinct, possibly exterminated through conflict with a new population, or they may have interbred and become absorbed into the modern human gene pool. A cultural discontinuity manifested itself around the same time. Whereas Neanderthals had produced simple, generalized, multipurpose tools from local materials, we—*Homo sapiens sapiens*—began to produce a great assortment of tools, many of which were specialized, from stone, bone, and antler: needles and sewn clothing, rope and nets, lamps, musical instruments, barbed weapons, bows and arrows, fish hooks, spear throwers, and more elaborate houses and shelters with fireplaces. Humans began to conduct long-distance trade of shells and flints through exchange over hundreds of miles, and they produced art, tracked the moon, and buried their dead. And yet, in terms of their basic social and economic way of life, they continued along the same path—they remained nomadic food-collectors.

Foraging for a Living

Prehistorians classify the period from 2 million years ago to the end of the last Ice Age at about 12,000 years ago as a single era. They label it the Paleolithic (from the Greek, *paleo,* "ancient"; *lithos,* "stone") or Old Stone Age. Food-collecting is its essential attribute, codified in the term *hunter-gatherer* society. Paleolithic tools aided in hunting or scavenging animals and for collecting and processing plant and animal

Fig. 1.2. "*H. erectus* Utilizing a Prairie Fire," by Jay H. Matternes. Control of fire became a fundamental technology in the human odyssey. Undoubtedly, members of the genus *Homo* first used wildfires before learning to control them.

food, and it is now understood that Paleolithic technology developed in the service of a basic food-collecting economy.

Paleolithic food-collecting bespeaks a subsistence economy and a communal society. Seasonal and migratory food-collecting produced little surplus and thus permitted little social ranking or dominance and no coercive institutions (or, indeed, any institutions) of the kind needed in stratified societies to store, tax, and redistribute surplus food. The record indicates that Paleolithic societies were essentially egalitarian, although grades of power and status may have existed within groups. People lived in small bands or groups of families, generally numbering fewer than 100. Much circumstantial evidence suggests that a division of labor based on gender governed the pattern of food collection. Although one has to allow for sexually ambiguous roles and individual exceptions, males generally attended to hunting and scavenging animals, while females most likely went about gleaning plants, seeds, and eggs as food and medicines. Men and women together contributed to the survival of the group, with women's work often providing the majority of calories. *Homo sapiens sapiens* lived longer than Neanderthals, it would seem; more true elders thus added experience and knowledge in those groups. Paleolithic bands may have converged seasonally into larger clans or macrobands for celebrations, acquiring mates, or other collective activities, and they probably ingested hallucinatory plants. Except as located in a handful of favored spots where year-round hunting or fishing might have been possible, Paleolithic food-collectors were nomadic, following the migrations of animals and the seasonal growth of plants. In some instances Paleolithic groups engaged in great seasonal moves to the sea or mountains. In the Upper Paleolithic (around 30,000 years ago) spear-throwers and the bow and arrow entered the weapons arsenal, and the dog (wolf) became domesticated, possibly as an aid in hunting.

Ice Age art is the most heralded example of the cultural flowering produced after anatomically modern humans appeared on the scene. Earlier human groups may have made beautified objects of perishable materials, but several late Upper Paleolithic cultures in Europe (30,000 to 10,000 years ago) produced enduring and justly renowned paintings and sculptures in hundreds of sites, often in hard-to-reach galleries and recesses of caves. Artists and artisans also created jewelry and portable adornments, and decorated small objects with animal motifs and other embellishments. No one has yet fully decoded what purposes cave paintings fulfilled; anthropologists have suggested hunting rituals, initiations, magical beliefs, and sexual symbolism. The many "Venus" statuettes with exaggerated feminine features, characteristic of the Paleolithic, have been interpreted in terms of fertility rituals and divination of one sort or another. By the same token, they may represent ideals of feminine beauty. But we should not overlook the technical dimension of Ice Age art, from pigments and painting techniques to ladders and scaffolding. The great cave paintings of Europe are the better known, but literally and figuratively Paleolithic peoples the world over left their artistic handprints.

Neanderthals had already begun to care for their old and invalid, and by 100,000 years ago they ceremonially buried some of their dead. Centers of mortuary and burial activity may have existed, and one can speak of a "cult of the dead" beginning in the Middle Paleolithic (100,000–50,000 years ago). Intentionally burying the dead is a distinctly human activity, and burials represent a major cultural landmark in human prehistory. They bespeak self-consciousness and effective social and group cohesion, and they suggest the beginning of symbolic thought.

It may be enlightening to speculate about the mental or spiritual world of Paleolithic peoples. What we have already seen and said of Paleolithic burials and cave art strongly suggests that Paleolithic populations, at least toward the end of the era, developed what we would call religious or spiritual attitudes. They may well have believed the natural world was filled with various gods or deities or that objects and places, such as stones or groves, were themselves alive. Religious beliefs and practices—however we might conceive them—formed a social technology, as it were, that knitted communities together and strengthened their effectiveness.

Fig. 1.3. Paleolithic art. In the late Paleolithic era food-collecting populations of *Homo sapiens* began to create art in many parts of the world. In southwestern Europe they adorned the walls of caves with naturalistic representations of animals.

For anatomically modern humans the Paleolithic way of life continued unabated and essentially unchanged for 30,000 years, a phenomenally long and stable cultural era, especially compared to the rapid pace of change in the periods that followed. Paleolithic peoples doubtless lived relatively unchanging lives involving great continuity with their own past. Well fed on a varied diet that included significant amounts of meat, not having to work too hard, cozy in fur and hide, comfortable by a warm fire, who can deny that our Paleolithic ancestors often enjoyed the good life?

Over the entire 2 million years of the Paleolithic, beginning with the first species of *Homo*, population density remained astonishingly low, perhaps no more than one person per square mile, and the rate of population increase, even in the late (or Upper) Paleolithic, may have been only one-five-hundredth of what it has been for modern populations over the past few centuries. The very low rate of population increase derives from several factors acting singly or in combination to restrict fertility rates: late weaning of infants (since nursing has somewhat of a contraceptive effect), low body fat, a mobile lifestyle, and infanticide. Nevertheless, humankind slowly but surely fanned out over the earth and, as long as suitable food-collecting habitats could be found, humanity had no need to alter its basic lifestyle. Food-collecting groups simply budded off from parent populations and founded new communities. Paleolithic peoples spread through Africa, Asia, Europe, and Australia, while waves of hunters and gatherers reached North America by at least 12,000 years ago, if not well before, ultimately spreading the Paleolithic mode of existence to the southernmost tip of South America. After many millennia of slow expansion, Paleolithic humans "filled up" the world with food-collectors. Only then, it seems, did population pressure against collectible resources trigger a revolutionary change from food-collecting to food-producing in the form of horticulture or herding.

Is Knowledge Science?

The extraordinary endurance of Paleolithic society and mode of existence depended on human mastery of an interlocked set of technologies and practices. It is sometimes said that Paleolithic peoples needed and possessed "science" as a source of the knowledge that underpinned their practical activities. It is all too easy to assume that in making and using fire, for example, Stone Age peoples practiced at least a rude form of "chemistry." In fact, however, while both science and technology involve "knowledge systems," the knowledge possessed by food-collectors cannot reasonably be considered theoretical or derivative of science or theories of nature. Although evidence of something akin to science appears in late Paleolithic "astronomy," it evidently played no role in the practice of Paleolithic crafts. To discover the origins and character of that science we need to understand why it did not impact technology.

Practical knowledge embodied in the crafts is different from knowledge deriving from some abstract understanding of a phenomenon. To change a car tire, one needs direct instruction or hands-on experience, not any special knowledge of mechanics or the strength of materials. By rubbing sticks together or sparking flint into dry kindling, a scout can build a fire without knowing the oxygen theory (or any other theory) of combustion. And conversely, knowledge of theory alone does not enable one to make a fire. It seems fair to say that Paleolithic peoples applied practical skills rather than any theoretical or scientific knowledge to practice their crafts. More than that, Paleolithic peoples may have had explanations for fire without it being meaningful to speak about Paleolithic "chemistry"—for example, if they somehow thought they were invoking a fire god or a spirit of fire in their actions. A major conclusion about Paleolithic technology follows from all this: to whatever small extent we may be able to speak about "science" in the Paleolithic, Paleolithic technologies clearly were prior to and independent of any such knowledge.

The record (or rather the absence of one) indicates that Paleolithic peoples did not self-consciously pursue "science" or deliberate inquiries into nature. Does the Paleolithic period nevertheless offer anything of note for the history of science? On the most rudimentary level one can recognize the extensive "knowledge of nature" possessed by Paleolithic peoples and gained directly from experience. They had to be keen observers since their very existence depended on what they knew of the plant and animal worlds around them. And, like surviving food-collectors observed by anthropologists, they may have developed taxonomies and natural histories to categorize and comprehend their observations.

Even more noteworthy, the archaeological record for the late Paleolithic era, beginning around 40,000 years ago, offers striking evidence of activities that look a lot like science. That evidence appears in the form of thousands of engraved fragments of reindeer and mammoth bones that seem to have recorded observations of the moon. An "unbroken line" of such artifacts stretches over tens of thousands of years. The engraved mammoth tusk from Gontzi in Ukraine is an example of such lunar records, which may have been kept at all major habitation sites. Pictured in figure 1.4, it dates from around 15,000 years ago.

We can only speculate, of course, but, as Paleolithic peoples lived close to nature, the waxing and waning moon would naturally present itself as a significant object of interest with its obvious rhythms and periods. One can easily imagine our intelligent forebears following those rhythms and beginning to record in one fashion or another the sequence and intervals of full and new moon. Moreover, the Gontzi bone and others like it could have served as a means of reckoning time. Although we cannot go so far as to say that Paleolithic peoples possessed a calendar, we can surmise that knowledge of the moon's periods would be useful in time-reckoning. For example, dispersed groups might have come together seasonally and would have needed to keep track of the intervening months. We need not envision a continuous tradition of such lunar records, for the process may have been invented and reinvented hundreds of times over: a simple counter fashioned over the course of a few months and discarded. The artifacts in question evidence the active observation and recording of natural phenomena over time. That activity indicates only a rudimentary approach to theoretical knowledge, but its results seem more abstract than knowledge gained from direct experience and different from what Paleolithic peoples otherwise embodied in their crafts.

Leaving the Garden

This picture of humankind's childhood, which has emerged from the research of archaeologists, paleoanthropologists, and prehistorians, raises several puzzling questions about the dynamics of social change. How can we explain the steadfast durability of a food-collecting social system for 2 million years including more than 200,000 years populated by our own species? How can the relative lack of technological innovation be accounted for? Why, after anatomically modern humans flourished culturally in the Paleolithic 40,000 to 30,000 years ago, did they continue to live as food-collectors, making stone tools and following a nomadic way of life? And why did the pace of change accel-

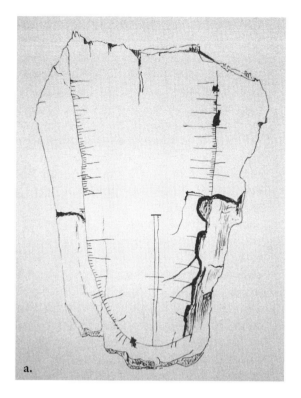

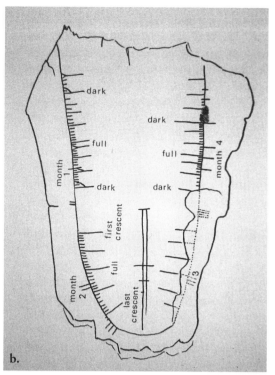

Fig. 1.4. Paleolithic lunar observations. a) An engraved mammoth tusk from Gontzi, Ukraine, that has been interpreted as a record of lunar cycles. Thousands of these artifacts have been found stretching back 30,000 years. This one dates from approximately 15,000 years ago. b) A diagrammatic rendition of the artifact showing cycles of four lunar months aligned with the engraved markings.

erate 15,000 years ago, as food-collecting finally gave way to food-producing, first in the form of gardening (horticulture) and animal husbandry in the Neolithic era and later, after another technological revolution in the form of intensified farming (agriculture) under the control and management of the political state?

Different explanations have been offered to explain the social and economic transformations that occurred at the end of the Paleolithic. It may have been set in motion by climate change and the retreat of the glaciers at the end of the last Ice Age about 10,000–12,000 years ago. The extinction of many large-bodied animals occurred then, restricting the food supply, and other animal-migration patterns shifted northward, probably leaving some human groups behind. Humans themselves probably overhunted large game, self-destructively changing their living conditions. Another line of argument that has recently gained credibility postulates that the food-collecting mode of life persisted as long as the population of hunters and gatherers remained small enough to exploit the resources of their habitats with reasonable ease. Since population increased slowly and since suitable habitats were numerous on a global scale, 2 million years passed before hunter-gatherers reached the "carrying capacities" of accessible environments through the increase of their own numbers and a resulting broadening of foraging activity. This account also explains the low rate of technological innovation prior to the late Paleolithic era: small populations blessed with ample resources were served well by their techniques and refined skills. Although Paleolithic peoples would have known that seeds grow and that gardening is possible (and occasionally practiced it), they had no compelling incentive to revolutionize their way of life. Only when increasing population density that could no longer be readily relieved by migration finally upset the balance between needs and resources were plant and animal husbandry taken up as a new way of life.

Our ancestors did not give up their Paleolithic existence willingly. By abandoning, under pressure of ecological degradation, a nomadic lifestyle of food-collecting, and adopting a mode of food-producing—by "progressing" from hunting and gathering to gardening and stock-raising—only then did humankind reluctantly fall out of the Garden of Eden into the Neolithic era.

CHAPTER 2

The Reign of the Farmer

At the end of the last Ice Age, around 12,000 years ago, the Neolithic Revolution began to unfold. This Revolution, first and foremost a socioeconomic and technological transformation, involved a shift from food-gathering to food-producing. It originated in a few regions before eventually spreading around the globe. In habitats suitable only as pasture it led to pastoral nomadism or herding animal flocks; in others it led to farming and settled village life. Thus arose the Neolithic or New Stone Age.

Growing Your Own

A surprising but grand fact of prehistory: Neolithic communities based on domesticated plants and animals arose independently several times in different parts of the world after 10,000 B.C.—the Near East, India, Africa, North Asia, Southeast Asia, and Central and South America. The physical separation of the world's hemispheres—the Old World and the New World—decisively argues against simple diffusion of Neolithic techniques, as do the separate domestications of wheat, rice, corn, and potatoes in different regions. On the time scale of prehistory the transformation appears to have been relatively abrupt, but in fact the process occurred gradually. Nonetheless, the Neolithic revolution radically altered the lives of the peoples affected and, indirectly, the conditions of their habitats. Although different interpretations exist concerning the origin of the Neolithic, no one disputes its world-transforming effects.

The Neolithic was the outcome of a cascading series of events and processes. In the case of gardening—low-intensity farming—we now know that in various locales around the world human groups settled down in permanent villages, yet continued to practice hunting, gathering, and a Paleolithic economy before the full transition to a Neolithic mode of production. These settled groups lived by complex foraging in limited territories, intensified plant collection, and exploitation of a broad spectrum of secondary or tertiary food sources, such as nuts and seafood. They also lived in houses, and in this sense early sedentary humans were themselves a domesticated species. (The English word "domestic" derives from the Latin word *domus,* meaning "house." Humans thus domesticated themselves as they domesticated plants or animals!) But the inexorable pressure of population against dwindling collectible resources, along with the greater nutritional value of wild and domesticated cereal grains, ultimately led to increasing dependence on farming and a more complete food-producing way of life.

In most places in the world people continued a Paleolithic existence after the appearance of Neolithic settlements 12,000 years ago. They were blissfully unpressured to take up a new Neolithic mode of food-producing, and as a cultural and economic mode of existence even today a few surviving groups follow a Paleolithic lifestyle. As a period in prehistory, the Neolithic has an arc of its own that covers developments from the first simple horticulturists and pastoralists to complex late Neolithic groups living in "towns." In retrospect, especially compared to the extreme length of the Paleolithic period, the Neolithic of prehistory lasted just a moment before civilization in Mesopotamia and Egypt began to usher in further transformations around 5,000 years ago. But even in its diminished time frame the Neolithic spread geographically and persisted in particular locales over thousands of years from roughly 12,000 to 5,000 years ago, when the Neolithic first gave way to civilization in the Near East. To those experiencing it, Neolithic life must have proceeded over generations at a leisurely seasonal pace.

Two alternative paths toward food production led out of the Paleolithic: one from gathering to cereal horticulture (gardening), and then to plow agriculture; the other from hunting to herding and pastoral nomadism. A distinct geography governed these Neolithic alternatives: in climates with sufficient atmospheric or surface water, horticulture and settled villages arose; in grasslands too arid for farming, nomadic people and herds of animals retained a nomadic way of life. Of these very different paths, one led historically to nomadic societies such as the Mongols and the Bedouins. The other, especially in the form that combined farming and domestication of animals, led to the great agrarian civilizations and eventually to industrialization.

Opportunistic and even systematic hunting and gathering persisted alongside food-producing, but where Neolithic settlements arose the basic economy shifted to raising crops on small cleared plots. Gardening contrasts with intensified agriculture using irrigation, plows, and draft animals which later developed in the first civilizations in the Near East. Early Neolithic peoples did not use the plow but, where necessary, cleared land using large stone axes and adzes; they cultivated their plots using hoes or digging sticks. In many areas of the world, especially tropical and subtropical ones, swidden, or "slash and burn," agriculture developed where plots were cultivated for a few years and then abandoned to replenish themselves before being cultivated again. The Neolithic tool kit continued to contain small chipped stones, used in sickles, for example, but was augmented by larger, often polished implements such as axes, grinding stones, and mortars and pestles found at all Neolithic sites. Animal antlers also proved useful as picks and digging sticks. And grain had to be collected, threshed, winnowed, stored, and ground, all of which required an elaborate set of technologies and social practices.

Human populations around the world independently domesticated and began cultivating a variety of plants: several wheats, barleys, rye, peas, lentils, and flax in Southwest Asia; millet and sorghum in Africa; millet and soybeans in North China; rice and beans in Southeast Asia; maize (corn) in Mesoamerica; potatoes, quinoa, beans, and manioc in South America. Domestication constitutes a process (not an act) that involves taming, breeding, genetic selection, and occasionally introducing plants into new ecological settings. In the case of wheat, for example, wild wheat is brittle, with seeds easily scattered by the wind and animals, a trait that enables the plant to survive under natural conditions. Domesticated wheat retains its seeds, which simplifies harvesting but which leaves the plant dependent on the farmer for its propagation. Humans changed the plant's genes; the plant changed humanity. And, with humans raising the grain, the rat, the mouse, and the house sparrow "self-domesticated" and joined the Neolithic ark.

The domestication of animals developed out of intimate and long-standing human contact with wild species. Logically, at least, there is a clear succession from hunting and following herds to corralling, herding, taming, and breeding. The living example of the Sami (Lapp) people who follow and exploit semiwild reindeer herds illustrates how the shift from hunting to husbandry and pastoral nomadism may have occurred. As with plant culture, the domestication of animals involved human selection from wild types, selective slaughtering, selective breeding, and what Darwin later called "unconscious selection" from among flocks and herds. Humans in the Old World domesticated cattle, goats, sheep, pigs, chickens, and, later, horses. In the New World Andean communities domesticated only llamas and the guinea pig; peoples in the Americas thus experienced a comparative deficiency of animal protein in the diet.

Animals are valuable to humans in diverse ways. Some of them convert inedible plants to meat, and meat contains more complex proteins than plants. Animals provide food on the hoof, food that keeps from spoiling until needed. Animals produce valuable secondary products that were increasingly exploited as the Neolithic unfolded in the Old World. Cattle, sheep, pigs, and the rest are "animal factories" that produce more cattle, sheep, and pigs. Chickens lay eggs, and cows, sheep,

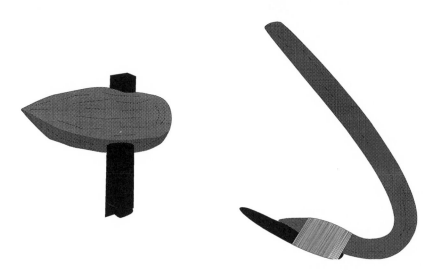

Fig. 2.1. Neolithic tools. Neolithic horticulture required larger tools for clearing and cultivating plots and for harvesting and processing grains.

goats, and horses produce milk. Treated and storable milk products in yogurts, cheeses, and brewed beverages sustained the great herding societies of Asia and pastoralists everywhere. Manure later became another valuable animal product as fertilizer and fuel. Animal hides provided raw material for leather and a variety of products, and sheep, of course, produced fleece. (Wool was first woven into fabric on Neolithic looms.) Animals provided traction and transportation. The Neolithic maintained the close dependence on plants and animals that humankind had developed over the previous 2 million years. But the technologies of exploiting them and the social system sustained by those technologies had changed radically.

After a few thousand years of the Neolithic in the Near East, mixed economies that combined the technologies of horticulture and animal husbandry made their appearance. Late Neolithic groups in the Old World apparently kept animals for traction and used wheeled carts on roads and pathways that have been favorably compared to those of medieval Europe. The historical route to intensified agriculture and to civilization was through this mixed Neolithic farming. If biology and evolution were partly responsible for the character of our first mode of existence in the Paleolithic, then the Neolithic Revolution represents a change of historical direction initiated by humans themselves in response to their changing environment.

Complementing the many techniques and skills involved in farming and husbandry, several ancillary technologies arose as part of the shift to the Neolithic. First among these novelties was textiles, an innovation independently arrived at in various parts of the Old and New Worlds. Recent findings show that some Paleolithic groups occasionally practiced techniques of weaving, perhaps in basketry, but only in the Neolithic did the need for cloth and storage vessels expand to the point where textile technologies flourished. The production of textiles involves several interconnected sets of technologies: shearing sheep or growing and harvesting flax or cotton, processing the raw material, spinning thread (an ever-present part of women's lives until the Industrial Revolution 10,000 years later), constructing looms, dyeing, and weaving the cloth. In considering the advent of textile production in the Neolithic, one cannot overlook design considerations and the symbolic and informational role of dress in all societies.

Pottery, which also originated independently in multiple centers around the world, is another new technology that formed a key part of the Neolithic Revolution. If only inadvertently, Paleolithic peoples had produced fired-clay ceramics, but nothing in the Paleolithic economy called for a further development of the technique. Pottery almost certainly arose in response to the need for a storage technology: jars or vessels to store and carry the surplus products of the first agrarian societies. Neolithic communities used plasters and mortars in building construction, and pottery may have arisen out of plastering techniques applied to baskets. Eventually, "manufacturing centers" and small-scale transport of ceramics developed. Pottery is a "pyrotechnology," for the secret of pottery is that water is driven from the clay when it is "fired," turning it into an artificial stone. Neolithic kilns produced temperatures upwards of 900°C. Later, in the Bronze and Iron Ages, the Neolithic pyrotechnology of pottery made metallurgy possible.

In Neolithic settings, hundreds if not thousands of techniques and technologies large and small melded to produce the new mode of life. Neolithic peoples built permanent structures in wood, mud brick, and stone, all of which testify to expert craft skills. They twisted rope and practiced lapidary crafts, and Neolithic peoples even developed metallurgy of a sort, using naturally occurring raw copper. The technology of cold metalworking produced useful tools. The now-famous "Ice man," the extraordinary frozen mummy exposed in 1991 by a retreating glacier in the Alps, was first thought to belong to a Bronze Age culture because of the fine copper axe he was carrying when he perished. As it turns out, he lived in Europe around 3300 B.C., evidently a prosperous Neolithic farmer with a superior cold-forged metal tool.

The Neolithic was also a social revolution and produced a radical change in lifeways. Decentralized and self-sufficient settled villages, consisting of a dozen to two dozen houses, with several hundred inhabitants became the norm among Neolithic groups. Compared to the smaller bands of the Paleolithic, village life supported collections of families united into tribes. The Neolithic house doubtless became the center of social organization; production took place on a household basis. The imaginative suggestion has been made that living inside houses forced Neolithic peoples to deal in new ways with issues concerning public space, privacy, and hospitality. Neolithic peoples may have used hallucinatory drugs, and they began to experiment with fermented beverages. Although a sexual division of labor probably persisted in the Neolithic, horticultural societies, by deemphasizing hunting, may have embodied greater gender equality. A comparatively sedentary lifestyle, a diet higher in carbohydrates, and earlier weaning increased fertility, while freedom from the burden of carrying infants from camp to camp enabled women to bear and care for more children. And one suspects that the economic value of children—in tending animals or helping in the garden, for example—was greater in Neolithic times than in the Paleolithic. At least with regard to Europe, some archaeologists have made compelling claims for the existence of cults devoted to Neolithic goddesses and goddess worship. There were doubtless shamans, or medicine "men," some of whom may also have been women. Neolithic societies remained patriarchal, but males were not as dominant as they would become with the advent of civilization.

In the early Neolithic, little or no occupational specialization differentiated individuals who earned their bread solely through craft expertise. This circumstance changed by the later Neolithic, as greater food surpluses and increased exchange led to more complex and wealthier settlements with full-time potters, weavers, masons, toolmakers, priests, and chiefs. Social stratification kept pace with the growth of surplus production. By the late Neolithic low-level hierarchal societies, tribal chiefdoms, or what anthropologists call "big men" societies appeared. These societies were based on kinship, ranking, and the power to accumulate and redistribute goods sometimes in great redistributive feasts. Leaders now controlled the resources of 5,000 to 20,000 people. They were not yet kings, however, because they retained relatively little for themselves and because Neolithic societies were incapable of producing truly great wealth.

Compared to the Paleolithic economy and lifestyle, one could argue that the standard of living actually became depressed in the transition to the Neolithic in that low-intensity horticulture required more labor, produced a less varied and nutritious diet, and allowed less leisure than Paleolithic hunting and gathering in its heyday. But—and this was the primary advantage—Neolithic economies produced more food and could therefore support more people and larger population densities (estimated at a hundredfold more per square mile) than Paleolithic foraging.

Populations expanded and the Neolithic economy spread rapidly to fill niches suited for them. By 3000 B.C. thousands of agrarian villages dotted the Near East, usually within a day's walk of one another. Wealthier and more complex social structures developed, regional crossroads and trading centers arose, and by the late Neolithic real towns had emerged. The classic example is the especially rich Neolithic town of Jericho, which by 7350 B.C. already had become a well-watered, brick-walled city of 2,000 or more people tending flocks and plots in the surrounding hinterland. Jericho had a tower nine meters high and ten meters in diameter, and its celebrated walls were three meters thick, four meters high, and 700 meters in circumference. The walls were necessary because the surplus stored behind them attracted raiders. War-like clashes between Paleolithic peoples had undoubtedly occurred repeatedly over the millennia in disputes over territory, to capture females, or for cannibalistic or ritual purposes. But with the Neolithic, for the first time, humans produced surplus food and wealth worth stealing and hence worth protecting. Paleolithic groups were forced to adapt to the Neolithic economies burgeoning around them. Thieving was one alternative; joining in a settled way of life was another. In the long run, Neolithic peoples marginalized hunter-gatherers and drove them virtually to extinction. Idealized memories of the foraging lifestyle left their mark in "Garden of Eden" or "happy hunting grounds" legends in many societies.

Blessed or cursed with a new economic mode of living, humans gained greater control over nature and began to make more of an impact on their environments. The ecological consequences of the Neolithic dictated that the domestic replace the wild, and where it occurred the Neolithic Revolution proved irreversible—a return to the Paleolithic was impossible because Paleolithic habitats had been transformed and the Paleolithic lifestyle was no longer sustainable.

Moonshine

The Neolithic Revolution was a techno-economic process that occurred without the aid or input of any independent "science." In assessing the connection between technology and science in the Neolithic, pottery provides an example exactly analogous to making fire in the Paleolithic. Potters made pots simply because pots were needed and because they acquired the necessary craft knowledge and skills. Neolithic potters possessed practical knowledge of the behavior of clay and of fire, and, although they may have had explanations for the phenomena of their crafts, they toiled without any systematic science of materials or the self-conscious application of theory to practice. It would denigrate Neolithic crafts to suppose that they could have developed only with the aid of higher learning.

Can anything, then, be said of science in the Neolithic? In one area, with regard to what can be called Neolithic astronomy, we stand on strong ground in speaking about knowledge in a field of science. Indeed, considerable evidence makes plain that many, and probably most, Neolithic peoples systematically observed the heavens, particularly the patterns of motion of the sun and moon and that they regularly created astronomically aligned monuments that served as seasonal calendars. In the case of Neolithic astronomy, we are dealing not with the prehistory of science, but with science in prehistory.

The famous monument of Stonehenge on the Salisbury Plain in southwest England provides the most dramatic and best-understood case in point. Stonehenge, it has now been determined by radiocarbon

Fig. 2.2. Jericho. Neolithic farming produced a surplus that needed to be stored and defended. Even in its early phases, the Neolithic settlement of Jericho surrounded itself with massive walls and towers, as shown in this archaeological dig.

dating, was built intermittently in three major phases by different groups over a 1,600-year period from 3100 B.C. to 1500 B.C., by which time the Bronze Age finally washed across the Salisbury Plain. The word "Stonehenge" means "hanging stone," and transporting, working, and erecting the huge stones represents a formidable technological achievement on the part of the Neolithic peoples of prehistoric Britain.

A huge amount of labor went into building Stonehenge—estimates range to 30 million man-hours, equivalent to an annual productive labor of 10,000 people. In order to create a circular ditch and an embankment 350 feet in diameter 3,500 cubic yards of earth were excavated. Outside the sanctuary the first builders of Stonehenge erected the so-called Heel Stone, estimated to weigh 35 tons. Eighty-two "bluestones" weighing approximately 5 tons apiece were brought to the site (mostly over water) from Wales, an incredible 240 kilometers (150 miles) away. Each of the 30 uprights of the outer stone circle of Stonehenge weighed in the neighborhood of 25 tons, and the 30 lintels running around the top of the ring weighed 7 tons apiece. More impressive still, inside the stone circle stood the five great trilithons or three-stone behemoths. The average trilithon upright weighs 30 tons and the largest probably weighs over 50 tons. (By contrast, the stones that went into building the pyramids in Egypt weighed on the order of 5 tons.) The great monoliths were transported 40 kilometers (25 miles) overland from Marlborough Downs, although the suggestion has been made that ancient glaciers may have been responsible for moving them at least part way to Stonehenge. The architects of Stonehenge appear to have laid out the monument on a true circle, and in so doing they may have used some practical geometry and a standard measure, the so-called megalithic yard.

The labor was probably seasonal, taking place over generations. A stored food surplus was required to feed workers, and some relatively centralized authority was needed to collect and distribute food and to supervise construction. Neolithic farming and ranching communities appeared on the Salisbury Plain by the fourth millennium B.C. and evidently reached the required level of productivity. Although Neolithic farming never attained the levels of intensification later achieved by civilized societies, Stonehenge and the other megalithic ("large stone") structures show that even comparatively low-intensity agriculture can produce sufficient surpluses to account for monumental building.

Recognition that Stonehenge is an astronomical device has been confirmed only in our day. As literate peoples encountered Stonehenge over the centuries, any number of wild interpretations emerged as to who built it and why. Geoffrey of Monmouth in his twelfth-century *History of Kings of Britain* has Merlin from King Arthur's court magically transporting the stones from Wales. Other authors have postulated that the Romans or the Danes built Stonehenge. A still-current fantasy holds that the Druids built and used Stonehenge as a ceremonial center. (In

Fig. 2.3. Stonehenge. Neolithic and early Bronze Age tribes in Britain built and rebuilt the famous monument at Stonehenge as a regional ceremonial center and as an "observatory" to track the seasons of the year.

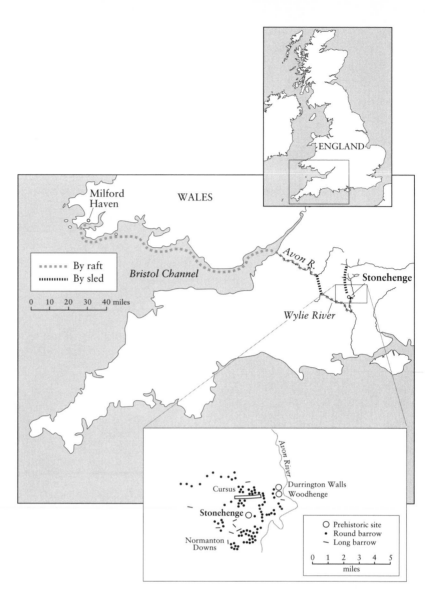

Map 2.1. The Salisbury plain. Stonehenge was set among a cluster of Neolithic sites, indicating the relative wealth and resources of the region. Some of the smaller stones that went into making Stonehenge were transported 150 miles by rollers and raft from Western Wales; some of the largest stones came from 25 miles north of the site.

Milford Haven

WALES

ENGLAND

By raft
By sled

Bristol Channel

Avon R.

Stonehenge

0 10 20 30 40 miles

Wylie River

Avon River

Cursus

Durrington Walls
Woodhenge

Stonehenge

Normanton
Downs

○ Prehistoric site
• Round barrow
⌒ Long barrow

0 1 2 3 4 5
miles

fact, the Celtic Iron Age Druids and their culture only appeared a thousand years after Stonehenge was completed.) Even in the 1950s, when the possibility became clear that Neolithic peoples from the Salisbury Plain themselves were responsible for Stonehenge, there was considerable resistance to the idea that "howling barbarians" might have been capable of building such an impressive monument, and some supposed that itinerant contractors from the Near East built it. All scholars now agree that Stonehenge was a major ceremonial center and cult site built by the people of the Salisbury Plain. Its astronomical uses indicate that it functioned as a Neolithic religious center for the worship of the sun and the moon and for establishing a regional calendar.

The English antiquarian William Stuckeley (1687–1765) was the first modern to write about the solar alignment of Stonehenge in 1740. The sun rises every day at a different point on the horizon; that point moves back and forth along the horizon over the course of a year, and each year at mid-summer the sun, viewed from the center of the sanctuary at Stonehenge, rises at its most southerly point, which is precisely where the builders placed the Heel Stone. The monument's primary astronomical orientation toward the midsummer sunrise is confirmed annually and has not been disputed since Stuckeley.

In the 1960s, however, controversy erupted over claims for Stonehenge as a sophisticated Neolithic astronomical "observatory" and "computer." The matter remains disputed today, but wide agreement exists on at least some larger astronomical significance for Stonehenge, especially with regard to tracking cyclical movements of the sun and the moon. The monument seems to have been built to mark the extreme and mean points of seasonal movement of both heavenly bodies along the horizon as they rise and set. Thus, the monument at Stonehenge marks not only the sun's rise at the summer solstice, but the rise of the sun at winter solstice and at the fall and spring equinoxes. It also indicates the sun's settings at these times, and it tracks the more complicated movements of the moon back and forth along the horizon, marking four different extremes for lunar motion.

The construction of Stonehenge required sustained observations of

Fig. 2.4. Midsummer sunrise at Stonehenge. On the morning of the summer solstice (June 21) the sun rises along the main axis of Stonehenge and sits atop the Heel Stone.

Fig. 2.5. Neolithic society on Easter Island. A society based on low-intensity agriculture flourished here for hundreds of years before it was extinguished by ecological ruin. During its heyday it produced megalithic sculptures called *moai* comparable in scale to Stonehenge and other monumental public works that are typical of Neolithic societies.

the sun and the moon over a period of decades and mastery of horizon astronomy. The monument embodied such observations, even in its earliest phases. The ruins testify to detailed knowledge of heavenly movements and to a widespread practice of "ritual astronomy." We have no access to what megalithic Europeans thought they were doing; their "theories" of the sun and the moon, if any, may have been utterly fantastic, and we would probably label their explanations more religious than naturalistic or scientific. Still, megalithic monuments embody a scientific approach in that they reflect understanding of regularities of celestial motions and they bespeak long-term systematic interest in and observations of nature. Although religious elders, hereditary experts, or priestly keepers of knowledge doubtless tended Stonehenge, it probably goes too far to suggest that megalithic monuments provide evidence for a class of professional astronomers or for astronomical research of the sort that later appeared in the first civilizations. Stonehenge may better be thought of as a celestial orrery or clock that kept track of the major motions of the major celestial bodies and possibly some stars. In addition, Stonehenge certainly functioned as a seasonal calendar, accurate and reliable down to a day. As a calendar, Stonehenge kept track of the solar year and, even more, harmonized the annual motion of the sun with the more complicated periodic motion of the moon. It may even have been used to predict eclipses, although that possibility seems unlikely. In these telling ways—systematically observing the heavens, mastering the clock-like movement of the sun and the moon, gaining intellectual control over the calendar—it is possible and even necessary to speak of Neolithic "astronomy" at Stonehenge. The further development of astronomy awaited the advent of writing and cohorts of full-time experts with the patronage of centralized bureaucratic governments. But long before those developments, Neolithic farmers systematically investigated the panorama of the heavens.

On the other side of the globe the remarkable giant statues of Easter Island (also known as Rapa Nui) provide mute testimony to the same forces at play. Easter Island is small and very isolated: a 46-square-mile speck of land 1,400 miles west of South America and 900 miles from the nearest inhabited Pacific island. Polynesian peoples reached Easter Island by sea sometime after A.D. 300 and prospered through cultivating sweet potatoes, harvesting in a subtropical palm forest, and fishing in an abundant sea. The economy was that of settled Paleolithic or simple Neolithic societies, but local resources were rich, and even at slow growth rates over a millennium the founding population inevitably expanded, reaching 7,000 to 9,000 at the peak of the culture around A.D. 1200 to 1500. (Some experts put the figure at over 20,000.)

Islanders carved and erected more than 250 of their monumental moai statues on giant ceremonial platforms facing the sea. Notably, the platforms possessed built-in astronomical orientations. Reminiscent of the works of the peoples of Stonehenge or the Olmecs of Central America, the average moai stood over 12 feet in height, weighed nearly 14 tons, and was transported up to six miles overland by gangs of 55 to 70 men; a few mammoth idols rose nearly 30 feet tall and weighed up to 90 tons. Hundreds more statues—some significantly larger still—remain unfinished in the quarry, where all activity seems to have stopped suddenly. Remote Easter Island became completely deforested because of the demand for firewood and construction material for sea-going canoes, without which islanders could not fish for their staple of porpoise and tuna. By 1500, with the elimination of the palm tree and the extinction of native bird populations, demographic pressures became devastatingly acute, and islanders intensified chicken-raising and resorted to cannibalism and eating rats. The population quickly crashed to perhaps one-tenth its former size, the sad remnant "discovered" by Europeans in 1722. Only 100 souls lived there in 1887. The wealth of

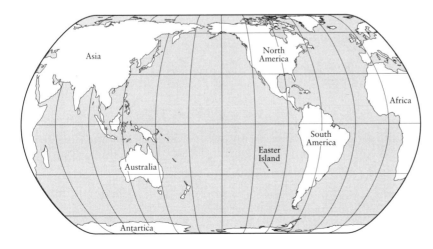

Map 2.2. Easter Island. This isolated speck of land in the South Pacific lies 1,400 miles off the coast of South America and 900 miles from the nearest inhabited island to the west. Polynesian seafarers, probably navigating by star charts and taking advantage of their knowledge of wind and current changes, arrived at Easter Island around A.D. 300. Europeans "discovered" the island in 1722.

the pristine island had provided rich resources where a human society evolved in a typically Neolithic (or settled Paleolithic) pattern. But human appetites and the island's narrow ecological limits doomed the continuation of the stone-working, heaven-gazing, and wood-burning culture that evolved there.

In general, through observation of the sun and the moon Neolithic peoples around the world established markers, usually horizon markers, that monitored the periodic motion of these bodies across the sky, tracked the year and the seasons, and provided information of great value to communities of farmers. In some cases the devices they created to reckon the year and predict the seasons became quite elaborate and costly and were possible only because of the surplus wealth produced in favored places.

Before Stonehenge and long before the settlement and ruination of Easter Island, in certain constricted environments growing populations pressed against even enlarged Neolithic resources, setting the stage in Egypt, Mesopotamia, and elsewhere for a great technological transformation of the human way of life—the advent of urban civilization.

CHAPTER 3

Pharaohs and Engineers

Neolithic societies never reached the complexity of kingdoms. They never built large cities, or large enclosed structures like palaces or temples; they had no need for writing to keep records; and they never established a tradition of higher learning or institutionalized science. These features arose only when Neolithic societies coalesced into civilizations—a second great transformation in human social evolution.

This revolution is often referred to as the Urban Revolution. Whatever its name, changes that began around 6,000 years ago in the Near East ushered in the first civilizations, replete with all the social and historical consequences accompanying cities, high population densities, centralized political and economic authority, the origin and organization of regional states, the development of complex and stratified societies, monumental architecture, and the beginnings of writing and higher learning. The transition was another techno-economic revolution, this time arising out of the need for intensified agricultural production to sustain increasingly large populations that pressed against the carrying capacities of their habitats. As an episode in human history and the history of technology, the Urban Revolution proved to be unrivaled in its consequences until the Industrial Revolution that took root in eighteenth-century Europe.

A new mode of intensified agriculture, distinct from Neolithic horticulture or pasturage, provided the underpinnings of the first civilizations. In that mode simple gardening was superseded by field agriculture based on large-scale water-management networks constructed and maintained as public works by conscripted labor gangs (the corvée) under the supervision of state-employed engineers. In the Old World the ox-drawn scratch plow replaced the hoe and digging stick. And subsistence-level farming gave way to the production of large surpluses of cereals (estimated at a minimum of 50 percent above Neolithic levels) that could be taxed, stored, and redistributed. Centralized political authorities dominated by a pharaoh or king came into being to manage these complex systems of agricultural production. Along with hydraulically intensified agriculture (generally artificial irrigation) and a centralized state authority, the Urban Revolution sustained much larger populations, urban centers, coercive institutions in the form of armies, tax collectors, and police, expanded trade, palaces and temples, a priestly class, religious institutions, and higher learning. In such bureaucratically organized societies, cadres of learned scribes developed mathematics, medicine, and astronomy.

Taming the Rivers

The Urban Revolution unfolded independently in multiple centers across the Old and New Worlds. The same remarkable pattern of Neolithic settlements coalescing into centralized kingdoms based on intensified agriculture occurs at least six times in six different sites around the globe: in Mesopotamia after 3500 B.C., in Egypt after 3400 B.C., in the Indus River Valley after 2500 B.C., along the Yellow River (Hwang-Ho) in China after 1800 B.C., in Mesoamerica at about 500 B.C., and in South America after 300 B.C. The origin and development of these civilizations were essentially independent and not the result of diffusion from a single center, and hence they are known as the pristine civilizations.

Why did civilization arise independently and repeatedly on a worldwide scale after the fourth millennium B.C. in those particular sites? Several explanations have been proposed. The precise processes involved in the leap to civilization are research questions actively debated by archaeologists and anthropologists, but many scholars emphasize the importance of hydrology and ecology, and they recognize that intensified agriculture, abetted by large-scale hydraulic engineering projects, was a key element in the formation of large, highly centralized bureaucratic states. The fact alone that pristine civilizations arose in hydrologically distressed regions—that is, where too little or too much water required hydraulic engineering for the successful practice of intensified agriculture—gives credence to what is called the hydraulic hypothesis, linking the rise of civilization with the technology of large-scale hydraulic systems. Under a hot, semitropical sun, irrigation agriculture is extraordinarily productive and yields that can literally fuel large populations become possible. Silt-laden rivers provide water for irrigation and, especially when controlled artificially, they enrich the soils around them. Irrigation agriculture and flood control required hydraulic engineering works and some level of communal action to build and maintain them and to distribute water when and where needed: marshes had to be drained; dams, dikes, canals, sluices, conduits, terraces, catchments, and embankments had to be built; and ditches had to be kept free of debris. Water disputes had to be settled by some authority, and grain surpluses had to be stored, guarded, and redistributed. The interacting effects of the geographical setting and the techniques of hydraulic agriculture reinforced trends toward an authoritarian state.

Along these lines, the notion of "environmental circumscription" provides the key explanatory concept: civilizations arose in prehistoric river valleys and flood plains that were environmentally restricted agricultural zones beyond which intensive farming was impossible or impractical. In these constricted habitats, like the Nile River Valley, expanding Neolithic populations soon pressed against the limits imposed by desert, cataracts, and sea, leading to pressures to intensify food production. Warfare became chronic and developed beyond raiding to involve conquest and subjugation since, in a habitat already filled, the losers could no longer bud off and form a new agricultural community. Whereas previously in both the Paleolithic and Neolithic, defeated groups could generally move on to a new locale, in environmentally restricted areas such as the Nile River Valley agriculturalists had nowhere to go. Victors not only took over land and smaller irrigation works but subjugated and dominated defeated groups, sparing their lives in return for their labor as slaves and peasants in maintaining systems of intensified farming. Once this process started, the historical momentum favoring confederating and centralizing forces was irreversible. Neolithic communities thus became increasingly stratified, culminating in a dominant elite in command of an agricultural underclass as regional powers subsumed local ones. Time and again civilization and the state emerged wherever these ecological and demographic conditions occurred.

Further research will doubtless amplify this picture, but for now a common pattern with common characteristics seems apparent. History is too easily thought of as a sequence of unique events—what has been lampooned as "one damned thing after another." But the recurrent rise of civilizations in the Near East, in the Far East, and in the New World testifies to significant regularities in the historical record.

The model described above admirably fits the first human civilization arising on the flood plain between the Tigris and the Euphrates Rivers in present-day Iraq. This was ancient Mesopotamia, the land "between the rivers." By 4000 B.C. Neolithic villages filled the Mesopotamian plain. Local authorities drained marshes in the lower delta and, later, installed extensive irrigation works on the flood plain upriver. Great walled cities such as Uruk, Ur, and Sumer, with populations between 50,000 and 200,000, arose after 3500 B.C., and the dynastic civilization of the Sumerians developed fully by 2500 B.C. Possibly because of the shifting and unpredictable courses and flood patterns of the Tigris and Euphrates, no single kingdom or polity dominated Mesopotamia as in Egypt, but rather a series of city-states along with empires based on them rose and fell over the succeeding millennia.

Mesopotamian civilization shows a great deal of continuity over

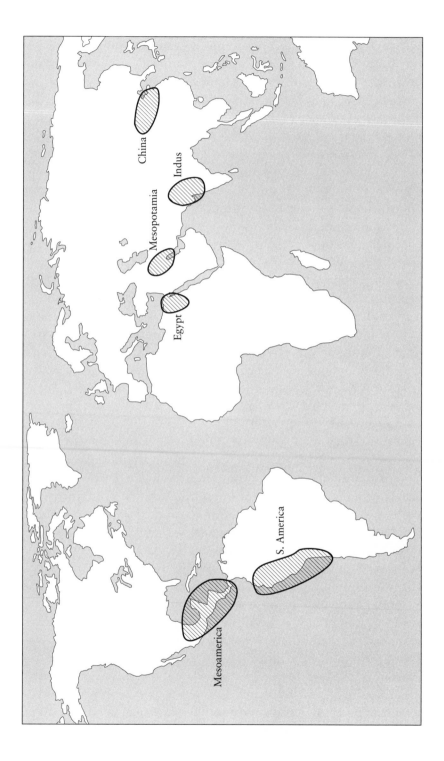

thousands of years, even though different groups, from different portions of Mesopotamia, took their turns at cultural, political, and military ascendance. When the Babylonians of central Mesopotamia became the dominant power, they absorbed a good deal of Sumerian culture and adapted Sumerian script for writing their own language. When Assyria (a kingdom in northern Mesopotamia) began to control the region, it similarly absorbed much of Babylonian culture.

All of these civilizations were based on irrigation agriculture. Main canals were upward of 75 feet wide and ran for several miles, with hundreds of connecting channels. All Mesopotamian civilizations developed centralized political authority and complex bureaucracies to collect, store, and redistribute agricultural surpluses. All are characterized by monumental building, including most notably great brick temple complexes and pyramids known as ziggurats. For example, Ur-Nammu's ziggurat of Third Dynasty Ur (dating to approximately 2000 B.C.) formed part of a larger complex measuring 400 by 200 yards. Nebuchadnezzar's tower (600 B.C.) rose over 90 meters (270 feet) and was, according to tradition, the basis of the biblical story of the Tower of Babel. Mesopotamian civilization also developed writing, mathematics, and a very sophisticated and mature astronomy.

Ancient Egypt illustrates a similar route to civilization. The Nile River Valley is a circumscribed strip of green hemmed in by a sea of desert to the east and west, mountains to the south, and the Mediterranean to the north; it forms a narrow ribbon 12–25 miles wide and hundreds of miles long. Neolithic settlements proliferated along the Nile, and already in the sixth millennium B.C. kingdoms emerged; seven predynastic kingdoms have been identified down to roughly 3400–3200 B.C. (Egyptologists agree about the order of events, but they differ by centuries on dating, especially in the early dynasties and Old Kingdom Egypt.) Sometime in that period King Menes united the two kingdoms of Upper and Lower Egypt, thus becoming the first Egyptian pharaoh of what we know as the first dynasty. And, according to tradition, Menes also organized hydraulic works, having embanked the Nile at Thebes. The explosive growth of Egyptian civilization followed. Based on managing the annual flooding of the Nile, Egypt manifested all the earmarks of high civilization, including large-scale building in the great pyramids at Giza, which were early creations of Egyptian civilization. Centralized authority grew correspondingly at an early date; 20,000 soldiers came to compose the Egyptian army; the pharaohs became legal heirs to all property in Egypt and controlled absolutely their 2.5 million subject-tenants; bureaucracy, writing, mathematics, elementary astronomy, expanded crafts, and all the other complexities of civilization displayed themselves in turn.

Less is known of civilization in the Indus River Valley, but the outlines of its historical development are plain. Neolithic settlements appeared along the Indus by 7000 B.C. Civilization may have arisen in-

Map 3.1. The earliest civilizations. The transition from Neolithic horticulture to intensified agriculture occurred independently in several regions of the Old and New Worlds. Increasing population in ecologically confined habitats apparently led to new technologies to increase food production. *(opposite)*

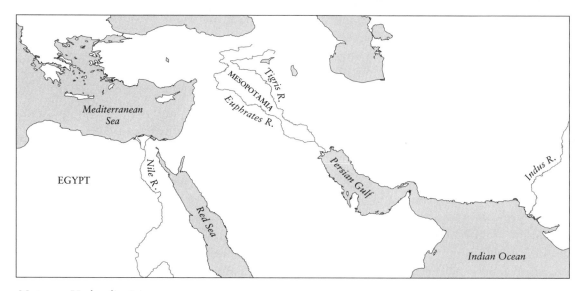

Map 3.2. Hydraulic civilizations. The first civilizations formed in ancient Mesopotamia (modern Iraq) on the flood plain of the Euphrates and Tigris Rivers, astride the Nile River in Egypt, and along the Indus River. The agricultural benefits of annual flooding were intensified by hydraulic management.

digenously or some of its incipient features may possibly have arrived with settlers or traders from Mesopotamia. One way or another, the alluvial flood plain of the Indus River Valley provided the indispensable setting for Indus civilization, and irrigation agriculture the requisite means. The cities of Mohenjo-daro and Harappa in modern-day Pakistan date to 2300 B.C. Harappan civilization, as it is known, thereafter spread inland and along the coast of the Arabian Sea. Peoples of the Indus River Valley farmed the arid plains, and they built embankments to protect cities against erratic, silt-laden floods. Indicative of strong central government, Harappan towns were rigidly planned walled communities with laid-out streets and blocks, towers, granaries, and sewers, and all the trappings of civilization. At the center of Mohenjo-daro, for example, stood an enclosed citadel (200 × 400 yards) with its 40-foot-high brick mound. Within, the Great Bath held a manmade pool 12 meters long, 7 meters wide, and almost 3 meters deep, and archaeologists have identified what may be priestly residences and an assembly hall. The population of Mohenjo-daro has been estimated at 40,000. Harappan metallurgists used copper, bronze, gold, silver, tin, and other metals; potters produced glazed pots; and writing and higher learning developed. Limited evidence suggests that even at an early period authoritarian regimes with a strong priestly-bureaucratic-military class already held command. But after 1750 B.C. the original urban culture of the Indus declined, probably because of climate and ecological factors, including the changing course of the Indus River.

In China a similar pattern repeated itself along the Yellow River (the Hwang-Ho). By 2500 B.C. thousands of late Neolithic villages spread out along the river, and as irrigation agriculture began to be practiced, kingdoms arose. Yü the Great, the putative founder of the semimythical first dynasty (Hsia), is legendary in China as the ruler who "controlled the waters." The Shang (Yin) dynasty (1520–1030 B.C.), which marks the documented beginning of Chinese civilization, made itself master of the Yellow River plain by dint of extensive irrigation works. Later, engineers brought irrigation techniques to the more southern Yangtze River. Rice cultivation spread northward from south China and also involved hydraulic control. One of the roles of government throughout Chinese history was to build and maintain waterworks; as a result, dikes, dams, canals, and artificial lakes (such as the 165-acre Lake Quebei) proliferated across China. Deliberate government policies of water conservancy and agricultural improvement also involved drainage. To effect these installations, massive corvée labor was extracted from the peasantry.

The early Chinese built cities with protective walls, palaces, and ceremonial centers. Their society became highly stratified; Chinese emperors functioned as high priests, and great emphasis was placed on royal burials that included the emperor's entourage, sacrificed by the hundreds to accompany him. China was first unified in 221 B.C., and unprecedented authority became centralized in the emperor, backed administratively by an elaborate and formidable bureaucracy associated with royal courts. The population of China under the control of the emperor has been estimated at 60 million at the beginning of the Christian Era. The early Chinese state built granaries and maintained standing armies. Sophisticated bronze metallurgy was also practiced, with the bronze tripod the symbol of administrative power invested in officials. As for monumental building, in addition to hydraulic works, the Great Wall of China has been hailed as the largest building project in history. Construction of the first 1,250 miles of the Great Wall (on the divide between steppe and arable land) began in the fourth and third centuries B.C. and was finished in 221–207 B.C., coincident with the first unification of China. (In later historical times the total length of Chinese defensive walls extended to over 3,000 miles.) The Grand Canal (originally built in A.D. 581–618), the interior waterway stretching 1,100 miles from Hangchow to Beijing, deserves mention as another example of monumental building associated with Chinese civilization. On the order of 5.5 million people labored on the project in which 2 million workers may have perished. No less characteristically, writing, mathematics, and astronomy came to be part of Chinese civilization.

Swamps and Deserts

The separate and independent rise of civilizations in the Old and New Worlds represents a great experiment in human social and cultural development. Despite departures in the New World, notably the absence of cattle, the wheel, and the plow, the independent appearance of civilization in the Western Hemisphere and the deep parallels among pristine civilizations in regions where water management was necessary lend support to the hydraulic hypothesis and the view that regularities in history derive from the material and technical bases of human existence.

Recent findings have confirmed that humans entered the Americas and hunted and gathered their way to southern Chile by at least 12,500 years ago. In Central (or Meso-)America, Paleolithic hunter-gatherers gave way to fully settled Neolithic villages by 1500 B.C. Increasingly complex Neolithic settlements filled the humid lowlands and coastal regions of Central America by 1000 B.C. Olmec culture flourished from 1150 to 600 B.C. inland along rivers flowing into the Gulf of Mexico and is sometimes said to be the first American "civilization." But in fact the Olmecs seem to have been at a high Neolithic stage comparable to the megalithic culture at Stonehenge. Olmec "towns" held populations of fewer than 1,000. Nonetheless, they built ceremonial centers with burial mounds, and they are known for colossal Olmec stone heads, some over 20 tons in weight and transported 100 miles, according to one report. They developed a calendar and, suggestive of the origins of true civilization, hieroglyphic writing. The Olmecs declined after 600 B.C., but they provided cultural models that later, more fully formed American civilizations built upon.

Founded around 500 B.C., the first true city in the New World was at Monte Albán looking down on the semiarid Oaxaca Valley in Central Mexico. Small-scale irrigation agriculture was practiced in the valley, and Monte Albán was a planned city that possibly represented the confederation or consolidation of three regional powers into what became Zapotec civilization. Engineers leveled the top of the mountain for a large astronomically oriented acropolis, stone temples, pyramids, and a ball court. Two miles of stone walls encircled the city; 15,000 people lived there by 200 B.C., 25,000 by the eighth century A.D. Before its subsequent decline, Zapotec scribes wrote with hieroglyphs and possessed a complex calendar.

Coexisting with Monte Albán but an order of magnitude larger, the huge city of Teotihuacán arose in the dry Teotihuacán Valley near modern Mexico City after 200 B.C. Estimates for the population of the city at its peak in the period A.D. 300–700 range from 125,000 to 200,000, making it the largest and most powerful urban center in Mesoamerica; it was the fifth largest city in the world in A.D. 500, and it remained one of the world's largest urban centers for several hundred years. Oriented astronomically, the planned town of Teotihuacán covered eight square miles, and the main avenue ran for over three miles. The largest structure was the gigantic Temple of the Sun, a huge stepped pyramid nearly 200 feet high, 35 million cubic feet in volume, with a temple on top. There were 600 other pyramids and temples in Teotihuacán and several thousand apartment complexes. As in other early civilizations, hydraulic works and irrigation agriculture made

Teotihuacán possible. In addition to farming land in the seasonally flooded upper valley, Teotihuacános built canals and installed extensive, permanent irrigation works along the San Juan River in the lower valley. Teotihuacán itself was well supplied with water by the river, canals, and reservoirs. Control over a highly developed obsidian trade also increased the prosperity of the city. What archaeologists have identified as a gigantic royal palace and a major bureaucratic/administrative center testify both to extreme social and economic stratification and to centralization of power into royal/priestly authority. At its height the civilization of Teotihuacán dominated the whole Valley of Mexico.

Contemporaneous with civilization in the dry valleys of central Mexico, Mayan civilization rose in the wet lowlands of the Yucatan and flourished for a thousand years between 100 B.C. and the ninth century A.D. Until the 1970s the archaeology of Mayan civilization seemed to discredit any link between civilization and the taming of waters. But an interpretative revolution in Mayan studies followed from the discoveries of extensive Mayan engineering installations covering 741 acres at Pulltrouser Swamp in modern Belize. The problem for lowland Mayan agriculture was not too little water, but too much, a problem the Maya overcame by farming raised fields (3 feet high, 15–30 feet wide, and 325 feet long at Pulltrouser) with canals and drainage channels in between. The works drained water from fields, the muck in canals served as fertilizer, and the system overall proved capable of producing surpluses sufficient to support large populations. And it required collective effort to build and maintain. The distinctive Mayan form of intensified wetland agriculture now reveals the hydraulic underpinnings of Mayan civilization.

The largest Mayan city was Tikal, which had a population of 77,000 before its collapse about A.D. 800. Population densities during the Maya Classic Period are estimated to have been 10 to 15 times greater than that supported in the remaining jungles of Central America today. Monumental building dominated Mayan cities, especially temple platforms and large stepped pyramids, similar to ziggurats, with a stairway leading to a temple on top. Political authority was centralized in noble classes and Mayan kings. And the Maya developed the most sophisticated mathematical, calendrical, and astronomical systems of any civilization in the Americas.

In the rise of civilization in South America, the pattern repeats itself yet again. Collectively covering millions of acres, Peruvian irrigation systems represent the largest archaeological artifact in the Western Hemisphere. The many short rivers flowing from the Andes Mountains to the Pacific across an arid coastal plain are now seen to form the ecological equivalent of the Nile River. Early village settlement arose in more than sixty of these extremely dry coastal valleys, and increasingly elaborate and well-engineered irrigation systems became

Fig. 3.1. Teotihuacán. Cities and monumental building are defining features of all civilizations. Here, the huge Temple of the Sun dominates the ancient Mesoamerican city of Teotihuacán.

essential to support the civilizations that developed there. One of the irrigation canals of the Chimu people, for example, ran 44 miles; their capital at Chan-Chan covered nearly seven square miles. In joining existing irrigation systems, Moche civilization expanded out of the Moche River Valley after 100 B.C., ultimately occupying 250 miles of desert coastline and up to 50 miles inland. The Moche urban center at Pampa Grande had a population of 10,000, and the Huaca del Sol pyramid, made of 147 million adobe bricks, stood 135 feet high. Moche civilization endured for nine centuries.

In southern Peru another center of civilization arose in the highlands around Lake Titicaca. There, based on the cultivation of potatoes, a fecund agricultural system of raised and ridged fields similar to Mayan wet farming fueled a line of civilizations. One report puts the population of the mountain city of Tiwanaku at 40,000–120,000 at the city's zenith between A.D. 375 and 675. The succeeding Incas installed irrigation works and practiced water management on a larger scale than their predecessors, and militarily the Incas were the first to unite the productive resources of the coastal plains and the mountain highlands. At its peak in the fifteenth century A.D., the Inca empire extended 2,700 miles and included 6 to 8 million people (some say 10 million). Monumental building is well represented in the Inca capital of Cuzco with its exquisite mortarless masonry and water supply and drainage systems, in remote Machu Picchu with its steeply terraced fields, and no less in the incredible system of roads that united the Inca empire. Two road systems—one coastal, one in the mountains—ran for 2,200 miles each, and all together the Incas built 19,000 miles of path and road, a huge engineering achievement accomplished without metal tools. The state maintained an elaborate system of grain-storage facilities and redistribution mechanisms. The Inca emperor was the sacred focus of an absolutist state rivaling ancient Egypt in despotism, and like the Egyptian pharaohs, dead Inca emperors in Peru were mummified and worshiped.

Thus, time and again the Urban Revolution produced civilizations that depended on large-scale hydraulic engineering, and it repeatedly transformed human existence from Neolithic roots. The similarities of ancient American civilizations and those of the Old World have often been noticed and sometimes attributed to diffusion from the Old World to the New. But rather than invoking exotic contact across space and time to explain these parallels, would it not be less remarkable simply to say that similar material, historical, and cultural conditions produced similar civilizations?

Men of Metal

Based on the new technologies of irrigation and field agriculture, the worldwide rise of urban civilization marks a fundamental and irreversible turning point in the history of technology and in human affairs generally. A cascade of ancillary technologies accompanied the rise of civilization, including, at least in the Old World, bronze metallurgy. The mastery of bronze (copper alloyed with tin) still lends its name to the new civilization as the Bronze Age. Metals offer several advantages over stone as tools and weapons, and in the long run metals replaced stone. Metalworking embodies a complicated set of technologies, including mining ore, smelting, and hammering or casting the product into useful tools and objects; and bronze metallurgy requires furnaces with bellows to raise temperatures to 1100°C. In the New World, bronze did not replace the digging stick, stone hammers, chisels, or the obsidian blade for tools, but highly expert gold and silver metallurgy developed nonetheless for decorative and ornamental purposes. The sophisticated gold craftsmanship of pre-Columbian Indians in Peru is justly renowned, and Chimu metallurgists apparently used techniques amounting to chemical electroplating of gold.

Control over mineral resources thus became significant in the early civilizations. Sinai copper mines proved of great importance to Egyptian pharaohs; tin for making bronze had to be transported over long distances throughout the Near East; and, as mentioned, an extensive obsidian trade developed in Mesoamerica. Increased trade and expanded economic activity stand out among the earmarks of early civilizations. Occupational specialization and a sharpened division of labor likewise characterized civilized life from the outset. Craft production was no longer exclusively part-time or carried on as a household system of production, but rather became the business of specialized crafts whose practitioners earned their daily bread primarily in exchange for the practice of their craft skills. Certain "industrial" quarters of early cities were apparently given over to certain crafts and craft specialists. Among the new technologies of the Bronze Age, one might also mention brewing beer from bread, which became a noteworthy activity in Mesopotamia, where the famous Hammurabi Code regulated beer parlors in detail. Likewise in Inca Peru, ceremonial consumption of intoxicating beverages amounted to a redistribution of state-owned vegetable protein.

As a feature of the rise of state-level civilizations, humans began to exploit new sources of energy and power to do work. The muscle power of the ox (a castrated bull) was applied to pull the plow, and the horse was domesticated and entered humanity's service. The Hittites of second millennium B.C. Anatolia first harnessed the horse and the ass to a wheeled cart, thus creating the chariot and transforming warfare throughout the Near East. In the first millennium B.C. the camel began to provide essential transport. So, too, did the llama in South America and the elephant in India and South Asia. Wind power became a new energy source tapped for the first time with the rise of civilization. The Nile River especially, with the current flowing north and the prevailing winds blowing south, became a highway for sailboats and a factor contributing to the unity of ancient Egypt. Boats also came to ply the waters between Mesopotamia and the Indus River Valley. Slavery arose coincident with civilization, and the corvée, while less coercive than slavery, fits into this same category of the human use of human beings.

Pyramids

Monumental architecture in the form of pyramids, temples, and palaces is diagnostic of high civilization and is remarkable in the history of technology, not only as a set of extraordinary technical accomplishments, but also as indicative of the institution and practice of architecture and the developed crafts and trades associated with engineering. The Egyptian pyramids provide the classic example of monumental building by an early civilization. The case is well documented, and it encapsulates the themes raised thus far regarding agriculture, civilization, and the Urban Revolution.

Consider first the sheer immensity of the Great Pyramid at Giza. Built on the west bank of the Nile during the zenith of the pyramid building era between 2789 and 2767 B.C. (or possibly 2589–2566 B.C.) by Khufu (Cheops), the first pharaoh of the Fourth Dynasty, the Great Pyramid is the largest solid-stone structure ever built: it consists of an unbelievable 94 million cubic feet of masonry, made up of 2.3 million blocks averaging 2.5 tons apiece, with a total weight of 6 million tons; it covers 13.5 acres, in 210 courses of stone, and stands 485 feet high and 763 feet on a side; chambers, buttresses, and passageways lie within. Sheathed with polished stone, the scale of the construction—not to mention the beauty of the finished structure—has not been surpassed in the nearly five millennia of human history since the Great Pyramid was built.

The architects and engineers who built the Great Pyramid and the others like it commanded some elementary and some not-so-elementary practical mathematics. Design and material requirements demanded such expertise, as did the very exact north-south and east-west alignment. Ancient Egyptian engineers and architects understood the mathematics and appreciated the elegance of perfect pyramids, but the Egyptian pyramids (and monumental building generally) need to be seen primarily as stupendous engineering achievements.

According to a report by the fifth-century B.C. Greek historian Herodotus, 100,000 people toiled for twenty years to build the Great Pyramid; perhaps 4,000–5,000 craftsmen worked at the site year round. The techniques of pyramid construction are now well understood, and excepting the possible use of a cantilevered machine to lift stones, no categorically new building methods developed compared to what one finds in Neolithic building techniques. Simple tools and practical procedures carried the day but, characteristic of the new powers of civilization, more people, by orders of magnitude, were deployed and construction completed that much faster than at Neolithic sites.

Such an extraordinary monument did not suddenly appear in the Egyptian desert. Rather, the Great Pyramid culminates a clear progression of pyramid building coincident with the growth and expansion of the Egyptian agrarian state.

Several fanciful theories have been put forward to explain why the Great Pyramid and preceding and succeeding pyramids were built, but the function of these structures as tombs for pharaohs seems irrefutable, even if it may not have been their only purpose. A problem exists, however: at some periods at least, the number of new pyramids exceeded the number of pharaohs; and several pyramids were built simultaneously by a single pharaoh. Moreover, most of the truly monumental pyramids came into being in just over a century in the late Third and early Fourth Dynasties. According to one account, in four generations over 112 years between 2834 and 2722 B.C., six pharaohs built thirteen pyramids. Clearly, something more than burying the dead is needed to explain the extraordinary sociocultural phenomenon of the Egyptian pyramids.

One explanation of pyramid building from an engineering point of view attempts to explain the more or less continuous construction that took place on the west bank of the Nile during the heyday of pyramid building. In this interpretation, pyramid building was an activity pur-

Fig. 3.2. The Great Pyramid at Giza. An engineering marvel of the third millennium B.C., the Great Pyramid of Cheops (Khufu) at Giza culminated the tradition of pyramid building in Egyptian civilization. Some modern interpreters see it as a monumental exercise in political "state building." The Cheops pyramid is on the right.

sued in its own right as an exercise in statecraft. The sequence of the early pyramids comprised giant public-works projects designed to mobilize the population during the agricultural off-season and to reinforce the idea and reality of the state in ancient Egypt. More than one pyramid arose simultaneously because a labor pool—and surely an increasingly large labor pool—was available and because the geometry of pyramids dictates that fewer laborers are required near the top of a pyramid than at the bottom, thus permitting the transfer of labor to newly started projects. Monumental building was therefore a kind of institutional muscle-flexing by the early Egyptian state, somewhat akin to the arms industry today.

The engineering key to this argument comes from two particular pyramids. The first, the pyramid at Meidum, begun by the pharaoh Huni (Uni), who reigned for 24 years between 2837 and 2814 B.C., and continued by his son Sneferu, stood 80 feet high and ran 130 feet on side. It was to have been the first true pyramid with sheer, sloping sides and no visible steps. However, the pyramid at Meidum turned out to be an engineering disaster and a monumental structural failure, as the outer stone casing collapsed in rubble around the inner core of the pyramid. Designed with the evidently excessive slope of 54 degrees, the collapsed ruin may still be seen by the traveler.

The second pyramid at issue is the succeeding "Bent" pyramid at Dashur, also built by King Sneferu. It is a huge pyramid 335 feet high, 620 feet on a side, with a volume of 50 million cubic feet. Extraordinarily, the Bent pyramid is truly bent, angled, like Meidum, at 54 degrees on the lower half and 43 degrees on the top. One supposes that when the pyramid at Meidum failed, engineers reduced the slope of the Bent pyramid, still under construction, as a precaution. The next pyramid built by Sneferu, the Red pyramid, retained the safer slope of 43 degrees. (The Great Pyramid and later pyramids returned to increased elevations over 50 degrees, but used improved internal buttressing techniques.)

One does not have to follow every detail in order to accept the general point. The Egyptian pyramids were large state-run construction projects. A surplus of idle agricultural workers available seasonally for three months a year during the Nile floods provided the labor pool. (Agricultural productivity was thus not affected by the demand for labor for pyramid building.) Contrary to a once-common belief, forced slave labor did not build the pyramids, but labor was conscripted (like military conscription today) and organized in work gangs. Workers received food supplied by state granaries, and the completed pyramids served as tombs for departed pharaohs. Inevitably, elaborate theologies, priestly ceremonies, and ancillary technologies (such as mummifying) grew up around burying pharaohs. But in their construction the pyramids functioned primarily as gigantic public-works projects, the effect of which helped maintain the economy of irrigation agriculture in the Nile River Valley and bolstered centralizing political and social forces, notably the state. Indeed, the heyday of pyramid building was the heyday of political centralization in Old Kingdom Egypt. The pyramids were symbolic as well as literal exercises in state building.

Writing

One earmark of the earliest civilizations, already alluded to, was the elaboration and institutionalization of higher learning—writing, record-keeping, literature, and science. The fact that aspects of arithmetic, geometry, and astronomy originated in all of the earliest civilizations merits close attention, and it specifically suggests that such societies imposed a distinctive mark on the scientific traditions they fostered.

Fig. 3.3. The pyramid at Meidum. Built at a steep angle, the outer casing of the pyramid at Meidum collapsed around its central core during construction.

Fig. 3.4. The Bent pyramid. The lower portion of this pyramid rises at the same angle as the pyramid at Meidum, but ancient Egyptian engineers reduced the slope for the upper portion to ensure its stability. The Bent and Meidum pyramids were apparently constructed concurrently with engineers decreasing the angle of the Bent pyramid once they learned of the failure at Meidum.

Knowledge in the first civilizations was subordinated to utilitarian ends and provided useful services in recordkeeping, political administration, economic transactions, calendrical exactitude, architectural and engineering projects, agricultural management, medicine and healing, religion, and astrological prediction. Since higher learning was heavily skewed toward useful knowledge and its applications, in this sociological sense applied science, in fact, preceded pure science or abstract theoretical research later fostered by the Greeks.

State and temple authorities patronized the acquisition and application of knowledge by cadres of learned scribes. The early states all created and maintained bureaucracies and a bureaucratic civil service which, in some measure, dealt with knowledge of mathematics and the natural world. A number of bureaucratic institutions prevailed in Mesopotamian city-states which employed learned civil servants, court astrologers, and specialized calendar keepers. Similarly in ancient Egypt, expert knowledge was institutionalized in the "House of Life," a scriptorium and center of learning that primarily maintained ritual knowledge and customs, but that harbored magical, medical, astronomical, mathematical, and possibly other lore and expertise. Archival halls and temple libraries also existed, and the record speaks of Egyptian savants, hierarchies of court doctors, magicians, and learned priests.

Again and again, higher learning with practical applications was supported by state and temple authorities and deployed to maintain the state and its agricultural economy. Knowledge became the concern of cadres of professional experts employed in state institutions whose efforts were bent to the service of sustaining society rather than to any individualistic craving for discovery. An additional characteristic of this bureaucratic pattern of science is the fact that scribal experts were anonymous; not a single biography of the individuals who over hundreds of years contributed to science in the first civilizations has come down to us.

Another odd characteristic of the first scientific traditions seems to be a penchant to record knowledge in the form of lists rather than in any analytical system of theorems or generalizations. Science in the first civilizations was characteristically pursued with a notable lack of abstraction or generality and without any of the naturalistic theory or the goal of knowledge as an end in its own right that the Greeks later emphasized.

Writing and reckoning were first and foremost practical technologies with practical origins meeting the practical needs of early civilizations. Centralized authority and bureaucracies responsible for redistributing large surpluses required the recording of verbal and quantitative information. All early civilizations developed arithmetical systems and systems of permanent record keeping. The archaeological discovery of what amount to ancient Mesopotamian invoices—insignia sealed in clay—underscores the economic and utilitarian roots of writing and reckoning. Eighty-five percent of cuneiform tablets uncovered at Uruk (3000 B.C.), for example, represent economic records, and Egyptian temple and palace records are similar. Ultimately writing came to supplant oral traditions and the skills and techniques of human memory. While the vast majority of early written records concern economic, legal, commercial, votive/religious, and administrative affairs, a significant literary component also came into being.

The scribal art was highly valued everywhere, and its practitioners enjoyed high social status. Educated scribes made up a privileged caste patronized by palace or temple, and literacy offered a pathway to power. It led to employment in huge and varied bureaucracies and often to high status in government. The large bureaucracies of the hydraulic civilizations, many of which left continuous records over thousands of years, provided civil service careers for junior and senior administrators, as well as specialized posts in specialized institutions as accountants, astrologer/astronomers, mathematicians, doctors, engineers, and teachers. No wonder that novice scribes were the sons (and occasionally the daughters) of the elite.

Civilization brought with it the first schools, institutions where writing was formally taught. In Mesopotamia scribal schools known as the é-dubba or "tablet house" taught writing, mathematics, and later a literature of myths and sayings. Many Mesopotamian tablets record the countless writing and calculating exercises performed by generations of students in schools that operated in the same location teaching the same curriculum for a thousand years and longer. In Egypt, writing was institutionalized in scribal schools and other institutions that contained scriptoria and libraries, and student exercises form a large part of the written records that have survived.

Although writing and recordkeeping are characteristic features of all civilizations, writing systems have varied considerably. The earliest, the cuneiform system of writing on clay tablets, arose with Sumerian

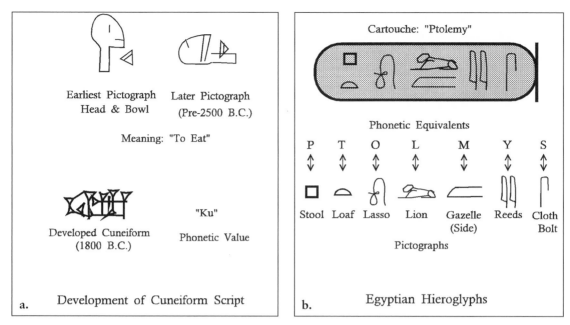

Earliest Pictograph
Head & Bowl

Later Pictograph
(Pre-2500 B.C.)

Meaning: "To Eat"

Developed Cuneiform
(1800 B.C.)

"Ku"

Phonetic Value

a.   Development of Cuneiform Script

Cartouche: "Ptolemy"

Phonetic Equivalents

P	T	O	L	M	Y	S

Stool	Loaf	Lasso	Lion	Gazelle (Side)	Reeds	Cloth Bolt

Pictographs

b.   Egyptian Hieroglyphs

Fig. 3.5a–b. Babylonian and Egyptian writing systems. Different civilizations developed different techniques for recording information in writing. Most began with representations called pictographs. Many later used signs to represent the sounds of a spoken language.

civilization in ancient Mesopotamia. Over the millennia of Mesopotamian civilization innumerable cuneiform clay tablets were dried or baked, stored, and catalogued in great libraries and archives, with tens of thousands ultimately preserved. Cuneiform—or wedge writing—is so called because Sumerian scribes used a wedge-shaped reed stylus to inscribe clay tablets. Sumerian scribes in the third millennium B.C. self-consciously developed a sophisticated system of 600–1,000 signs, where signs (called ideograms) represent the idea of a word or an action, as in "I ♥ my dog." Later, the number of Sumerian characters was reduced, but the scribal art remained very difficult to master and literacy remained restricted to a scribal profession. Cuneiform signs assumed sound (or phonographic) values at an early period and were written as syllables voicing the Sumerian language. Indeed, Old Babylonian (Akkadian), a different language from the original Sumerian, came to be written using Sumerian phonetic values. In other words, pictographs originally pictured things, whereas the signs later came to represent sounds of spoken languages. Sumerian continued to be taught in the é-dubba as a dead language after the eighteenth century B.C., similar to the way Latin was taught in European universities until the nineteenth and twentieth centuries. Sumerian and Babylonian languages had written grammars, and many tablets record word lists, bilingual lexicons, and bilingual texts.

Pictographic writing is known in Egypt from predynastic times, and the hieroglyphs ("sacred carvings") of ancient Egypt were used by the first dynasty, around 3000 B.C. The idea of writing may have passed from Mesopotamia, but specific Egyptian writing developed independently. Hieroglyphs are ideographic, but from an early period Egyptian writing incorporated phonographic elements voicing the Egyptian language. Six thousand formal Egyptian hieroglyphs have been identified, but pharaonic engravers and scribes commonly used only 700–800 across the millennia. Formal hieroglyphs were obviously not easy to write, so scribes developed simpler scripts (called hieratic and demotic) for the day-to-day maintenance of Egyptian civilization. (Among the technologies that made this possible was papyrus paper.) The last hieroglyphic inscription dates from A.D. 394, after which knowledge of ancient Egyptian writing was lost. Only the acclaimed Rosetta stone—an inscription dated to 196 B.C. with its text written in hieroglyphics, demotic, and Greek—discovered by Napoleon's soldiers in 1799 and deciphered by J.-F. Champollion in 1824—allows us to read again the records of the ancient Egyptian scribes. It should also be noted that purely phonetic alphabets where the sign stands only for a vowel or consonant sound—such as the Greek or Roman alphabets—are a late historical development of secondary civilizations, first appearing after 1100 B.C. with the Phoenicians.

Reckoning

Mathematical methods developed along with writing and out of the same practical needs. The ancient Greek historian Herodotus made the point when he placed the origins of geometry (or "earth measure") in Egypt and the need to resurvey fields after the Nile floods. Along these lines, with the agricultural surpluses generated by irrigation agriculture came the first money (in ancient Babylonia and in Shang dynasty China) and the first standardized weights and measures (in ancient Egypt, the Indus River Valley, and in early China). Although pure mathematics later became an abstract game played by mathematicians, the practical, economic, and craft roots of early mathematics remain visible in these applications.

As for the early civilizations developed its own system of mathematics. The ancient Sumerians and Babylonians evolved a sexigesimal or base-60 system (in contrast with our own decimal or base-10 system). Although not entirely consistent and initially lacking a zero, it was the first place-value system, where the "digits" represented powers of 60. Sexigesimal remnants can be found today in the 60-minute hour, the 60-second minute, and the 360 degrees of the circle. In contrast, Egyptian numbers resembled later Roman numerals with separate signs for the decimal numbers and no place value. Such a number system was more cumbersome and less efficient in handling the calculating requirements of Egyptian civilization.

As for mathematical operations, Babylonian mathematicians, using tables of numbers—multiples, reciprocals, squares, cubes, Pythagorean triplets, and the like—could perform many complex calculations, including recipe-like procedures that calculated compound interest and

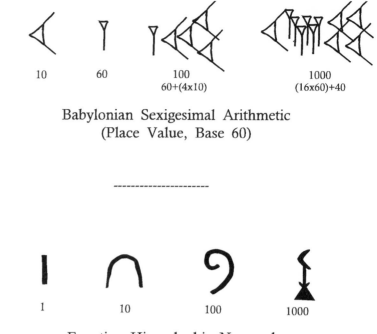

Fig. 3.6. Babylonian and Egyptian number systems. Different civilizations developed different numeral systems and reckoning methods. The Babylonian system was a base-60, place-value system with number signs for the values 1 and 10. Egyptian hieroglyphic numerals represented values of 10 in a manner reminiscent of later Roman numerals. No civilization was without a system to record numerical information.

Babylonian Sexigesimal Arithmetic
(Place Value, Base 60)

Egyptian Hieroglyphic Numerals

solved quadratic and cubic equations. In ancient Egypt, the "method of duplication," that is, the process of multiplication by doubling and redoubling numbers, was especially handy with a Roman-style number system. Egyptian mathematicians arrived at a superior estimate of the value of π (256/81 or 3.16 compared to the rough value of 3 of Babylonian mathematics and the Bible), and they developed tables that facilitated working with fractions.

In every early civilization the problems tackled by mathematicians reflect the practical and utilitarian direction of their interests. Engineering and supply problems predominated, generally solved by mathematical recipes involving little or no abstract understanding of numbers. The solution was usually arrived at recipe-style ("add 2 cups of sugar, 1 cup of milk, etc."), much like a computer program would handle the underlying equation ("square a, multiply $a \times b$, add a^2 and ab"). Although we do not know how the recipes were concocted, they were computationally sound and gave correct answers.

The Greeks had yet to invent abstract mathematics, but in a few restricted instances some very esoteric nonutilitarian "playfulness" becomes apparent in the achievements of the early scribes. In Babylonia, for example, mathematicians calculated the square root of 2 to the equivalent of six decimal places, beyond any conceivable need in engineering or reckoning. Similarly in China expert mathematicians computed π to the very high and, practically speaking, useless accuracy of seven decimal places. However, as interesting as they are, even these steps toward abstract mathematics developed in the context of broad programs of study directed at practical ends. In ancient Mesopotamia tables of exponential functions that would appear to be as abstract as an excessively accurate approximation of the square root of 2 were, in fact, used to calculate compound interest, and "quadratic equations" were solved in connection with other problems. Linear equations were solved to determine shares of inheritance and the division of fields. Lists of coefficients for building materials may have been used for the quick calculation of carrying loads. Coefficients for precious metals and for economic goods presumably had equally practical applications. And calculation of volumes reflected no idle interest in geometry but was applied in the construction of canals and other components of the infrastructure.

Time, the Gods, and the Heavens

All agricultural civilizations developed calendrical systems based on astronomical observations, and in several of the first civilizations we can identify what can only be called sophisticated astronomical research. The utility and necessity of accurate calendars in agrarian societies seems self-evident, not only for agricultural purposes, but also for regulating ritual activities. The commercial and economic role of the calendar in, for example, dating contracts and future transactions likewise seems clear.

In Mesopotamia a highly accurate calendar was in place by 1000 B.C., and by 300 B.C. Mesopotamian calendrical experts had created a mathematically abstract calendar valid for centuries ahead. Since they had adopted lunar calendars of 12 lunar months or 354 days, which is obviously out of sync with the solar year of 365¼ days, an extra lunar month occasionally had to be inserted (or intercalated) to keep lunar months and (seasonal) solar years in harmony; Babylonian astronomers inserted seven intercalary months over periods of 19 years. Ancient Egyptian priest/astronomers maintained two different lunar calendars, but a third solar/civil calendar governed official Egyptian life. That calendar consisted of 12 months of 30 days and five festival days. Each year the 365-day civil calendar thus deviated from the solar year by one-quarter day; and so over the long course of Egyptian history the civil year drifted backward and every 1460 years (4 times 365) completely circled the solar/agricultural year. The civil and solar calendars thus coincided in 2770 B.C. and again in 1310. This unwieldy calendric confusion is resolved when one remembers that the central event in Egypt—the annual, highly regular Nile flood—could be predicted independently from the seasonal first appearance of the star Sirius above the horizon.

Calendars, astronomy, astrology, meteorology, and magic formed part of a general pattern, repeated in Mesopotamia, Egypt, India, China, and the Americas. Despite our modern biases it is not possible or justifiable to separate astronomy from astrology or astronomers from astrologers and magicians in these early civilizations, for the enterprises formed an inseparable unity. In predicting the fate of crops, the outcome of military action, or the future affairs of the king, astrology and occult learning were universally seen as useful knowledge. Indeed, along with calendrical astronomy (which, after all, predicted the seasons), they constituted the first applied sciences.

Of all the ancient scientific traditions, Babylonian astronomy was the best developed, and it merits detailed attention. In ancient Babylonia a shift in divination from reading the entrails of animals to an astral religion may have encouraged the study of the heavens. Astronomical observations were recorded as early as 2000 B.C., and continuous observations date from 747 B.C. By the fifth century B.C. Babylonian astronomers could track the principal heavenly bodies indefinitely into the future. Mesopotamian astronomers fully mastered solstices, equinoxes, and the cycles of the sun and moon. In particular, later Babylonian astronomy understood and could predict solar and lunar eclipses and eclipse magnitudes. Astronomers computed and extrapolated the risings, settings, and visibility of planets, especially Venus as a morning and evening star. The legacy of Babylonian astronomy and the sexigesimal system was great, not only for our measure of the circle in degrees, but also for the seven-day week and the identification of the planets. Indeed, many technical procedures of Babylonian astronomy were handed down and adopted by later Greek and Hellenistic astronomers. What needs emphasis here is the *research* conducted by Babylonian astronomers. Obviously, they observed the heavens, no doubt with sighting instruments, and kept accurate records. We now know that they did much more than observe and keep records; they also conducted systematic research to solve very specific scientific problems in astronomy.

It is instructive to examine the "new moon problem" as a case in point. For calendrical and religious reasons Babylonian astronomers needed to know the length of the lunar month in days. The interval between full moons or new moons varies between 29 and 30 days (the average is 29.53 days). Which was it going to be in any given month? Several independent variables affect the outcome: the relative distance between the sun and moon in the heavens as seen from the earth (*AB* on the figure), the season of the year (α), and longer-term lunar cycles (*CD*). With these independent variables at play the reappearance of the new moon obviously becomes difficult to predict. Babylonian astronomers conducted research and mastered the "new moon problem" to the point of being able to create exact astronomical tables that reliably predicted when a new moon would be visible. The "new moon problem" indicates active scientific research by ancient Babylonian astronomers on a very specific problem (29 or 30 days?). This research was based on observation, mathematical analysis, and modeling of the phe-

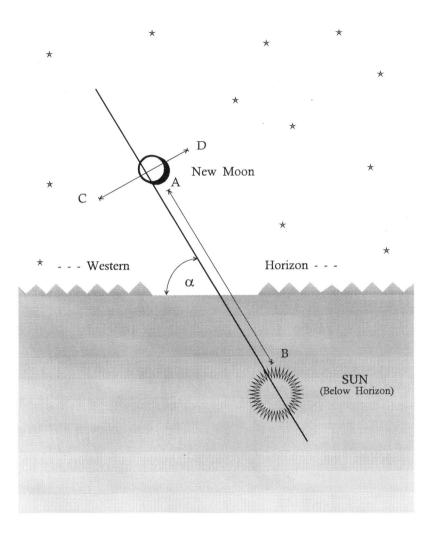

Fig. 3.7. The earliest scientific research. Ancient Babylonian astronomers systematically investigated variables determining the first appearance of the new moon each lunar month. More than simply observing the phenomena, Babylonian astronomers investigated patterns in the variations of these factors, reflecting a maturing science of astronomy.

nomena, and was theoretical insofar as more attention was paid to the abstract models of mathematical cycles than to what was visibly going on in the heavens.

Medicine and the social organization of medicine also formed a distinct feature of the bureaucratic model of state support for useful knowledge. Cadres of official medical practitioners appeared in every early state, and their practical and empirical knowledge of anatomy, surgery, and herbal medicines grew as a result of state support for medical learning. The Edwin Smith medical papyrus from the Egyptian New Kingdom (ca. 1200 B.C.) is often cited for its "rational," nontheistic approaches to medical cases.

Similarly, alchemy and alchemical expertise began to be patronized at an early date in the first civilizations; the roots of alchemy doubtless lay in the practice of ancient metallurgy, a case, if ever there was one, of technology giving rise to science. Alchemy, like astrology, offered the promise of utility, and the theme of state support for alchemy winds its way through all cultures until the modern era. The distinction that we draw between the rational and the pseudoscientific was not recognized. All of these investigations seemed to be fields of useful knowledge.

A cautious word needs to be added about the cosmologies and worldviews of the earliest civilizations. It seems safe to assume that these were all societies in which religion played a prominent role. For the most part their heavens were divine, magical, and inhabited by gods; heavenly bodies were often associated with sacred deities, and the heavens embodied myths and stories of gods. Thus in Egypt, the goddess Nut held up the sky, and deceased pharaohs became stars. In Babylonia the movement of the planets represented the movement of celestial gods. In ancient Mesoamerica, according to the Maya, the earth was a giant reptile floating in a pond. The Chinese held more organic and less pantheistic views of the cosmos. But none of the first civilizations developed any theoretical models of the cosmos as a whole, certainly no abstract, mechanical, or naturalistic ones. Little is recognizable in these cultures as independent naturalistic inquiries into the natural world or as a conception of "nature" to be studied abstractly.

The first civilizations tended to treat knowledge extensively, by drawing up encyclopedic tables and lists of words, numbers, gods, plants, animals, stones, cities, rulers, occupations, or scribes, sometimes indiscriminately. This manner of coping with and recording knowledge—what has been called the "science of lists"—may have been favored generally in societies that had not yet discovered formal logic and analytical thought. The laborious drudgery that went into them, intellectually unrewarding to the individuals who compiled the data, may have been possible only where the state patronized battalions of scribes as civil servants.

In sum, deriving from practical necessity, science repeatedly emerged part and parcel with civilization. Writing and arithmetic were new technologies applicable to the solution of many practical problems. Institutions and the institutionalized status of specialized experts underwritten by the state served the same utilitarian purposes. The evidence of advanced calendars, sophisticated astronomical puzzle-solving, and occasional mathematical "playfulness" make plain the high level of scientific accomplishment in the first civilizations. Lacking was the abstract dimension of theory that we recognize as a further hallmark of science. What has to be explained, therefore, is the origin of scientific theory and the pursuit of natural knowledge for its own sake, what came to be called natural philosophy—the philosophy of nature. If science in the form of mathematics and astronomy arose independently and many times over with the first civilizations, natural philosophy originated uniquely with the Greeks.

CHAPTER 4

Greeks Bearing Gifts

Ancient history displays a remarkable singularity in what has sometimes been termed the "Greek miracle." Just to the west of the Near
Eastern civilizations, around the shores of the Aegean Sea, Greek-
speaking peoples originated a unique civilization.

Given its proximity to Egypt and Mesopotamia, Greek civilization
derived some of its traits from its older neighbors. But those traits took
root in a habitat sharply different from the semiarid flood plains of
Egypt and Mesopotamia. Instead of a centralized kingdom, Greek civilization arose as a set of decentralized city-states, and it retained its
loose structure until Alexander the Great (356–323 B.C.) unified Greece
in the fourth century B.C. Its pre-imperial period, from 600 to 300 B.C.,
is known as the Hellenic era, while the period following Alexander's
conquests has been designated as the Hellenistic.

During the Hellenic period Greek science took an unprecedented
turn as natural philosophers, unsupported by the state and uncommit-
ted to any program of useful knowledge, developed a series of abstract
speculations about the natural world. Then, with Alexander's conquest of the wealthy districts of the East, Greek science entered its
Golden Age through a merger of its theoretical spirit with the bureaucratic pattern of institutional patronage.

Several features characterize Hellenic science. The most remarkable
was the Greek invention of scientific theory—"natural philosophy" or
the philosophy of nature. Early Greek speculations on the cosmos and
the disinterested Hellenic quest for abstract knowledge were unprecedented endeavors. They added a fundamental new element to the definition of science and shifted the direction of its history. In launching
their novel intellectual enterprise, early Greek natural philosophers
raised fundamental questions that proved highly influential and continue to be asked today.

A second notable feature of Hellenic science concerns its institutional
status. At least in the period down to Alexander the Great, state patron-

age for Greek science did not exist and, unlike the Near East, there were no scientific institutions. Some informal "schools"—intellectually very important ones—did appear in classical Greek culture, but these operated more in the vein of private associations or clubs rather than educational institutions. No public support or funding existed for schools of higher learning, libraries, or observatories, nor did scientists or natural philosophers receive public employment. Quite unlike his state-sponsored counterpart, the Greek natural philosopher was an independent operator. Although we know little of their private lives, it appears that early natural philosophers either possessed independent wealth or earned a living as private teachers, doctors, or engineers since there was no social role for natural philosophers or scientists as such. Hellenic science thus floated in a sociological vacuum to the point where the utterly impractical and apparently meaningless private investigations of its practitioners sometimes excited animosity and ridicule.

In the East, knowledge had been turned to practical ends and purposes. But in Hellenic Greece a distinctive ideology stressed the philosophical dimension of knowledge and a detachment from any social or economic objectives. In an influential passage in his *Republic* (c. 390 B.C.), for example, Plato mocks the idea that one should study geometry or astronomy in pursuit of practical benefits for agriculture, military affairs, navigation, or the calendar. Plato insisted on separating the pursuit of natural knowledge from the lesser activities of the crafts and technology. In this regard it might be said that the Greeks undertook natural philosophy as play or recreation or to fulfill higher goals concerning the life of reason and philosophic contemplation. By contrast, no comparable disinterested intellectual endeavor had been evident in the scientific cultures of the ancient hydraulic civilizations. Finally in this connection, whereas a utilitarian pattern appeared in each of the pristine civilizations, Hellenic natural philosophy appeared once, in Hellas, the result of a singular set of historical circumstances. In sum, Hellenic natural knowledge represents a new sort of science and scientific activity—self-consciously theoretical inquiries into nature.

Recent research, while not taking away from the glories of early Greek natural philosophy, has tended to set the Greek scientific enterprise in a larger, more pluralistic cultural context. It used to be thought, for example, that science and rationality arose almost miraculously from the dark world of religion and myth prevailing before the Hellenic. Today, historians emphasize that ancient Greece was not culturally insulated from the East or from the "barbarian" world beyond Greece itself. In particular, recent interpretations stress the influence of Egyptian civilization on the development of Hellenic culture around the Aegean Sea. Within the Hellenic world the continuation of popular beliefs in magic, folklore, alchemy, astrology, and religious mysticism of one variety or another represented intellectual competition to relatively secularized scientific knowledge.

Roots

The appearance of Greek science and natural philosophy may seem less surprising than it once did, but the question remains of how to account for the rise of natural philosophy in ancient Greece. Greece was a so-called secondary civilization, arising on the periphery of Egypt and Mesopotamia, but ecologically and economically very different from the principal centers of civilization in the Near East and elsewhere. (See Map 4.1.) Whereas these pristine civilizations arose on the basis of hydraulic agriculture, food production and farming in the Greek city-states depended almost wholly on seasonal rainfall and runoff from mountain snow. The Greeks did not disdain waterworks, as research has shown, but these remained small scale since Greece lacked a great river and a large, productive flood plain. Furthermore, Neolithic deforestation and erosion had already degraded the ecology and productive capabilities of Greece to the extent that only comparatively low population densities could be supported. The spawning of scores of Greek colonies by a constant flow of emigrants around the Mediterranean in the eighth through sixth centuries B.C. testifies to these ecological and cultural pressures. Classical Greece could not feed itself and depended on grain imports from abroad. The relatively poor agrarian economy of ancient Greece sustained itself on goat and sheep husbandry and on cultivating olive trees and grapevines which flourish on marginal soils by tapping subsurface water. The secondary products of wine and olive oil gave the Greeks something to trade and, as a result, Hellenic civilization acquired a maritime, mercantile, and outward-looking cast.

Just as the mountains of Greece compartmentalized the land in separate valleys, Hellenic civilization was politically decentralized and fragmented into small, independent city-states. The government of a city-state in a region with a limited and eroded agricultural base could never concentrate enormous wealth like that of an Egyptian pharaoh to patronize a pervasive bureaucracy that bent every social and cultural activity toward the interests of the state.

The Greeks are famous for the level of their political debate about law and justice and for their analysis of kingdoms, aristocracies, democracies, tyrannies, and the like. A small step separates rational debate about political constitutions from inquiring into the constitution of nature—and vice versa, as the later history of science was to show. These political debates may indeed have provided one route to the origins of Greek science. It may be impossible to reach an understanding of exactly *why* a new scientific culture came into being in the unique habitat of Hellas. (If Ionia and Athens had remained as bereft of science as, say, Corinth and Sparta, would there be any grounds for surprise?) But once a scientific culture arose in ancient Greece it was shaped by a society that attached no social value to scientific research or in-

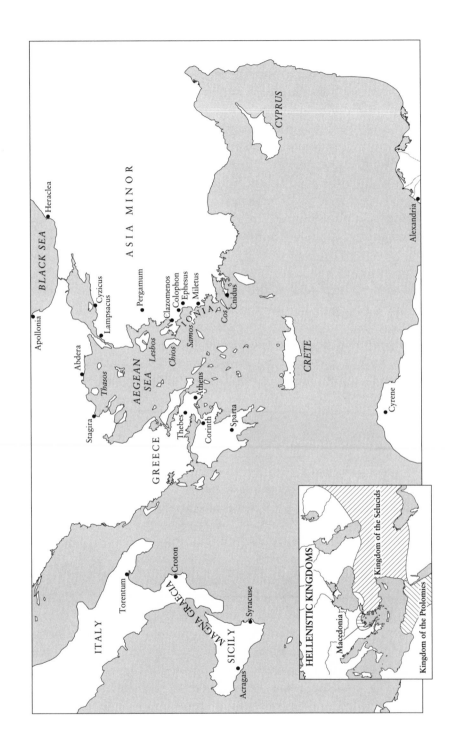

BLACK SEA

Heraclea

ASIA MINOR

CYPRUS

Apollonia

Cyzicus
Lampsacus

Pergamum

Abdera

Alexandria

Clazomenos
Colophon
Ephesus
Miletus

Stagira

Thasos

Lesbos

Chios

IONIA

Samos

Cos

Cnidus

AEGEAN
SEA

Athens

CRETE

GREECE

Thebes

Corinth
Sparta

Cyrene

ITALY

Torentum

MAGNA GRAECIA

Croton

Syracuse

SICILY

Acragas

HELLENISTIC KINGDOMS

Macedonia

Kingdom of the Selucids

Kingdom of the Ptolomies

struction and that provided no public support for schools of higher learning.

Greek science did not originate in Greece, but in Asia Minor on the (then-) fertile Mediterranean coast of present-day Turkey, at first in the city of Miletus and later in several other cities of the region known as Ionia. In the seventh century B.C. Ionia was the center of Greek civilization while the Greek mainland was decidedly the province. Lying on the eastern shores of the Aegean, it had more fertile land and received more rainfall than mainland Greece. Ionia remained more urbanized and economically superior to Greece proper for two centuries. Not surprisingly, the majority of the first natural philosophers hailed from Ionia.

The Ionians and the entire collection of early Greek natural philosophers are known as the pre-Socratics, that is, thinkers active in the formative period of Greek philosophical and scientific thought before Socrates (470?–399 B.C.). (See Table 4.1.) Greek natural philosophy is usually said to begin with Thales of Miletus, who lived from about 625 to about 545 B.C. Thales is a test case for historical interpretation, for we have nothing from Thales himself and are wholly dependent on secondary reports. Our view of Thales is thus refracted through both the biases of ancient commentators and our own interpretative frames. We do know that he came from Miletus, a vibrant trading city on the Ionian coast of Asia Minor, and that he was later crowned as one of the seven "wise men" of archaic Greece, along with his contemporary, the lawgiver Solon. Thales was probably rich, and he probably traveled to Egypt, from where he is said to have brought geometry to the Greek-speaking world. As Plato reports, maliciously perhaps, Thales and his philosophy earned a reputation for unworldliness: "A servant-girl is said to have mocked Thales for falling into a well while he was observing the stars and gazing upwards, declaring that he was eager to know things in the sky, but that what was behind him and just by his feet escaped his notice." By the same token, according to Aristotle, Thales exploited his knowledge of nature through an astute scientific observation of a forthcoming harvest in order to corner the market on olive presses and thus to demonstrate that philosophers could be rich and useful, if those were their concerns. Thales allegedly also applied his acute scientific knowledge in wartime to help King Croesus ford a river in 547 B.C. In the end, the social role of wise man or magus probably befits Thales better than that of the "first scientist," which he is often called, if by "scientist" one has more modern social models in mind.

That we know Thales's name and these details about his life unexpectedly reveals something significant about his natural philosophy and about the subsequent development of science. Thales's claims about nature were just that, *his* claims, made on his own authority as an individual (with or without other support). Put another way, in the tradition stemming from Greek science, ideas are the intellectual property

Map 4.1. The world of ancient Greece. Greek civilization originated as a cluster of small city-states around the Aegean Sea. Greek science first arose in towns along the Ionian coast of Asia Minor. After the conquests of Alexander the Great in the fourth century B.C. the Greek world stretched from Egypt to the borders of China, forming the largest empire in the ancient world. After Alexander's death in 323, his empire *(inset)* collapsed into three states: Macedonian Greece, Ptolemaic Egypt, and the Seleucid Kingdom in Mesopotamia. *(opposite)*

Table 4.1
The Pre-Socratic Natural Philosophers

The Milesians	
Thales	fl. 585 B.C.
Anaximander	fl. 555 B.C.
Anaximenes	fl. 535 B.C.
Empedocles of Acragas	fl. 445 B.C.
The Pythagoreans	
Pythagoras of Samos	fl. 525 B.C.
Philosophers of Change	
Heraclitus of Ephesus	fl. 500 B.C.
Parmenides of Elea	fl. 480 B.C.
The Atomists	
Leucippus of Miletus	fl. 435 B.C.
Democritus of Abdera	fl. 410 B.C.
Socrates of Athens	470?–399 B.C.
Plato of Athens	428–347 B.C.
Aristotle of Stagira	384–322 B.C.

of individuals (or, less often, close-knit groups) who take responsibility and are assigned credit (sometimes by naming laws after them) for their contributions. This circumstance is in sharp contrast with the anonymity of scientists in the ancient bureaucratic kingdoms and, in fact, in all pre-Greek civilizations.

Thales made claims about nature, including his idea that the south-blowing Etesian winds cause the Nile flood. Another theory of his held that the earth floats on water like a log or a ship and that the earth quakes when rocked by some movement of the water. Only a hundred years after Thales, Herodotus savagely attacked these ideas, and to the modern scientific mind they may seem oddly primitive notions. But they are nonetheless extraordinary in several important regards. For one, the explanations offered by Thales are entirely general; they seek to account for *all* earthquakes and *all* Nile floods, and not only a single case. In a related way, Thales invokes no gods or supernatural entities in his explanations; to use the stock phrase, he "leaves the gods out." Thus, "hail ruined my olive crop" not as punishment because I offended Zeus or Hera in a particular instance, but accidentally because in all instances—mine unfortunately included—hail results from natural processes that involve the freezing of water in the atmosphere. Note that a feature of Greek natural philosophy—the "discovery of nature"—required objectifying and demystifying nature, in order that theories might be proposed regarding nature in the first place. That is, "nature" had to be defined as an entity to be investigated; the concept may appear self-evident to us, but it was not necessarily so to our scientific forebears. "Naturalistic" explanations first posit the phenomenon in question to be a regular part of some external nature and hence a natural phenomenon, then account for the phenomenon also in terms of nature. Thus for the Nile, naturally occurring winds are invoked to explain the natural phenomenon of flooding. Interestingly in the case of earthquakes, Thales employs analogies to what we see in the world (ships, floating logs) in his explanation. It is far from the case, however, that Thales or (most of) his successors were atheists or irreligious; in fact, Thales also taught that the world is divine and "full of gods" and that the magnet possesses a "soul." There is no contradiction, however, for despite whatever homage is due the gods, Thales sets the natural world off both as somehow separate from the divine and as something comprehensible by the powers of human intellect.

Thales is known for his view that the world is composed of a primordial watery substrate. This deceptively simple pronouncement represents the first attempt to say something about the material "stuff" making up the world around us. It marks the beginning of matter theory—that line of scientific theorizing concerned with the makeup of the physical world below the level of ordinary perception. In asking about the material basis of things early in the sixth century B.C., Thales became the founding father of the first of the "schools" of Greek natural philosophy mentioned above, the Milesians. This Milesian school and its tradition of matter theory are an important element of pre-Socratic thought, but consideration of the intellectual dynamic driving the Milesians reveals another feature of the enterprise of early Greek science: the rise of science as rational debate. In a word, the Milesian philosophers disagreed, and they used reason, logic, and observation to attack the ideas of others and to bolster their own propositions.

Thales's notion that water is the primary substance had its problems, notably to explain how water could give rise to fire, its opposite, water and fire being mutually destructive, as in fire boiling away water or water quenching fire. Anaximander of Miletus (fl. 555 B.C.) dealt with this problem in the generation following Thales by rejecting water as the underlying agent and by putting forth the much vaguer notion of some "boundless" or formless initial state (the *Apeiron*) out of which duality and the world grew. By allowing duality to emerge from unity, as it were, Anaximander's "boundless" explained hot and cold, which Thales could not, but the concept of the "boundless" remained forbiddingly abstract and metaphysical. The next Milesian, Anaximenes, responded to this difficulty and to the same general question around 535 B.C. His answer was to suggest air (or the "pneuma") as the primeval element. More down to earth, this suggestion would also seem to suffer from the problem of opposites that troubled Thales's water theory, except that Anaximenes posited two conflicting *forces* in the universe, rarefaction and condensation, which variously condensed air into liquids and solids and rarefied it into fire. The tradition of the Milesian school culminated a century later with the thought of Empedocles (fl. 445 B.C.), who as an adult lived in Greek Italy. In a theory that remained influential for 2,000 years Empedocles postulated four primary elements—earth, air, fire, and water—and the attracting and repelling forces of (what else?) Love and Strife.

The pluralistic and abstract character of natural knowledge among the early Greeks is no better illustrated than by another pre-Socratic "school," the cult of the Pythagoreans. The Pythagoreans, centered in Italy, formed an organized religious brotherhood and sect, and individual innovator-adepts submerged their contributions to the collectivity by giving credit to their founding guru, Pythagoras (fl. 525 B.C.), originally from the island of Samos off the Ionian coast. The Pythagoreans embodied a certain "orientalism" reminiscent of the master's sixth-century Persian contemporary, Zoroaster.

The Pythagoreans are famed for introducing mathematics into natural philosophy. Their mathematics was not the crude arithmetic of the marketplace or the practical geometrical procedures of the surveyor or architect, or even the exact mathematical tools of Babylonian astronomers. Rather, the Pythagoreans elevated mathematics to the level of the abstract and the theoretical, and they made the concept of number central to their view of nature. In its way, number was the Pythagorean response to the Milesian question about the material stuff of the world. In focusing on number, the Pythagoreans introduced potent notions of idealism into natural philosophy and science—the idea that some more perfect reality accessible through intellectual understanding underlies the observed world of appearances. Put crudely, the real world contains no perfect triangles, no absolutely straight lines, or numerical abstractions; such entities exist only in the realm of pure mathematics. That the Pythagoreans and their intellectual successors thought that such mathematical perfection somehow constitutes the world (or even that it is useful to think so) inaugurated a whole new way of thinking about nature, and it launched the great tradition of mathematical idealism that has been so powerful a current in scientific thought since then.

Pythagoras is supposed to have achieved the profound insight of mathematical order in the universe in considering musical strings and the tones they sound; half the length producing the octave above, one-third producing the higher fifth tone, and so on. Based on this unexpected correlation between small integers and the real world, Pythagoras and his followers extended their mathematical investigations. Some of their results, such as their classification of odd and even numbers, seem unexceptional to us; others, such as a sacred triangle (the *Tetratkys*) representing the sum of the numbers 1, 2, 3, and 4 (= 10), or their association of the institution of marriage with the number 5 in joining the 2 of femaleness with the 3 of maleness, reflect what we would all too easily consider a bizarre numerology.

Of course, Pythagoras is credited with the discovery of the theorem in geometry that bears his name. It says that for any right triangle (to use the algebraic formulation) $a^2 + b^2 = c^2$, where c is the hypotenuse of the triangle and a and b are the legs. Lurking in the Pythagorean theorem is a corollary that says that not all line lengths can be expressed as ratios or fractions of other unit lengths. Some pairs of lines (like a leg and the diagonal of a square) are incommensurable—that is, their ratio cannot be expressed by any pair of integers. To the Pythagoreans the square root of 2 was "alogon," the unutterable. The discovery of irrationality was subversive of the Pythagorean commitment to integers and the program of investigating mathematical harmonies in the world, and, supposedly, knowledge of the irrational was therefore held as the innermost secret of the Pythagorean cult.

The more fundamental point to be made about these discoveries is the role of mathematical proof in demonstrating their certainty. The invention of deductive reasoning and proof, wherein even the most skeptical auditor is forced step by step to the inevitable Q.E.D. ("thus proven") at the end, was a remarkable innovation in the histories of mathematics, logic, and science. The Egyptians knew of Pythagorean triplets (whole numbers obeying the Pythagorean theorem, as in 3-4-5 right triangles), and the Babylonians prepared tables listing them. But no one until the Pythagoreans saw in them a theorem to be proved. Rigorous mathematical demonstrations did not appear full-blown with the Pythagoreans, and the process of developing an axiomatic and deductive plane geometry continued until Euclid compiled his *Elements* around 300 B.C. Nevertheless, to the early Pythagoreans goes the credit for studying mathematics as natural philosophy, for turning Greek mathematics away from practical arithmetic to pure arithmetic and geometry, and for developing the proof as a means and model for justifying claims to knowledge.

The different traditions represented by the Milesians, the Pythagoreans, and their successors make plain that Greek natural philosophy in the pre-Socratic period lacked an agreed-upon unity and was fragmented into different schools of thought. In this connection two other major groups of pre-Socratic natural philosophers need to be mentioned at least briefly: the atomists and the so-called philosophers of change. The atomists, notably Leucippus of Miletus (fl. 435 B.C.) and Democritus of Abdera (fl. 410 B.C.), responded in their way to the Milesian challenge of a century earlier by imagining that the world is composed of atoms, the least reducible, indivisible particles of matter. These theorists supposed that differences in the shape, position, motion, and arrangement of atoms in the void are the root cause of the differences we see in objects around us. Ancient atomism faced a grave difficulty in explaining how random atoms could assume any coherent or lasting pattern in nature other than by cosmic accident, and atomist philosophy thereby earned a reputation for atheism. Some atomist demonstrations designed to illustrate the corporeality of air (a bottle of air held underwater) may be viewed as early scientific experiments,

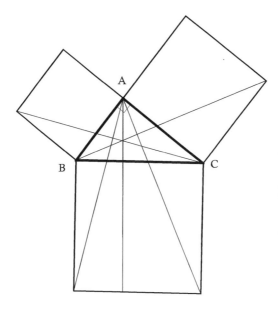

Fig. 4.1. The Pythagorean theorem. Although Pythagorean triplets (like 3-4-5) were recorded by Babylonian scribes, the Pythagorean theorem ($AB^2 + AC^2 = BC^2$) was first proved in Euclid's *Elements*. When the nineteenth-century philosopher Schopenhauer saw the diagram, he remarked, "That's not a proof, it's a mousetrap."

but ones whose purposes were to illustrate and not to test. Atomism attracted a small following, notably in the person of the Roman poet Lucretius, but the movement was decidedly a minor branch of thought until its revival in seventeenth-century Europe and the emergence of modern atomic theories in the nineteenth century. Indeed the attention usually given to ancient atomism reflects *our* interests more than the ancients'.

The pre-Socratics did not limit their inquiries to the inanimate world around them, but also initiated natural philosophical investigations of the living world. Alcmaeon of Croton (fl. 500 B.C.), for example, reportedly undertook anatomical research and dissected merely for the sake of learning.

Heraclitus of Ephesus (fl. 500 B.C.) and Parmenides of Elea (fl. 480 B.C.) are labeled the "philosophers of change" because they initiated a great debate over the nature of what we experience as change in the world. Heraclitus held that change is perpetual, that everything flows, that the same river is never crossed twice. Parmenides countered with the radical notion that nothing changes, that change is nothing but an illusion, despite the apparent evidence of our senses. The debate was important because it made explaining change central to natural philosophy. While the Milesians and the Pythagoreans do not seem to have considered the question, after Parmenides it was unavoidable: not simply the world, but apparent *flux* in the world is what natural philosophy needs to explain. The Heraclitean-Parmenidean debate also raised fundamental questions about the senses and about how we can know things. In part these questions involved the psychology of perception (e.g., the stick that seems to bend in water, the red of a red apple) and the general reliability of the senses. On another level they dealt with whether and, if so, how knowledge can be based on the senses or indeed on anything at all. The consequence for natural science was that thenceforth not only did every claim to knowledge about nature formally have to be buttressed by its own internal evidence and reasoning, but it had also to be accompanied (either implicitly or explicitly) by a separate rationale as to why *any* evidence or reasoning might support any such claims.

Whereas science in the ancient bureaucratic kingdoms had been patronized by the state and, accordingly, held to strict standards of usefulness, the work of these Greek natural philosophers was its polar opposite—theoretical, abstract, and whimsical. There was, however, one field of Greek higher learning that was more akin to the social pattern of the ancient East, the Hippocratic medical tradition that arose in the Hellenic period—the collective body of medical literature credited to the great fifth-century physician Hippocrates of Cos (fl. 425 B.C.). In the Hippocratic tradition, with its emphasis on reason, cautious observation, medical prognostication, and natural healing, one finds a good deal of natural knowledge and scientific thinking akin to that pursued by the natural philosophers. For example, articulating a view that remained influential into the nineteenth century of our era, Hippocratic theorists correlated the four elements (earth, air, fire, and water) with four bodily humors (blood, phlegm, yellow bile, and black bile), and then argued that health represents a balance between and among the humors. By the same token, the skepticism of Hippocratic medicine—the doubt that certain knowledge is even possible—set it apart from most of the speculations of natural philosophy. Ancient medicine remained more tied to practice and craft than natural philosophy, and "scientific physicians," such as they were, competed alongside many "schools" and diverse forms of healing arts that included magic, incantations, and dream cures.

Around the Greek world clearly identifiable medical institutions could be found, notably in the temples and cult centers devoted to Asclepius, the deified physician and supposed offspring of Apollo. Asclepieions, or healing centers, appeared in Cos, Epidauros, Athens, and elsewhere. Medical practice was not regulated in antiquity, and doctors were often itinerant. Medicine was a highly specialized trade, and practitioners could become wealthy. City-states contracted with doctors in wartime, but by and large Hippocratic and other doctors operated independently of the political state or any government bureaucracy.

Worlds of Pure Thought

Although early Greek natural philosophers initiated abstract inquiries into nature there was no unity to their endeavors, and nothing like sustained scientific research is evident in their traditions. That changed in the fourth century B.C. with the two great intellectual syntheses of Plato and Aristotle.

Before Plato there was no consensus in Greek cosmology or astronomical theory. Instead, the pre-Socratic tradition was notorious for the diversity of the models proposed. In the sixth century B.C. Anaximander of Miletus had hypothesized that the earth is a disk, floating naturally in space with humans living on its flat surface. There are wheels of fire in the heavens and the luminous heavenly bodies we see are really holes in the fire wheels; the wheel of the stars is closest to the earth, the wheel of the sun farthest; eclipses result from holes becoming blocked; and certain mathematical proportions govern the location of the heavenly wheels. This cosmological model is remarkable for just that, being a model—some simplified simulation of the real thing, a likeness we might construct. Anaximander's view is more sophisticated than Egyptian and Mesopotamian cosmologies as well as the succeeding model of Anaximines (who held that the earth is a table held up by air), in that Anaximander can account for what supports the earth, that is, the earth positioned in the middle of nowhere. The model of the Pythagoreans displaced the earth from the center of the cosmos and held that it (and probably the sun) went around some vague central fire and an even more mysterious counter-Earth. The mechanical and vaguely mathematical character of these models made them distinctly Greek inventions, but their advocates did not pursue any of their details.

The case of Plato of Athens (428–347 B.C.) and his geometrical astronomy carries us past the founding era of the pre-Socratics and lands us solidly in classical fourth-century Greece. Plato was a pupil of Socrates, the fifth-century master who "called philosophy down from the skies." In his youth, Socrates is said to have been interested in natural philosophy, but he concluded that nothing certain was to be learned in the study of nature, and he focused his attentions instead on examining the human experience and the good life. But he offended the political authorities and was sentenced to death. After his execution in 399 B.C., the mantle of philosophy passed to Plato, who seemingly felt better prepared to make direct statements about the natural world. Plato formalized the enterprises of philosophy and natural philosophy by establishing a private school, his Academy at Athens (which survived for 800 years). Significantly, inscribed over the portals of the Academy was the motto, "Let no one enter who is ignorant of geometry."

Geometry was important to Plato and his philosophy as a form of intellectual discipline and as a model for all that was metaphysically abstract and perfect. Geometry was also key to Plato's matter theory, as he identified the fundamental elements of earth, air, fire, water, and an extra aether with the five so-called perfect solids, three-dimensional polyhedra each with identical regular polygonal faces, which geometers had proven could be only five in number. But Plato himself was a philosopher, not a serious geometer or mathematician. Nor was he an astronomer. He did not observe the heavens, and he disdained those who did. Nevertheless, in his *Timaeus* Plato presents a fairly complex model of the heavens, involving a central earth linked mechanically along a common axis to a series of spinning shells or spheres that carry around the various heavenly bodies. A mystical part of Plato's cosmology and a common philosophical opinion for centuries held that the heavens were alive and divine. Although the cosmology was influential, in most respects it was no advance over the previous models of the pre-Socratics. In one crucial particular, however, Plato exerted a profound and lasting effect on astronomy and the history of science: he set Greek astronomers to solving problems.

Plato believed that the heavenly bodies revolve in circles around a stationary earth. He held this opinion not because he observed that the sun, moon, planets, fixed stars, and everything in the heavens move in circular arcs across the sky once every 24 hours, which sensory evidence confirms. Nor did his belief that the heavens were essentially unchanging apart from their motion rest only on the reported experience of previous generations. Rather, Plato held his views concerning celestial motion on first principles. Because of their majesty and virtually divine status Plato believed that the heavens represent an embodiment of the eternal, transcendent, and perfect world of pure Form. Plato's world of the Forms constitutes an unchanging ideal reality, of which our temporalized world is only a pale and imperfect reflection. Therefore, circular motion was the only motion suitable to the heavens because the circle is a figure of constant curvature with no beginning or end. Because they faithfully mirrored the perfection of the world of the Forms, Plato likewise concluded that the heavens must necessarily move uniformly; uniform motion does not speed up or slow down, betraying the imperfection of change, but remains constant and undeviating. Uniform circular motion of heavenly spheres was not questioned thereafter in antiquity.

While most motions in the heavens do seem to be circular, some motions are plainly not circular and equally plainly not uniform. The daily movement of the stars, the annual trip of the sun around the heavens, and the monthly revolution of the moon are apparently circular, but other movements in the heavens are not, notably the movement of the planets or "wandering stars" as observed over a period of months. Relative to the background of the fixed stars, the planets slow in their courses, stop, move backwards, stop again, and move forward again, sweeping out great, noncircular loops in the sky. This was the great problem of the "stations and retrogradations" of the planets that Plato had uppermost in mind when, to use the famous phrase, he enjoined astronomers to "save the phenomena" with circles. Explaining the stations and retrogradations of the planets was the central problem in astronomy for nearly 2,000 years from Plato's era until after Copernicus in the sixteenth century A.D.

Planetary motions presented difficulties, Plato believing the planets

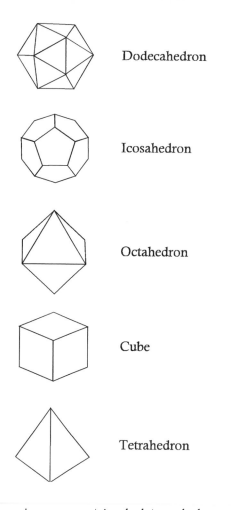

Dodecahedron

Icosahedron

Octahedron

Cube

Tetrahedron

Fig. 4.2. The Platonic solids. Plato knew that there cannot be more than these five regular shapes (each with congruent equilateral polygonal faces), and he correlated these geometrical forms with the elements.

move in one way (circularly), and observation showing they move in another way (loopingly); there was an obvious conflict to be worked out, an area for research. But note the crucial converse: there is nothing at all problematical about the observed stations and retrogradations unless, like Plato and his followers, one thought the planets ought to move otherwise than they appear, in this case with uniform circular motion. The astronomical paradigm initiated by Plato represents more than the onset of some straightforward "research" into self-evident phenomena; Plato's prior philosophical (theoretical) commitments to Forms and circles made manifest the phenomena to be investigated. Thus Plato defined a problem in natural philosophy where none existed before. But the import of Plato's paradigm in astronomy goes further: he also defined for theorists and astronomers what constituted appropriate or acceptable solutions to the problem of the planets, that is, models that use uniform circular motion to produce apparently nonuniform appearances. Nothing else qualified as a solution to the puzzle.

Fourth-century astronomers took up the problem and formed a small but distinct tradition of research in astronomy and cosmology. Plato's student Eudoxus of Cnidus (fl. 365 B.C.) was the first to respond. He proposed a model for the heavens which consisted of twenty-seven nested (homocentric) celestial spheres variously revolving around a central earth. The Eudoxean model made the universe into something akin to a grand cosmic onion. Some of the spheres were deployed to explain the apparent motion of the stars, sun, and moon, and each retrograding planet was assigned a system of four rotating spheres: one to account for daily motion, one for periodic motion through the heavens, and two, moving oppositely and tracing out a figure-8-like path of stations and retrogradations, known as the "hippopede." The model "worked," but there were problems with it. The observed inequality of the four seasons (they are not all the same length in days) was one, and to account for it a younger contemporary of Eudoxus, Callipus of Cyzicus (fl. 330 B.C.), improved on the model by adding an extra sphere for the sun and raising the number of spheres to thirty-five in all. But the model was still imperfect, notably in not being able to explain how the universe could function mechanically with all those spheres spinning just below and above each other at different rates and inclinations. In the next generation Aristotle (384–322 B.C.) tried his hand at this issue in technical astronomy, and, by inserting a number of counteracting spheres, he increased their number to fifty-five or fifty-six.

The Eudoxean model of homocentric spheres and the small research tradition associated with it hardly survived the Hellenic era, much less antiquity. In the final analysis the intellectual and conceptual problems afflicting Eudoxus's approach proved fatal. Those problems included difficulties explaining why the seasons are not the same number of days, why Venus varies in brightness, and why Venus, Mercury, and the sun should always stay close to one another. By the second century B.C. astronomers were actively considering alternatives to homocentrism, and the culmination of ancient astronomy in the work of Claudius Ptolemy (fl. A.D. 150) 500 years later shows only the vaguest relation to what Plato, Eudoxus, and their colleagues had in mind with their spinning sets of spheres.

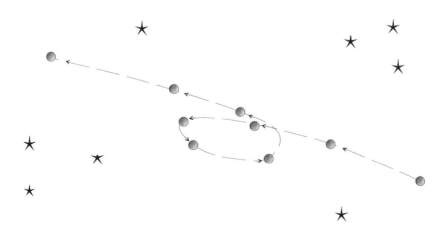

Fig. 4.3. Retrograde motion of Mars. To an observer on Earth, over a period of months Mars appears to reverse its direction as seen against the background of the fixed stars and then reverse it again to resume its forward trajectory. Accounting for these loops in terms of uniform circular motion defined a key problem that occupied astronomers for 2,000 years.

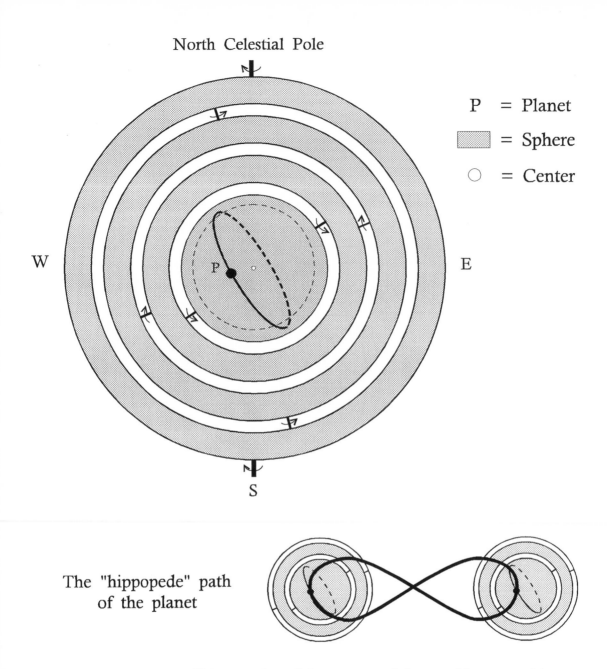

North Celestial Pole

P = Planet

▨ = Sphere

○ = Center

W

E

S

The "hippopede" path
of the planet

This research tradition was nonetheless notable in several key respects. For one, the case makes evident how much scientific research at this level depends on consensus among scientific practitioners. It makes no sense, in other words, for Eudoxus, Callipus, and Aristotle to have taken up the detailed investigations just described unless they agreed that Plato's approach was basically correct. The instance makes plain once again the community-based nature of the scientific enterprise in its Greek as well as its bureaucratic guise. In some larger sense groups, not individuals, practice science. Finally, like their counterparts among anonymous Babylonian astronomers and astrologers, Eudoxus, Callipus, and Aristotle did not simply know things about nature, they were not simply manipulating nature, and they were not simply theorizing about nature. They were checking nature out in detail, along lines established by their general philosophical, metaphysical, and theoretical commitments. The arsenal of techniques pertinent to the human inquiry into nature had expanded considerably from the first paleolithic lunar tallies.

Enter Aristotle

Aristotle marked a watershed in the history of science. His work, which encompassed logic, physics, cosmology, psychology, natural history, anatomy, metaphysics, ethics, and aesthetics, represents both the culmination of the Hellenic Enlightenment and the fountainhead of science and higher learning for the following 2,000 years. Aristotle dominated scientific traditions in late antiquity, in medieval Islam, and in early modern Europe where his science and his worldview defined scientific methodology and the research agenda up to just a few centuries ago.

Born in the town of Stagira in Thrace in northern Greece in 384 B.C., Aristotle came from a privileged family, his father being royal physician to the king of Macedonia. In his teens Aristotle went to Athens to study with Plato, and he remained in Athens as a member of the Academy for 20 years until Plato's death in 347. He then traveled around the Aegean until 343, when King Philip II of Macedonia called him to his court to tutor his son, Alexander, who became Alexander the Great. After Alexander assumed the throne and began his world conquest in 336, Aristotle returned to Athens, where he founded his own school, the Lyceum. After Alexander's early death in 323, Aristotle found it politically prudent to leave Athens; he died the following year at the age of 62. The extensive writings that we commonly regard as Aristotle's were compiled to some extent during his lifetime and to some extent by disciples during the first two centuries after his death. In any event, several entire books have survived, unlike the fragments of the natural philosophers who preceded him. Indeed, Aristotle's commentaries on their work tell us much of what we know of his predecessors.

From a sociological point of view, as with all Hellenic scientists, Aristotle's research was undirected by any state authority, and he had no institutional affiliation. The Lyceum, his own place of teaching—a grove on the outskirts of Athens—was not formally established as a school during his lifetime. He was thus in some measure a footloose professor, one of the leisured intellectuals to whom, in fact, he attributed the achievements of pure science. The substance of his studies reflected his sociological standing—utterly abstract and of no possible use in engineering, medicine, or statecraft. Although he recognized the distinction between pure and applied research, "speculative philosophers" and "[medical] practitioners," Aristotle confined his research to

Fig. 4.4. Eudoxus's system of homocentric spheres. In Eudoxus's "onion" system, Earth is at rest in the center of the universe, and each planet is nestled in a separate set of spheres that account for its daily and other periodic motions across the heavens. From the point of view of an observer on Earth, two of the spheres produce the apparent "hippopede" (or figure-eight) motion that resembles the stations and retrogradations of the planets. *(opposite)*

his private interests in the philosophy of nature. Even when he wrote on anatomy and biology, fields that might have lent themselves to useful studies applicable to the treatment of illness, he focused his objectives on the place of living beings in a rational cosmology. Similarly, his studies of the theory of motion, which remained influential until the seventeenth century, formed part of a program of purely theoretical research and were of no practical use in technical or economic applications.

Aristotle expressed himself in unambiguous terms concerning the relationship between science and technology. After humanity acquired the needed practical arts, leisured intellectuals cultivated pure science: "When everything [practical] had been already provided, those sciences were discovered which deal neither with the necessities nor with the enjoyment of life, and this took place earliest in regions where men had leisure." Curiosity provided the motivation for the development of pure science: "For men were first led to study [natural] philosophy, as indeed they are today, by wonder. . . . Thus, if they took to philosophy to escape ignorance, it is patent that they were pursuing science for the sake of knowledge itself, and not for utilitarian applications." Aristotle's opinions are thus consistent with modern studies that show the ratio of pure to applied orientations among known Hellenic scientists to have been roughly 4 to 1.

For the generations of natural philosophers who followed Aristotle the beauty and power of his achievement stemmed in large measure from the unity and universality of his worldview. He offered a comprehensive, coherent, and intellectually satisfying vision of the natural world and humanity's place in it, a vision that remains unequaled in scope and explanatory ambition.

Aristotle's physics, and indeed all of Aristotle's natural philosophy, is rightly said to represent the science of common sense. Unlike Plato's transcendentalism, Aristotle held that sensation and observation are valid—indeed, they represent the only route to knowledge. Time and again Aristotle's views conform with everyday observation and the commonplace experiences of the world we know (unlike modern science, which often contradicts plain observation and requires a reeducation of the senses before it can be accepted). Aristotle emphasized the sensible *qualities* of things, in opposition to the quantitative and transcendental approaches of the Pythagoreans or Plato's followers. Aristotle's natural philosophy was therefore more common sensical and scientifically promising.

Aristotle's theory of matter provides an easy entrée to his overall vision of the cosmos. He followed Empedocles and Plato in adhering to the four elements of earth, air, fire, and water. But unlike Plato, who believed the elements to be fashioned of abstract polyhedrons, Aristotle took them to be composed of pairs of even more fundamental *qualities:* hot, cold, wet, and dry, projected onto a theoretically quality-less "first matter" or *prima materia*. Thus, as Figure 4.5 illustrates, the qualities wet and cold make up the element water, hot and dry = fire, wet and hot = air, cold and dry = earth. Ordinary earth and all other composite bodies are mixtures of the pure elements, which are never found in an isolated state. And, again unlike Plato who found reality only in the transcendent world of the Forms, Aristotle held that the world we experience is materially real because objects in the world (such as tables and trees) are inseparable amalgamations of elemental matter and Form. Aristotle's matter theory is eminently rational and conformable to experience in, for example, explaining the boiling of water as a transformation of water into "air" by the substitution of the quality of hot for the quality of cold. In this case the application of fire replaces the hot and wet of air for the cold and wet of water. It should be noted that such a qualitative theory of the elements provides a theoretical basis for alchemy, in that qualities are projected onto a quality-less *prima materia* or "first matter" and it thus becomes theoretically possible to strip, say, lead of its qualities and substitute the qualities of gold. The theory as much as the authority of Aristotle thus legitimated the alchemical enterprise.

For Aristotle, the physics of motion—change of place—is only a special case of change or alteration in general, such as growth, fermentation, and decay. He associated a motion with each element according to its nature: earth and water, being heavy, naturally move to the center of the universe (that is, the earth); air and fire, being light, naturally move away from the center. Nothing else is required to explain this intrinsic motion, just as nothing else is required by modern physics to explain inertial motion. Accordingly, each element seeks a place in the universe, its so-called natural place: earth at the center layered with concentric shells of water, air, and fire. Thus, his theoretical analysis accords well with what we observe in nature, with lakes and oceans atop the earth, with bubbles of air rising in water, with the atmosphere atop the waters and the earth, and with fire seeming to rise in the air and meteors to spark in the sky. Indeed, theoretically, the reason concentric shells of earth, water, air, and fire that surround the cosmic center are not perfectly spherical is that the terrestrial region represents the realm of change, violence, imperfection, and corruption. On Earth things get jumbled up, unlike the perfect, unchanging, and incorruptible celestial regions. In support of these conjectures Aristotle alluded to experimental confirmation. If one attempts to submerge a bag or bladder of air in water one will sense resistance against displacing air from its natural place into the realm of water, and if the balloon is forcibly submerged and then released it will spontaneously return to the realm of air.

In Aristotle's scheme of the world, the earth we live on is essentially spherical and totally motionless at the center of the universe. If, in some extraordinary thought experiment, we could displace the earth from

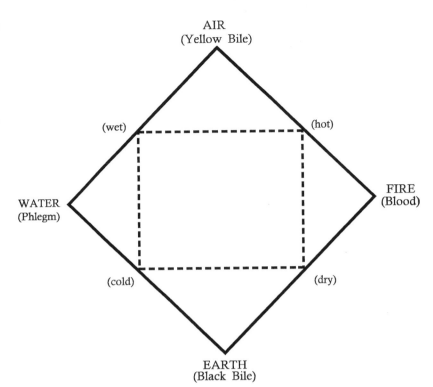

Fig. 4.5. The Aristotelian elements. In Aristotle's matter theory, pairs of qualities (hot-cold, wet-dry) define each of the four elements: earth, air, fire, and water. Substitute one quality for another, and the element changes. Each of the four elements also came to have an associated "humour," which connected Aristotle's views on matter with physiology and medical theory.

AIR
(Yellow Bile)

(wet)

(hot)

WATER
(Phlegm)

FIRE
(Blood)

(cold)

(dry)

EARTH
(Black Bile)

the center, it would naturally return to the center and reassemble there, just as stones fall back to the center through air and through water in order to return to their natural place. Thus, Aristotle's geocentric cosmology—the idea that the spherical earth remains unmoving at the center of the cosmos—is backed up by the authority of physics and confirms our sensory experience of the earth at rest and the heavens in motion. Aristotle confirmed the sphericity of the earth, for example, from the shadow it casts on the moon during lunar eclipses; and he offered common-sense arguments against a moving earth, such as the observation that a ball thrown straight up falls back to its point of origin and is not left behind as the earth turns underneath.

Since different natural motions occur in the region of the earth (either up or down) and the region of the heavens (always circular), Aristotle's cosmology makes a sharp distinction between the physics of the two regions. When terrestrial objects move naturally, that is, without the motion being started or sustained by a living or external mover, they move either up or down, toward or away from the center of the earth, depending on whether they are heavy or light. The terrestrial or sublunary realm is defined as the world below the orbit of the moon, wherein the four elements seek their natural place. The heavens above the moon are the celestial realm of a fifth element—the quintessence, Aristotle's aether. This element never combines with the other elements and, unlike them, is incorruptible and exists only in the pure state, separately in its own realm in the heavens. Aristotle associated a natural motion with the aether, too, not straight-line motion toward or away from the center, but perfect circles around the center. This seemingly metaphysical doctrine of the perfection of the celestial region is also based on naturalistic observations—heavenly objects appear in fact to be spherical and seem (at least in daily motion) to move in perfect circles around the earth. The enduring and unchanging face of the heavens that we observe from our world of flux and change is related to the unchangeable character of the aether. This dual physics, with separate laws of motion for terrestrial and celestial realms, was consistent with everyday experience and observation and remained intact until the seventeenth century when it was superseded by Newton's laws of motion and universal attraction which postulated a single physics for the whole cosmos.

In addition to the naturally occurring up or down motion of bodies composed of earth, water, fire, and air, nonspontaneous motion observed in the world around us, such as the flight of an arrow, requires explanation. Aristotle envisioned all such motion as forced or violent (as against natural) motion. He proclaimed that such motion always requires an external mover, someone or something to apply an outside force of some sort to cause the motion in question. Moreover, the mover must be in constant contact with the object. In the vast majority of instances Aristotelian movers can be easily identified and the principle apparently confirmed: the horse pulls the cart, the wind blows the sail, and the hand guides the pen. But paradoxical counterexamples exist: the arrow or the javelin in flight after it has lost contact with its mover. Where is the mover in those cases? (Aristotle himself said the medium somehow does the pushing.) In addition, for Aristotle the apparently

Fig. 4.6. The Aristotelian cosmos. According to Aristotle, each of the four elements has a "natural place" in the universe. In the terrestrial region (to the height of the moon), earth and water move "naturally" in straight lines downward toward the center of the cosmos (Earth), while air and fire "naturally" move in straight lines upward and away from the center. The sphere of the moon separates the terrestrial region, with its four elements (including fiery meteors and comets), from the celestial region—the realm of the fifth, "aetherial" element whose natural motion is circular. The stars and planets reside in the celestial region and take their circular motions from the aetherial spheres in which they are embedded.

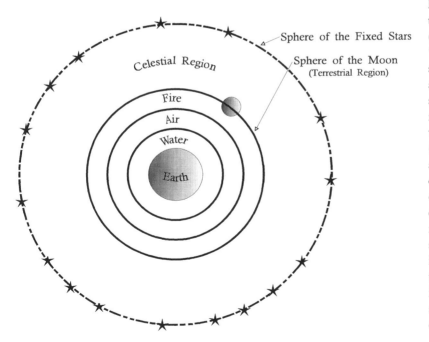

Sphere of the Fixed Stars

Sphere of the Moon
(Terrestrial Region)

Celestial Region

Fire

Air

Water

Earth

unmoved motion of animals or plants derives from the faculties of their souls—the animal or vegetable (or, in the case of human beings, the rational) souls they possess.

Except for the puzzling case of projectile motion Aristotle's theory appears to be consistent with at least casual observations of the physical world. Aristotle went beyond these general principles and postulated quantitative relationships among force, velocity, and resistance. His results were not self-evidently implausible. He gave the example of a boat dragged across a beach. Clearly, the boat will not move by itself; an external motive force is required. That force has to be sufficient to overcome the resistance of the friction between boat and sand before any motion can occur; and the speed with which the boat moves thereafter depends on how much force is applied beyond that minimum. The harder the haulers haul, the faster the boat will go; the greater the friction, the slower it will go. In the case of a falling body, the motive force is proportional to the *weight* of the body, so it follows that heavy bodies will fall downwards faster than light bodies (the more earthy matter a body has, the heavier it is, and the more easily it "divides the air" to descend to its natural place). This notion follows from Aristotle's principles and harmonizes with what we observe. For example, a heavy book falls faster than a light sheet of paper. Similarly, the same object falls more slowly in water than in air, and still slower in honey or in molten lead, where it may even float. In these and many other ways, Aristotle's notions hold for what we observe and experience. One can easily understand why his philosophy of nature prevailed for so long.

Another historically significant principle follows from Aristotle's law of motion, the idea that motion must take place in a medium of some appreciable density. In other words, motion in a vacuum is impossible. Motion in a vacuum implies motion without resistance; but if resistance tends toward zero, the velocity of a moving body becomes infinitely large, which implies that a body can move with infinite speed and can thus be at two places at the same time, an apparent absurdity completely inconsistent with all experience. A corollary of Aristotle's rejection of the vacuum was repudiation of atomism, denying the doctrine of empty space through which atoms supposedly moved. For Aristotle space must be completely filled. The power and comprehensiveness of Aristotle's conception of motion overcame the objections that were intermittently leveled against it. It would ultimately take a profound scientific revolution to overthrow Aristotelian views on motion in a medium and to replace them with an alternative doctrine. For two millennia Aristotle's views concerning the stuff of the world, the concept of place, and the principles of motion made great sense and were accordingly held and shared by those who studied natural philosophy in the Greek tradition.

It would be a mistake to overemphasize the physical sciences in analyzing Aristotle's thought, even though they were fundamental to his worldview. Aristotle was tremendously influential and highly skilled as an observational—one can almost say experimental—biologist and taxonomist. (We must remember, however, that the word *biology* did not come into existence until the nineteenth century of our era.) He conducted empirical investigations, carefully observing the development of the chick embryo, for example. Something like a third of his writings concern matters biological. Crucially, the model that Aristotle used to explain the all-important issue of change derives not from physics but from biology. The growth and development of living things provided a model of change for Aristotle with change embodying a process of becoming, of coming-into-being, the "actualization of that which is potential" in things, as in the classic example of the potential oak in the actual acorn. Growth or change merely brings out features that already exist potentially, thus avoiding the Parmenidian paradox of creating something from nothing. Furthermore, for Aristotle, the passing away of one form involves the coming-to-be of another, and therefore the cosmos must be eternal, with cycles of time repeating themselves *ad infinitum.*

In the details of his dealings with living things Aristotle became the pioneer of systematic taxonomy. He ranked life into a grand hierarchy, classifying animals into "bloodless" invertebrates and vertebrates with blood. His identification of three types of "soul" (nutritive, sensitive, and rational), corresponding to vegetable, animal, and the higher cognitive functions of humans, provided a link to anatomy and physiology, or considerations of how the body operates. Aristotle endorsed the concept of spontaneous generation, and he conceived reproduction as males contributing "form" and females only "matter" to the offspring. Over the ages, Aristotle proved as influential in the life sciences as he did in the physical sciences; in particular the later Greco-Roman physician and likewise influential theorist Galen began his work within the basic frame of reference that Aristotle had set down. Theophrastus of Eresus (371–286 B.C.), Aristotle's successor as head of the Lyceum in Athens, extended the range of the master's investigations into botany in work that remained a standard source until the eighteenth century.

Aristotle was not a dogmatic philosopher and his word was not taken as gospel. Rather, while his basic tenets were retained, his work provided a springboard for scientific research and for traditions of inquiry that unfolded over the succeeding centuries. Theophrastus directed a trenchant criticism at Aristotle's doctrine of fire as one of the elements. With regard to local motion, by highlighting the phenomenon of acceleration, Strato of Lampsacus, successor to Theophrastus at the Lyceum from 286 to 268 B.C., criticized Aristotle's lack of attention to the speeding up and slowing down of bodies as they begin or end their motion. The Byzantine natural philosopher John Philoponus later added to this ongoing debate over Aristotle's theories of motion, and thinkers in the

European Middle Ages intensified the controversies and eventually produced radical revisions of Aristotle's doctrines. This critical tradition evolved over a period of 2,000 years.

Aristotle's writings provided the basis of higher learning in the cultures of late antiquity, Islam, and the European Middle Ages. His cosmos remained at root theological, and, like Plato, he held the heavens to be animate and divine, set in motion by the Unmoved, or Prime Mover. To this extent Aristotle's philosophy could be harmonized with the theologies of Judaism, Christianity, and Islam, and ultimately theologians of all three faiths bent their efforts to squaring their religious doctrines with Aristotle's teachings. By the same token, many Byzantine, Muslim, and Christian scientists found their inspiration in attempts to understand nature—what they believed to be God's handiwork. With its notions of hierarchy and chains of being, much else in Aristotle resonated with later Christianity and the political interests of ruling political authorities, a circumstance that doubtless also helped insure the long-term success of his natural philosophy.

The intellectual legacy of Aristotle's studies shaped the history of scientific thought in the civilizations that inherited Greek learning. The clarity of his analyses and the cosmological comprehensiveness of his views set the standard for the scientific cultures following the Hellenic Enlightenment. The coincidence that Aristotle and his pupil Alexander the Great died within a year of one another (322 and 323 B.C., respectively) seems emblematic, for in their ways they both transformed the contemporary world. The world that immediately followed their deaths was far different—scientifically and politically—than the one they had inhabited.

After Alexander

During Aristotle's lifetime the separate Greek city-states began to coalesce. In Macedonia, the northern district of Greece, a local king, Philip II, gathered his forces, which included horse-mounted infantry and rock-throwing artillery, and began the conquest and unification of the Greek peninsula. When Philip was assassinated in 336 his son, Alexander, who became known to his contemporaries as "the Triumphant" and to us as "the Great," continued his father's expansionist course and forged the most extensive empire in the ancient world.

The empire of Alexander the Great lasted for only 11 years, from 334 when he defeated the Persians until his early death at the age of 33 in 323 B.C. At its height it encompassed not only the Hellenic world of fourth-century Greece, including Ionia and Macedonia, but extended from the great river valley civilization in Egypt, through the Mesopotamian heartland of the first civilizations on the flood plain of the Tigris and Euphrates Rivers, and on to the Indus River Valley in the east. After Alexander's death India returned to Indian control, and the empire collapsed into three kingdoms: Macedonia (including the Greek peninsula), Egypt, and the Seleucid Empire in Mesopotamia (see inset Map 4.1). Because circumstances in the extended Greek world became so reordered after Alexander, the subsequent period is designated the *Hellenistic* era to distinguish it from the pre-imperialist *Hellenic* civilization.

The onset of the Hellenistic also marks a break in the historical chronology of ancient science. Hellenic natural philosophy, with its unpatronized individualists, gave way to the more cosmopolitan world of the Hellenistic—the Golden Age of Greek science—and a new mode for the organization and social support of research. The distinctive character of Hellenistic and Greco-Roman science derives at least in part from the institutionalization of pure science and natural philosophy, most notably in the Museum and Library at Alexandria. To put the point another way, Hellenistic science represents the historical melding or hybridization of the tradition of Hellenic natural philosophy with patterns of state-supported science that had originated in the eastern kingdoms. Kings and emperors had patronized a bureaucratic science that leaned toward useful applications; Hellenic science was the work of unpatronized thinkers who tended toward abstract thought. Hellenistic science in the lands of the ancient Near East combined those disparate traditions. State support and patronage for scientific theory and abstract learning were potent novelties of Hellenistic culture and remained part of the historical pattern of science in all subsequent societies that inherited the Greek tradition.

The Museum and its associated Library at Alexandria exemplify this blend of the heretofore separate scientific traditions of East and West. The most important Hellenistic and Greco-Roman center of science developed in Egypt, where the resources of irrigation agriculture that for thousands of years had supported a great civilization continued munificently to underwrite science in the Hellenistic tradition.

Geographically at a cultural crossroads of East and West, Alexandria was a new town built as a port on the Mediterranean shore of the Nile delta during Alexander's lifetime. The first Greek king of Hellenistic Egypt, Ptolemaios Soter, began the tradition of royal patronage of science and learning, and it fell to his successor, Ptolemaios Philadelphus, to found the Museum at Alexandria formally around 280 B.C. With varying degrees of official support and patronage the Museum existed for 700 years—into the fifth century A.D., an existence roughly as long as the world's oldest universities today.

In essence, the Museum at Alexandria was a research institution—an ancient Institute for Advanced Study. The Museum, unlike its present namesake, did not display collections of objects (a function museums acquired only in the European Renaissance). It was, instead, a temple dedicated to the nine mythical Muses of culture—including Clio, the muse of history, and Urania, the muse of astronomy. There, subsidized members, in a combination of Greek and Oriental traditions, conducted their own research fully supported by state resources. In the royal precinct the Egyptian Ptolemies and their successors maintained splendid quarters for the Museum and its staff which included rooms, lecture halls, dissection studios, gardens, a zoo, an observatory, and possibly other facilities for research. The Ptolemies added a glorious library that soon contained 500,000 or more papyrus scrolls. Patronage always proved fickle in antiquity, depending on the largesse of individual kings and emperors, but at any one time the Museum harbored upwards of 100 scientists and literary scholars who received stipends from the state and meals from the Museum's kitchen while being allowed a Hellenic-style freedom of inquiry, freed even from the obligation to teach. No wonder the record indicates that stipendiaries were the targets of envious attacks as "rare birds" fed in gilded cages— such is the cultural ambiguity provoked by state support for pure learning. The later Roman emperors of Egypt kept up this extraordinary tradition of state support no less than their Hellenistic predecessors, making Alexandria the most significant center of science in the Hellenistic and Greco-Roman eras.

The motives of the Ptolemies and other Hellenistic and Greco-Roman patrons of science and learning are not clear. Doubtless they sought practical returns, and institutional pressure was at least indirectly brought to bear on the scientists at the Museum to bend their research toward useful applications. The fact that the Museum supported anatomical-medical research lends support to that conjecture. Similarly the zoo sheltered the war elephants of the king, the Library collected books on government and contemporary "political science," and academicians pursued geography and cartography. Applied military research may also have taken place at the Museum. Data suggest that Hellenistic scientists were somewhat more practically oriented than their earlier Hellenic counterparts. Beyond any immediate utility, however, it would seem that fame, glory, and prestige accruing to patrons were major motives for supporting the rare birds who roosted in the Museum. Whether the Ptolemies or their Roman successors got their money's worth depends on one's assessment of the relative values of the abstract and practical products of research.

The Hellenistic pattern of support for learning was not limited to Alexandria, and several cities in late antiquity came to boast of museums and libraries, including the library at Pergamum, a city that rivaled Alexandria as a center of state-supported science and scholarship. At Athens the institutional status of Plato's Academy and Aristotle's Lyceum is revealing in this regard. These schools, too, acquired a Hellenistic dimension. We saw that they began in the Hellenic era as informal, entirely private associations of masters and students devoted to the study of their founders' thought. They received legal status, notably as religious associations, but got no public support at the outset, remaining self-supporting as schools and communities of scholars. The formal institutional character of the Academy and the Lyceum became strengthened when, in the Alexandrian mode, the Roman emperors Antoninus Pius and Marcus Aurelius endowed imperial chairs in Athens and elsewhere in the second century A.D. The Lyceum in Athens and the Museum at Alexandria also shared contacts and personnel. The Lyceum continued to be active at least until the end of the second century A.D., and the Academy survived into the sixth century, nearly a thousand years after its founding. Still, the Lyceum and the Academy were primarily schools with teaching the key activity; research itself remained incidental, unlike the extraordinary case of the Alexandrian Museum where scholars received support for unfettered research.

Although literary and philological studies predominated at Alexandria, a historically unparalleled flourish of scientific activity also occurred there, especially during the first century of the Museum's existence, the third century B.C. A tradition of abstract, formal mathematics is the greatest and most enduring Alexandrian accomplishment. As exemplified by Euclid's geometry, Hellenistic mathematics was exceedingly formal and nonarithmetical, qualities that placed it far from the needs of artisans but squarely at the fountainhead of subsequent mathematical research. Euclid had probably studied at the Academy in Athens before he moved to Alexandria under the patronage of the Ptolemies. Apollonius of Perga (fl. 220–190 B.C.) did most of his work there, too; he was known for his mastery of the conic sections (which found its first application in Johannes Kepler's astronomical theories 1,800 years later). To this tradition belongs Archimedes of Syracuse (287–212 B.C.), probably the greatest mathematical genius of antiquity. Archimedes lived and died in Syracuse in Italy, but he traveled to Alexandria at one point and corresponded with the head of the Library, Eratosthenes of Cyrene (fl. 225 B.C.). Eratosthenes, himself a multifaceted man of science, performed a famous observation and calculation to determine the circumference of the earth, and persons educated in the Greek tradition did not believe the earth to be flat; Eratosthenes also inaugurated notable work in geography and cartography. The latter fields of research continued in Alexandria down through the astronomer Ptolemy 400 years later. Innovative anatomical research also took place at the Museum, seen notably in the work of Herophilus of Chalcedon (fl. 270 B.C.) and Erasistratus of Chios (fl. 260 B.C.). Alexandrian anatomists evidently conducted human dissections and possibly vivisections as well. Other Alexandrian scientists undertook substantial research in astronomy, optics, harmonics, acoustics, and mechanics.

In astronomy, the Eudoxean model of geocentric spheres was challenged early in the Hellenistic period. The reader will recall the research tradition that stemmed from Plato's legendary injunction to "save the phenomena"—particularly the problem of the stations and retrogradations of the planets—and Eudoxus's geocentric solution in terms of his onion-skin universe and its rotating and counter-rotating spheres. But the model of nested homocentric spheres, even as refined by Aristotle, faced serious difficulties, notably in accurately reproducing the retrograde motions of planets. And the unequal lengths of the seasons, difficult to explain if the sun moves uniformly at a constant distance from a central earth, was another technical problem undermining the Eudoxean approach. Already in the fourth century—the century of Plato and Aristotle—Heraclides of Pontus (fl. 330 B.C.) suggested that the apparent daily circling of the heavens could be accounted for by assuming that the heavens remained stationary as the earth spun on its axis once a day. The suggestion was generally considered implausible since it seemingly contradicted the direct sensory evidence that the earth is stationary.

Astronomical theory and cosmology posed questions that continued to excite the curiosity of many natural philosophers over subsequent centuries. One of those was Aristarchus of Samos (310–230 B.C.), an expert astronomer and mathematician and, it seems, an associate of the Museum. According to Archimedes, Aristarchus espoused a heliocentric, or sun-centered, cosmology, not unlike the system proposed by Copernicus nearly 2,000 years later. He placed the sun at the center and attributed two motions to the earth: a daily rotation on its axis (to account for the apparent daily circuit of the heavens) and an annual revolution around the sun (to account for the apparent path of the sun around the zodiac).

Aristarchus's heliocentrism was known but overwhelmingly rejected in antiquity, not for some anti-intellectual bias but rather for its essential implausibility. The heliocentric theory, which in its essentials we hold today, faced so many scientific objections at the time that only a zealot would subscribe to it. If the earth whirled on its axis and raced around the sun, surely everything not nailed down would go flying off the earth or be left behind in a wake of debris, a conclusion contradicted by the sensible evidence of birds flying with equal ease in all directions and bodies projected directly upwards returning to where they began. In addition, the motion of the earth postulated by Aristarchus's heliocentrism plainly violated Aristotle's physics of natural motion. Earthy and watery things that make up the earth naturally tend to the center of the cosmos—to require the earth to spin like celestial matter or move otherwise through space is to ask it to undertake motions that Aristotle and all of science declared impossible. If the earth was displaced from the center, its parts would simply return and reorder themselves there. Rational scientists would never accept a theory that flew in the face of everyday observations and that violated long-held doctrines that formed the basis of ongoing, productive research. Today, we also become suspicious of people who propose ideas that violate the laws of physics.

A highly technical but scientifically more telling point also counted strongly against Aristarchus and his sun-centered theory, the problem of stellar parallax. To state the problem simply, if the earth orbits the sun, then the relative position of the stars ought to change over the course of six months as the earthbound observer viewed the heavens from widely different positions. But no such change was observed, at least not until the nineteenth century. (To observe parallax the reader might hold a finger in front of his or her nose and watch it "move" as the left and right eyes are alternately opened and closed.) Archimedes gives us Aristarchus's response to the difficulty: Aristarchus compared the size of the earth's orbit to a grain of sand, meaning that the diameter of the earth's orbit around the sun is so small in relation to the distance to the fixed stars that the change of stellar position would be too small to be observed. This was an ingenious answer for why stellar parallax cannot be observed (the same answer, incidentally, that Copernicus later gave), but Aristarchus then faced the further objection that the size of the universe had to be expanded to extraordinary, (then) unbelievable proportions in order for heliocentrism to hold. The scientific problems facing the heliocentric hypothesis were formidable, and ancient astronomers stood on strong ground in repudiating it. Religious objections also arose against setting the corrupt and changeable earth in the divine and incorruptible heavens. Not surprisingly, Aristarchus was threatened with charges of impiety.

The difficult problem of accounting for planetary motions resulted in alternatives to the astronomies of Eudoxus and Aristarchus. Apollonius of Perga, the Alexandrian scientist mentioned above in connection with the conic sections, helped build an alternative means of "saving the phenomena" and preserving geocentrism. He developed two powerful mathematical tools that astronomers used to model observed motion in the heavens: *epicycles* and *eccentrics*. The epicycle model had planets orbiting on small circles which in turn moved on larger circles; the eccentric is simply an off-centered circle. Both the retrograde motion of the planets and the variable lengths of the seasons could be easily and accurately modeled using epicycles. By assigning different sizes, speeds, and directions to these circles, Hellenistic astronomers developed increasingly accurate models for heavenly motion.

Ancient astronomy culminated in the work of Claudius Ptolemy in the second century A.D. Ptolemy lived and worked under Roman governance in Alexandria. Building on his predecessors' use of epicycles and eccentrics Ptolemy composed a massive and highly technical manual of astronomy, the *Mathematical Syntaxis,* the celebrated *Almagest* (so named by later Muslim scholars). The *Almagest* is premised upon geocentrism and circular motion in the heavens and is extremely mathematical and geometrical in its approach. To his arsenal of epicycles and eccentrics, Ptolemy added a third tool, the so-called equant point, necessitated by the still-elusive harmony between planetary theory and observation. Viewed from the equant point an observer would see the

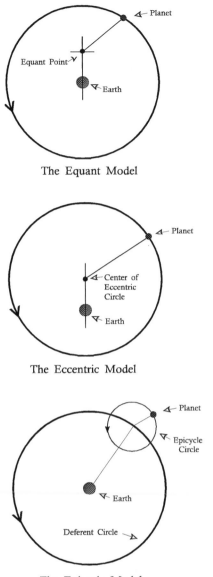

Fig. 4.7. Ptolemy's astronomical devices. To reconcile observed planetary positions with the doctrine of uniform circular motion Ptolemy employed epicycles, eccentric circles, and equant points. The epicycle model involves placing circles on circles; eccentrics are off-centered circles; the equant point is an imaginary point in space from which uniform circular motion is measured.

planet move with uniform circular motion, while it was in fact moving at a changing rate with respect to the earth. Ptolemy's equant violated the spirit, if not the letter, of Plato's injunction to "save the phenomena" using uniform circular motion, but the objection was abstruse even to astronomers and in no way undermined commitments to geocentrism. The equant proved a handy tool, and Ptolemy deployed it and other improvisations to create elaborate, if wholly abstract, mathematical constructions, celestial "Ferris Wheels" whose stately turnings charted the eternal and unchanging heavens. In theory a "Ptolemaic" system with appropriate epicycles, eccentrics, and equants can be designed today to match the accuracy of *any* observed orbit. Ptolemy's *Almagest* was a major scientific achievement. For 1,500 years it remained the bible of every astronomer whose work derived from Hellenistic sources.

Ptolemy also contributed to a Greek tradition of geometrical optics, notably incorporating experimental data into his study of refraction— the bending of light in different media. And his work in geography and cartography similarly proved influential. But one should not put too modern a spin on Ptolemy. For him, mathematical science was a form of philosophy and essentially an ethical and spiritual enterprise. He believed the heavens to be divine and, indeed, animate. For Ptolemy, the movement of the heavens self-evidently affected the sublunary world (through the tides or the seasons, for example). Thus, although Ptolemy distinguished between astrology and astronomy, he recognized the legitimacy of astrology and the effort to predict the future. In fact, he wrote a large and influential book on astrology, the *Tetrabiblos,* and, not least of his many accomplishments, he may fairly be said to have been the greatest astrologer of antiquity.

An upsurge of alchemy matched the strength of contemporary astrology. What came to be the foundational texts of a semisecret tradition were compiled in Hellenistic Alexandria and elsewhere. The tradition is labeled "Hermetic" because these compilations were attributed to its mythical founder, Hermes Trismegistus, a legendary Egyptian priest thought to be living around the time of Moses. This body of mystical work contained esoteric and supposedly divinely inspired doctrines pertaining to the secret workings of the universe. Although the idea and

Fig. 4.8. Ptolemy's model for Mercury. Ptolemy deployed epicycles, eccentrics, and equants in elaborate and often confusing combinations to solve problems of planetary motion. In the case of Mercury (pictured here) the planet orbits on an epicycle circle; the center of that circle revolves on a larger eccentric circle, the center of which moves in the opposite direction on an epicycle circle of its own. The required uniformity of the planet's motion is measured by the angle (α) swept out unvaryingly by a line joining the equant point and the center of the planet's epicycle circle. These techniques can be made to account for any observed trajectories. The intricate sets of solutions Ptolemy and his successors produced constituted gigantic "Ferris wheel" mechanisms moving the heavens.

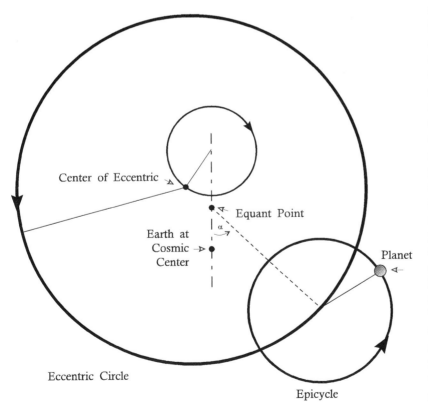

Center of Eccentric

Equant Point

Earth at
Cosmic
Center

α

Planet

Eccentric Circle

Epicycle

practice that base metals can be transmuted into gold and silver doubtless involved some amount of fraud in antiquity, the roots of alchemy lay in demonstrated metallurgical practice, and alchemical science, so to speak, evolved out of Bronze and Iron Age technologies involving metals. Alchemy offered the promise of utility, and in that sense it represents another early applied science, especially to the extent that rulers patronized it. But for serious practitioners, the alchemical quest for elixirs of immortality or the philosopher's stone that would transmute metals always entailed a spiritual dimension, wherein the alchemist sought to purify himself as much as he hoped to purify base metals. Ancient and medieval alchemy should not be thought of as pseudochemistry. Rather, alchemy needs to be understood on its own as a technically based applied science that combined substantial mystical and spiritual elements.

The impact of alchemy was small, and on the whole Hellenistic science at Alexandria and elsewhere in the ancient world was not applied to technology or, in general, pursued for utilitarian ends. Natural philosophy remained largely insular, as it had been previously in the Hellenic. It remained isolated, not in any direct way connected or applied to the predominant practical problems of the age. In addition, the ideology stemming from Plato and the pre-Socratics that held manual labor in contempt and that repudiated any practical or economic utility to science continued in the Hellenistic. Ideology thus reinforced the existing separation of *theoria* and *praxis*.

Mechanics itself, however, was subject to scientific analysis by Hellenistic scientists in theoretical treatises on mechanics and the mechanical arts. Archimedes, above all, mastered the mechanical principles of the simple machines: the lever, wedge, screw, pulley, and windlass; and in analyzing the balance (including the hydrostatic balance), the ancients articulated a theoretical and mathematical science of weight. The practical possibilities of this mechanical tradition are evident in the work of Ctesibius of Alexandria (fl. 270 B.C.), Philo of Byzantium (fl. 200 B.C.), and Hero of Alexandria (fl. 60 B.C.). Based on their knowledge of weight and pneumatics, these men designed ingenious mechanical devices—in the category of "wondrous machines" that could automatically open temple doors or pour libations, but whose purpose was to provoke awe and wonder, and not to contribute to economic progress. Hero, for example, contrived to spin a ball using fire and steam, but no one in antiquity conceived or followed up with a practical steam engine. In a word, the study of mechanics in Alexandria was, like its kindred sciences, almost completely detached from the wider world of technology in antiquity.

But not completely. The Archimedean screw, for example, was a machine that lifted water; it was invented in the third century B.C. purportedly by Archimedes, and it derived from this tradition of scientific mechanics. Archimedes, who died in 212 B.C. during the defense of his

native Syracuse against the Romans, became legendary for his technological wizardry with siege engines and war machinery. His published
work remained abstract and philosophical, but even discounting as
legend much that is credited to him, Archimedes probably did apply
himself to engineering technology and practical achievement. He supposedly used his knowledge of mechanics in wartime, and in this ca-
pacity he acted as an ancient engineer *(architecton)*, one of whose
domains was military engineering.

The case of the ancient torsion-spring catapult is revealing. Weapons
development was nothing new in antiquity, and, indeed, something like
a technological arms race took place among craftsmen and sponsoring
patrons to build the largest rowed warship. Philip II of Macedon and
Greek kings in Syracuse, Rhodes, and elsewhere supported programs
to develop and improve the catapult and a variety of ballistic machines.
Sophisticated engineering research in the form of systematic tests took
place at Alexandria to identify variables affecting the functioning of
catapults and to create the most effective and efficient machines. The
government sponsored this research, and scientists at Alexandria carried out some of it. Although the mechanical tradition at Alexandria
was less otherworldly than once thought, one must qualify the ways in
which catapult research represents applied science in antiquity. Overall, the tests seem to have been entirely empirical, that is, executed by
scientist-engineers perhaps, but without the application of any scientific theory or the exploitation of theoretical knowledge. After decades
of patient effort and recordkeeping the scientist-engineers at Alexandria created a practical and mathematically exact "catapult formula"
that involved extracting cube roots and that specified the optimal dimensions for any ballistic machine and its projectile. With this formula
Archimedes himself purportedly built the largest stone-throwing catapult on record. But the formula is simply a rule of thumb expressed in
mathematical terms. Development of the catapult is better thought of
as applied engineering research.

Some scientific instruments existed in antiquity—notably finely crafted
astronomical clockwork and other observing devices, where science
and technology combined in the service, not of warfare or the larger
economy, but of the scientific enterprise itself. As interesting and historically revealing as all these examples are, they do not belie the general point that ancient science on the whole had very little practical
import, was not as a rule directed to practical ends, and had no significant impact on ancient engineering as a whole.

Technology in antiquity needs to be seen as a domain separate from
ancient science, a robust world of farming, weaving, potting, building,
transporting, healing, governing, and like myriads of crafts and techniques great and small that made up and maintained Hellenistic and
Greco-Roman civilization. Hundreds of small new technologies and
technological refinements occurred in the 800 years of the Hellenistic

Fig. 4.9. The torsion-spring catapult. The Greeks under Philip of Macedon had begun to use ballistic artillery in the form of machines that could hurl large projectiles at an enemy. In some designs the action was produced by twisting and suddenly releasing bundles of elastic material. This large Roman model fired stones weighing 70 pounds. Hellenistic scientist-engineers experimented to improve the devices.

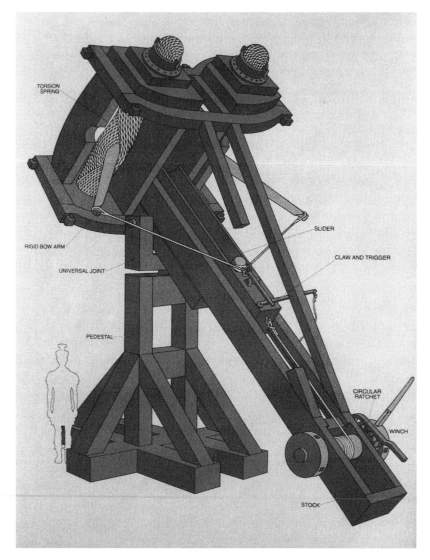

and Greco-Roman periods (such as a kickwheel added to the potter's wheel), but overall the technological bases of production did not change fundamentally during the period. Some industrial-style production occurred in a few fields like mining; and long-distance commercial movement of people and goods took place regularly. But most production remained craft-based and local, and artisans, traditionally secretive about knowledge of their skills, tended to monopolize practice without the benefit of writing, science, or natural philosophy.

While ancient science formed part of civilized life in towns, technology and engineering practice could be found everywhere in the ancient world, vigorously and expertly developed in great cities and towns, to be sure, but also in the countryside, where the practice of science and natural philosophy was notably absent. The engineer *(architecton)* was a recognized and employable social type in antiquity. A handful of individuals stood at the top rank of ancient engineering. The Roman Vitruvius, for example, worked as architect/engineer to the first Roman emperor, Augustus, at the turn of the first century A.D., and he contributed to an engineering literature. However, most engineers and, indeed, most artisans were anonymous practitioners plying trades at great remove socially, intellectually, and practically from the scientific world of Alexandria.

The Romans were the greatest technologists and engineers of the ancient world, and one can argue that Roman civilization itself represents one grand technological achievement. In the first centuries B.C. and A.D. Roman military and political power came to dominate the whole of the Mediterranean basin and much of the Hellenistic world that had arisen in the east. (Mesopotamia remained beyond the reach of Rome.) The Roman empire grew up around several technologies. Military and naval technologies created the disciplined Roman legion and the Roman navy. The extensive systems of Roman roads and aqueducts provided an essential infrastructure. The expertise and sophistication of the Romans in matters of formal law may also be thought of as a social technology of no small import in running the empire. Less lofty perhaps, but no less important as a building block of Roman civilization, the invention of cement was a key new technology introduced by the Romans, one that made stone construction much cheaper and easier, and it literally cemented the expansion of the Roman empire. The fact that Rome produced known engineers, a few of whom wrote books (an uncommon practice among engineers), such as Vitruvius and Frontinus (A.D. 35–103), likewise testifies to the significance of engineering and technology to Roman civilization and vice versa.

While Roman engineering flourished, there was little Roman science. Very little Greek science was ever translated into Latin. For the sake of tradition, Roman emperors patronized the Museum in faraway Alexandria, but the Romans did not value, indeed they spurned, science, mathematics, and Greek learning in general. Some privileged young Romans learned Greek and toured and studied in Greece. But Rome itself produced no Roman scientist or natural philosopher of the first or even the second rank. This circumstance has proved a puzzlement for those who see science and technology as always and necessarily linked. The temptation has been to overemphasize those exceptional Romans who did write on matters scientific. The notable Roman poet Lucretius (d. 55 B.C.), whose long poem *On the Nature of Things* advanced atomist notions, is one example. The great Roman compiler Pliny the Elder (A.D. 24–79), whose multivolume *Natural History* summarized as much of the natural world as he could document (the fabulous along with the commonplace), is another. For better or worse, Pliny's work remained the starting point for the study of natural history until the sixteenth century; that he devoted consider-

Fig. 4.10a–b. Roman building technology. Roman engineers were highly competent in the use of the wedge arch in the construction of buildings, bridges, and elevated aqueducts. This Roman aqueduct in Segovia, Spain, is an outstanding example. The invention of cement greatly facilitated Roman building techniques.

able space to the practical uses of animals and that he dedicated his *Natural History* to the emperor Titus suggests that, insofar as Roman science existed at all, familiar social forces were at play.

Greeks scholars lectured in Rome. Greek doctors, particularly, found employment in Rome, more for their clinical skills than for their theoretical knowledge. The illustrious scientist-physician Galen of Pergamum (ca. A.D. 130–200) was born, raised, and trained in Asia Minor and Alexandria, but climbed the ladder of medical success in Rome through gladiatorial and court circles, becoming court physician to the Roman emperor Marcus Aurelius. Galen produced a large and influential body of work in anatomy, physiology, and what we would call biology. He built on Aristotle and the Hippocratic corpus and he articulated rational and comprehensive accounts of the workings of the human body based on detailed anatomical analysis.

Galen's anatomy and physiology differ markedly from succeeding views in the European Renaissance and those held today, but that fact should not detract from the power and persuasiveness of his understanding of the human fabric. For Galen and his many successors, three different vital systems and as many *pneuma* operated in humans. He held that the liver and venous system absorbed nutrients and distributed a nourishing blood throughout the body. The brain and nerves distributed a psychic essence that permitted thought. The heart was the seat of innate heat which distributed a third, vital fluid through the arteries, thus enabling movement; the lungs regulated respiration and the cooling of the heart's innate heat. The circulation of the blood was a conceptual impossibility for Galenists because they believed that, except for a minor passageway in the heart where the nutrifying blood provided the raw material for the arterial spirit, veins and arteries were two entirely separate systems.

Galen was a prolific author, supposedly writing some 500 treatises, of which 83 survived antiquity. He remained the undisputed authority in anatomy and physiology into the early modern era. Galen exemplifies the continuing interaction between medicine and philosophy in the Hellenistic and Greco-Roman eras, but Galen was Greek, and the tradition out of which he emerged and to which he contributed was Hellenistic and not Roman. The phenomenal lack of any Roman tradition in mathematics or the natural sciences contrasts strongly not only with Roman engineering but also with the substantial record of Roman literary and artistic accomplishment in poetry, the theater, literature, history, and the fine arts. The names of Cicero, Virgil, Horace, and Suetonius alone suffice to indicate the extent to which literary and learned culture held a valued place in Roman civilization generally. The Roman case shows that a civilization of great social and technological complexity could thrive for centuries essentially without theoretical science or natural philosophy.

Decline and Fall

The causes of the marked decline of science and natural philosophy at the end of the Greco-Roman era have long been debated among historians of science. Not all agree even about the facts. Some claim the decline can be dated from 200 B.C. in the Hellenistic era; others say it only began after A.D. 200 in the Greco-Roman period. Certainly, not all scientific and natural philosophical activity came to a halt after the second century A.D. Still, ancient science seems to have run out of steam in late antiquity. Generally speaking, less overall activity took place, and the level of scientific originality declined as time went on. Intellectual labor was increasingly directed less toward discovering new knowledge than toward preserving old knowledge. This characteristic state of affairs gave rise to generations of compilers and commentators. Oribasius at Constantinople, for example, in the middle of the fourth century A.D., wrote a formidable medical compendium of seventy volumes. (It is notable, but hardly surprising in this regard, that medicine displayed greater historical continuity in antiquity than did ancient science or natural philosophy.) Whatever animated the pursuit of science seems to have disappeared. Eventually, the desire merely to preserve past knowledge fell off. Increasing skepticism arose about even the possibility of secure knowledge, and magic and popular superstitious beliefs gained ground. The substance and spirit of Greek scientific accom-

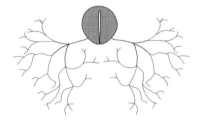

Brain/Nerves
distribute
"Animal Spirits"

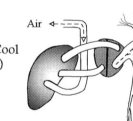

Air

(Lungs Cool
Heart)

Heart/Arteries
distribute
"Vital Spirits"

Fig. 4.11. Galenic physiology. Ancient physicians and students of anatomy separated the internal organs into three distinct subsystems governed by three different "spirits" functioning in the human body: a nutrifying venous spirit originating in the liver, a vivifying arterial spirit arising in the heart, and a psychic essence permeating the brain and the nerves.

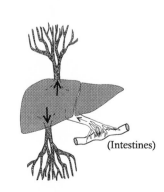

(Intestines)

Liver/Veins
distribute
"Nutrifying Spirits"

plishment in its Hellenic and Hellenistic modes gradually faded away in late antiquity.

Several explanations have been proposed to explain why. One possible explanation points to the lack of a clear social role for science and scientific careers. Science was weakly socialized and institutionalized in the ancient world, and it largely lacked an ideological or material basis of support in society. No employment was available for individuals in their capacities as scientists or as natural philosophers, and the historical separation of science and natural philosophy from *philosophy* itself in the Hellenistic further undercut any social role for the scientific enterprise.

A related explanation has to do with the ancient economy and the separation of science and technology in antiquity. That is, given slavery, the relative cheapness of human labor, and the ideology that natural knowledge should not be applied to practical ends, little incentive existed to employ scientists or to invest in applied science. In other words, by excluding the possible utility of natural knowledge, the social role and social support for science were undermined.

Historians have also made a strong case that the flourishing of various religious cults and sects in late antiquity did much to weaken the authority and vitality of ancient scientific traditions. To varying degrees the many cults of late antiquity were anti-intellectual in their approach, and they represented intellectual and spiritual competition with traditional knowledge of nature. The cult of the Greek fertility goddess Demeter and the cult of the Egyptian goddess Isis attracted wide followings. Popular among officials of the Roman empire, Mithraism, a late oriental mystery cult worshiping the Persian god of light, Mithras, embodied arcane and secret astrological and astronomical knowledge. And growing out of messianic Judaism, the most successful new cult was Christianity.

Official toleration of Christians in A.D. 313, the emperor Constantine's conversion to Christianity in 337, and the declaration in 391 of Christianity as the official state religion of the Roman Empire marked the extraordinary social and institutional success of the Christian church. Experts debate the effect of Christianity on ancient science, but with its heavy theological orientation, its devotional stress on the religious life, and its emphasis on revelation, the afterlife, and the second coming of Christ, the early Christian church and the church leaders who fashioned it displayed greater or lesser degrees of hostility, skepticism, ambivalence, and/or indifference toward pagan culture in general and toward science and inquiries into nature in particular. To cite only one example, Saint Augustine (A.D. 354–430) railed against natural philosophy and "those whom the Greeks call *physici*." On a more mundane level the church became firmly institutionalized in ancient civilization and a formidable institutional presence at every level of society. Church bureaucracy and administration offered employment and careers, which had the effect of draining talent, intellectual and otherwise, which previously might have been recruited for the Museum at Alexandria or for science in general.

Historians of technology have asked why no industrial revolution developed in antiquity. The simple answer seems to be that there was no need, that contemporary modes of production and the slave-based economy of the day satisfactorily maintained the status quo. The capitalist idea of profit as a desirable end to pursue was completely foreign to the contemporary mentality. So, too, was the idea that technology on a large scale could or should be harnessed to those ends. An industrial revolution was literally unthinkable in antiquity.

Alexandria and its intellectual infrastructure suffered many blows from the late third century onward. Much of the town was destroyed in 270–75 in Roman efforts to reconquer the city after its momentary capture by Syrian and Arab invaders. Christian vigilantes may well have burned books in the fourth century, and in 415 with the murder by Christian fanatics of the pagan Hypatia, the first known female mathematician and the last known stipendiary of the Museum, the centuries-old Museum itself came to an end. Later, the initial Islamic conquerors effected further depredations on what remained of the Library at Alexandria. Elsewhere, in A.D. 529 the Christian Byzantine emperor Justinian ordered the closing of the Platonic Academy in Athens.

The Roman Empire split into its western and eastern divisions in the fourth century A.D. (see Map 5.1 of the Byzantine Empire). In A.D. 330, Constantine the Great transferred the capital of the empire from Rome to Constantinople, modern-day Istanbul. Waves of barbarian tribes pressed in on the western empire from Europe. Visigoth invaders sacked Rome for the first time in A.D. 410. Other Germans deposed the last Roman emperor in Italy in A.D. 476, a date that marks the traditional end of the Roman Empire. While the latinized West Roman Empire fell, the Hellenized East Roman Empire—the Greek-speaking Byzantine Empire—centered in Constantinople, endured, indeed flourished. But, the fortunes of Byzantium declined in the seventh century as its glory and granaries contracted before the armed might of ascendant Islamic Arabs. Pouring out of Arabia after A.D. 632, the followers of the Prophet Mohammed conquered Syria and Mesopotamia. They captured Egypt and Alexandria in A.D. 642 and pressed in on Constantinople itself by the end of the century. Science and civilization would continue to develop in Muslim Spain, in eastern regions, and throughout the Islamic world, but by the seventh century A.D. the era of Greek science in antiquity had clearly come to an end.

The Roman West, which included much of Europe, had always been underdeveloped compared to the East. Decline, intellectual and otherwise, at the end of antiquity affected the West much more than the East, where greater continuity prevailed. Indeed, the words *disruption* and *discontinuity* aptly describe "western civilization" at the end of Greco-Roman antiquity. The population of Italy, for example, dropped by 50 percent between A.D. 200 and 600. An era had ended, and surely to contemporaries no promise of renewal seemed to be forthcoming. The late Roman author and senator Boethius (A.D. 480–524) knew that he stood at a historical crossroads, and his case is a poignant one on that account. Boethius was exceptionally well educated and fully the inheritor of the classical tradition of Greek and Latin antiquity that stretched back a millennium to Plato, Aristotle, and the pre-Socratics. Yet he held office and attended not a Roman emperor, but the Ostrogoth king in Rome, Theodoric. Imprisoned for many years by Theodoric, Boethius made every effort to pass on to the succeeding age as much of antiquity's accumulated knowledge as he could. He wrote short elementary handbooks on arithmetic, geometry, astronomy, mechanics, physics, and music. In addition, he translated some of Aristotle's logical treatises, some Euclid, and perhaps Archimedes and Ptolemy. In prison, he also wrote his immortal meditations, *On the Consolation of Philosophy,* which proved small consolation indeed. Theodoric had Boethius executed in 524.

Historians interested in the European Middle Ages and in the history of science in the Middle Ages often point to Boethius and his compeers to indicate the extent to which knowledge from classical antiquity passed directly into the stream of European history and culture. Cassiodorus (488–575), a Roman like Boethius, who influenced the early monastic movement, is regularly cited in this connection, as are the later learned churchmen Isidore of Seville (560–636) and the Venerable Bede (d. 735). There is much of interest about these men and their circumstances, but the Latin West inherited the merest crumbs of Greek science. From a world perspective, what needs to be emphasized is the utterly sorry state of learning among the Christian barbarians of Europe and the Latin West in the early Middle Ages. After the fall of Rome literacy itself virtually disappeared, and knowledge of Greek for all intents and purposes vanished. Isidore of Seville apparently thought the sun illuminated the stars. Two eleventh-century European scholars, Regimbold of Cologne and Radolf of Liège, could not fathom the sense of the elementary proposition from geometry that "the interior angles of a triangle equal two right angles." The terms "feet," "square feet," and "cubic feet" had no meaning for them.

How and why the scientific traditions of Greek antiquity took hold in western Europe centuries later require separate explanations and a return to the world stage.

PART II

Thinking and Doing among the World's Peoples

After the Hellenistic merger of the institutional patterns of the ancient Near East with the abstract intellectual approach of the Hellenic Greeks, scientific traditions took root in a series of Near Eastern empires during late antiquity and the Middle Ages: in Byzantium, in Sassanid Persia, and in the vast region that formed the Islamic conquest. At the same time, indigenous traditions of scientific research developed independently in China, India, and Central and South America. Monarchs in control of rich agricultural lands patronized experts in calendrical astronomy, astrology, mathematics, and medicine in the hope and expectation that these fields of research would produce useful knowledge. In the meantime, science and the traditional crafts continued everywhere as essentially separate enterprises. Part 2 investigates these developments.

CHAPTER 5

The Enduring East

Byzantine Orthodoxy

After Rome fell in A.D. 476, the empire's eastern districts with their capital at Constantinople gradually metamorphosed into the Greek-speaking Byzantine Empire (see Map 5.1). A Christian state, headed by an
emperor and governed by elaborate and scheming bureaucracies (hence
"byzantine"), the Byzantine Empire endured for a thousand years before being overrun by the Ottoman Turks in 1453. In possession of the
Egyptian breadbasket the empire flourished and wealthy emperors continued to patronize many old institutions of higher learning.

Science in the Byzantine Empire remains to be studied by historians
in greater detail. Byzantine civilization is often criticized as anti-intellectual and stifled by a mystical Christianity that was imposed as a state
religion. That the emperor Justinian (r. 527–565) closed Plato's still-
functioning Academy at Athens along with other schools in A.D. 529
is commonly seen as evidence of the state's repressive posture toward
science. Yet, to dismiss Byzantium from the history of science would
be to overlook continuations of Hellenistic traditions and the ways,
quite typical of eastern bureaucratic civilizations, in which science and
useful knowledge became institutionalized in society.

Even after the Justinian closures, state schools and church schools
provided instruction in the mathematical sciences (the quadrivium:
arithmetic, geometry, astronomy, and music), the physical sciences, and
medicine; libraries existed as centers of learning. The true hospital, as
an institution of in-patient medical treatment (and Christian mercy),
was a notable Byzantine innovation. It was, like the hospital today,
primarily a center of medical technology, not science. As hospitals
arose throughout the Byzantine Empire through the largesse of government, church, and aristocratic patrons, in some measure they also
became centers of medical research. Byzantine medicine fully assimilated the medical and physiological teachings of Galen and Hippoc-

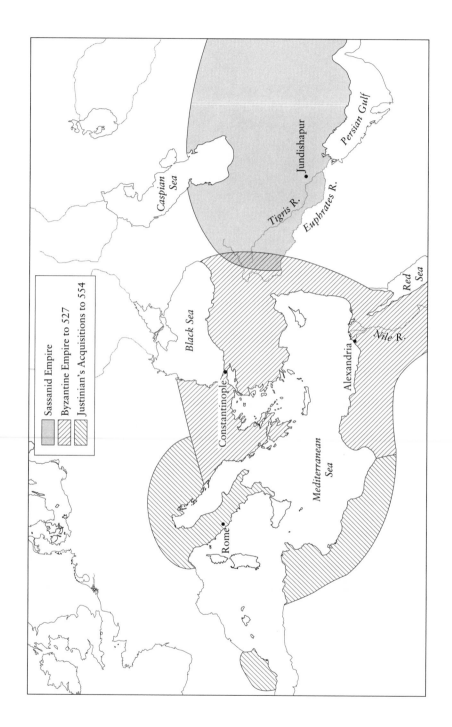

Sassanid Empire

Byzantine Empire to 527

Justinian's Acquisitions to 554

Caspian Sea

Persian Gulf

Jundishapur

Tigris R.

Euphrates R.

Black Sea

Red Sea

Nile R.

Alexandria

Constantinople

Mediterranean Sea

Rome

rates, while some hospitals maintained libraries and teaching programs and even fostered some original investigations and innovative techniques. Learned Byzantine doctors turned out influential medical and pharmacological tracts, albeit with much repetition of Greek knowledge. Veterinary medicine was a notable aspect of scientifico-medical activity in Byzantine civilization, one heavily supported by monarchs who had an interest in the well-being of their war horses since cavalry and the cavalry charge formed the basis of the Byzantine army and military tactics. As a result, Byzantine veterinarians produced many veterinary manuals, occasionally on a high level of originality.

In the fields of the exact sciences the Greek-speaking Byzantine scholars inherited much Greek learning from antiquity. They knew their Aristotle, Euclid, and Ptolemy, and Byzantine astronomers and mathematicians themselves sometimes produced sophisticated tracts based on earlier Greek and contemporary Persian sources. In addition to calendar work, a strong element of astrology, reflecting that venerable and inextinguishable desire to know the future, characterized Byzantine astronomy. Experts likewise studied music and mathematical music theory, perhaps for liturgical purposes. And finally, Byzantine alchemy and alchemical mineralogy cannot be overlooked as areas of both considerable research activity and of perceived practical utility.

The most notable natural philosopher of the early Byzantine era was John Philoponus. Philoponus lived and worked in Alexandria under Byzantine governance during the middle years of the sixth century A.D., and he launched the most sweeping attack on Aristotelian physics prior to the Scientific Revolution in Europe. In various commentaries he developed trenchant critiques of Aristotle and several aspects of Aristotelian physical theory. In his ingenious analysis of projectile motion, for example—motion Aristotle had lamely explained by invoking the ambient air as the required mover—Philoponus suggested that the thrower endowed the projectile with a certain power to move itself. Philoponus's views in turn sparked critical responses from other commentators, and because he so focused the debate on specific problems in Aristotle's writings on natural philosophy, Philoponus was later influential among Islamic and European natural philosophers when they came to review Aristotle's work. Although he wrote in Latin and studied the science of the Greeks, his career and his accomplishments are landmarks in the tradition of Byzantine science.

A full social history of Byzantine science would display the subject in a more favorable light than does intellectual history alone, which stresses originality and pure theory. Such a social history would pay close attention to intellectually unambitious medical tracts, to treatises published by veterinary surgeons retained by Byzantine monarchs, to the many farmers' manuals and herbals produced under Byzantine governance, as well as to astrology and alchemy. In a society where bureaucratic centralization was extreme, support came precisely for encyclo-

Map 5.1. Byzantium and Sassanid Persia. In late antiquity two derivative civilizations took root in the Middle East—the Byzantine Empire centered in Constantinople and Sassanid Persia in the heartland of ancient Mesopotamia. Both assimilated ancient Greek science and became centers of learning. *(opposite)*

pedists, translators, and writers of manuals on subjects useful and mundane. And it is precisely the kind of work that historians intent on detecting theoretical novelty tend to neglect.

The loss of Egypt and the productive resources of the Nile Valley to invading Arabs in the seventh century was a severe setback to the economy and society of Byzantium. Yet a reduced Byzantine civilization maintained itself, its cities, its institutions, and its science for hundreds of years. Inevitably, however, decline set in after the year 1000, as Byzantium faced challenges from the Turks, from the Venetians, and from friendly and not-so-friendly European Christians on crusade. In 1204 Crusaders pillaged Constantinople and occupied it until 1261. Finally, in 1453 the city and the empire fell to the Turks.

Although Byzantium never became a center of significant original science, it did not repudiate the tradition of secular Greek learning. Indeed, it tolerated and even preserved that tradition alongside the official state religion of Christianity.

Mesopotamia Revisited

In the heartland of ancient Mesopotamia the Sassanid dynasty created a typical Near Eastern system of scientific institutions along with a typical Near Eastern economy based on hydraulic agriculture and the restoration and maintenance of the old irrigation systems. The Sassanid dynasty was founded in A.D. 224, and was characterized by a strong central government and a bureaucratic caste that included scribes, astrologers, doctors, poets, and musicians. By the sixth century the royal residence at Jundishapur, northeast of present-day Basra, had become a cultural crossroads where many different learned traditions mingled: Persian, Christian, Greek, Hindu, Jewish, and Syrian. Persian cultural life became enriched when Nestorian Christians—bringing Greek learning with them—fled Byzantium after their center at Edessa in Syria was shut down in 489. A significant translation effort, centered in Jundishapur, rendered Greek texts into Syriac, the local language. Texts deemed to contain useful knowledge were generally chosen for translation—mainly the medical arts, but also scientific subjects, including Aristotle's logical tracts, mathematics, and astronomy. Jundishapur also became the site of a hospital and medical school, enriched by the presence of Indian medical masters; later taken over by Arab-Islamic caliphs, the medical school at Jundishapur continued to flourish until the eleventh century. Persian government authorities also sponsored astronomical and astrological investigations. While recent reappraisals have tempered its importance, Jundishapur was nonetheless a cosmopolitan intellectual center and a center of scientific patronage for several centuries before Persia fell to the forces of Islam in A.D. 642.

Sassanid civilization illustrates once again that centralization of authority to manage a hydraulic agricultural economy fostered scientific institutions. Its culture in some measure hybridized the institutional and intellectual traditions of the ancient Oriental kingdoms and those of classical, Hellenic Greece, and it produced state-dominated institutions, in some of which a Greek tradition of pure science found a niche. Once again, enormous wealth in the form of large agricultural surpluses generated by intensified irrigation agriculture made such institutional patronage possible. The case also confirms that prior to the development of modern science in western Europe Greek scientific influence flourished predominantly in the East.

Under the Banner of Islam

The Middle East produced still another scientific civilization, this time under the aegis of Islam. The flight of the Prophet Mohammed from Mecca in A.D. 622 marks the traditional beginning of the Muslim era. The word *Islam* means submission to the will of God, and Muslims (or Moslems) are those who submit. Arabs are the peoples of Arabia, and out of the Arabian desert and a nomadic pastoral society of the seventh century the faith of Islam spread to many different peoples east and west. Within three decades Islamic armies conquered Arabia, Egypt, and Mesopotamia—replacing Persian power and severely reducing the Byzantine Empire. In slightly more than a century they established an Islamic commonwealth stretching from Portugal to Central Asia. A unified sociocultural domain, Islam prospered as a great world civilization, and its scientific culture flourished for at least five centuries.

The success of Islam depended as much on its faithful farmers as on its soldiers. The former took over established flood plains in Mesopotamia and Egypt, and in what amounted to an agricultural revolution they adapted new and more diversified food crops to the Mediterranean ecosystem: rice, sugar cane, cotton, melons, citrus fruits, and other products. With rebuilt and enlarged systems of irrigation, Islamic farming extended the growing season and increased productivity. That Islamic scientists turned out an uninterrupted series of treatises on agriculture and irrigation is one indication of the importance of these endeavors. So, too, are the specialized treatises on camels, horses, bees, and falcons, all animals of note for Islamic farmers and Islamic rulers.

The effects of such improved agricultural productivity were typical: unprecedented population increases, urbanization, social stratification, political centralization, and state patronage of higher learning. Baghdad, founded in 762 on the Tigris, became the largest city in the world in the 930s with a population of 1.1 million. Córdoba in southwestern Spain reached a population close to 1,000,000 under Islamic rule, and several other Islamic cities had populations between 100,000 and 500,000 during a period when the largest European cities had populations numbering fewer than 50,000.

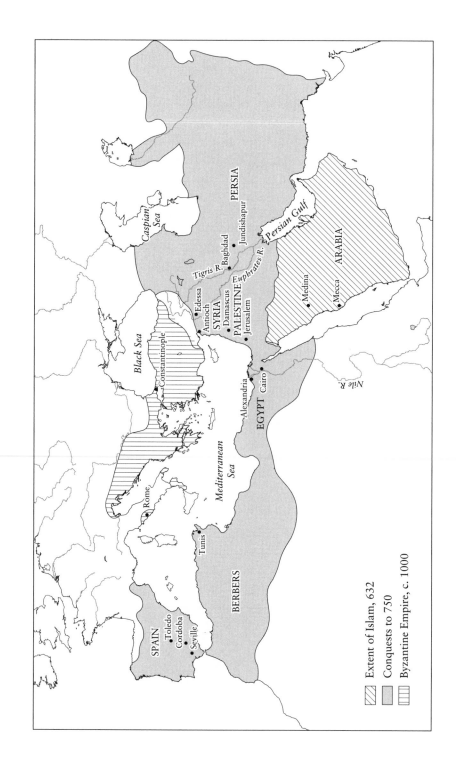

Caspian Sea

PERSIA

Jundishapur

Persian Gulf

Baghdad

Tigris R.

Euphrates R.

ARABIA

Edessa

Antioch

SYRIA

Damascus

PALESTINE

Jerusalem

Medina

Mecca

Black Sea

Constantinople

Alexandria

Cairo

EGYPT

Nile R.

Rome

Mediterranean
Sea

Tunis

BERBERS

SPAIN

Toledo

Cordoba

Seville

Extent of Islam, 632

Conquests to 750

Byzantine Empire, c. 1000

Islam was and is based on literacy and the holy book of the Quran (Koran); and, although policy vacillated, Islam showed itself tolerant toward Christians and Jews, also "people of the book." Thus, in contrast to the barbarian farmers of Europe who pillaged and destroyed the high civilizations they encountered, the nomadic and pastoral Arabs established conquest empires by maintaining and assimilating the high cultures they encountered. Early Islamic rulers encouraged the mastery of foreign cultural traditions, including notably Greek philosophy and science, perhaps in order to bolster the logical and rhetorical position of their new religion in the face of more highly developed religions and critical intellectual traditions. The result was another hybrid society, the cultural "hellenization" of Islam and its typically bureaucratized institutions, funded by wealthy monarchs and patrons who encouraged useful knowledge along with a dash of natural philosophy.

Medieval Islam became the principal heir to ancient Greek science, and Islamic civilization remained the world leader in virtually every field of science from at least A.D. 800–1300. The sheer level of scientific activity makes the point, as the number of Islamic scientists during the four centuries after the Prophet matched the number of Greek scientists during the four centuries following Thales. Islamic scientists established the first truly international scientific community, stretching from Iberia to Central Asia. Yet, despite considerable scholarly attention, medieval Islamic science is sometimes still dismissed as a conduit passively "transmitting" ancient Greek science to the European Middle Ages. A moment's thought, however, shows how ahistorical it is to evaluate the history of Islamic science only or even largely as a link to European science, or even to subsume Islamic science into the "Western tradition." Medieval Islam and its science must be judged on their own terms, and those terms are as much Eastern as Western.

Only a small fraction of Islamic scientific texts have been published. Most remain unstudied and in manuscript. Scholarly emphasis to date has been on classic texts, on the "internal" history of scientific ideas, on biographies, and on "precursor-itis," or identifying Arabic scientists who were precursors of ideas that were of later importance to European science. The institutional aspects of Islamic science are only beginning to be studied with scholarly rigor, and nothing like a full historical survey exists for the Islamic case.

Furthermore, the field divides into two divergent interpretative schools. One school argues for a "marginality" thesis, holding that the secular, rational sciences inherited from Greek civilization—known in Islam as the "foreign" (aw'il) sciences—never became integrated into Islamic culture, remaining only on the cultural margins, tolerated at best, but never a fundamental part of Islamic society. The "assimilationist" school, on the other hand, contends that the foreign sciences became woven into the fabric of Islamic life. Neither interpretation quite fits the facts, but the presentation favored here leans toward the

Map 5.2. Islam. Following the birth of Mohammedanism in the seventh century the Islamic conquest stretched from the Atlantic Ocean almost to the borders of China. In capturing Egypt and the resources of the Nile, the forces of Islam dealt a severe blow to Byzantine civilization. *(opposite)*

assimilationists, especially in tracing the institutional basis of Islamic science and in recognizing a similarity between the social function of science in Islam and in other Eastern civilizations.

Islamic scientific culture originated through the effort to master the learning of more established civilizations, and that first required the translation of documents into Arabic. Given the early conquest of Jundishapur, Persian and Indian influences, rather than Greek, were more influential in the early stages of Islamic civilization. Already in the 760s, for example, an Indian mission reached Baghdad to teach Indian science and philosophy and to aid in translations of Indian astronomical and mathematical texts from Sanskrit into Arabic. Later, Muslim men of science traveled to India to study with Indian masters.

In the following century, however, the translation movement came to focus on Greek scientific works. The governing caliph in Baghdad, Al-Ma'mun, founded the House of Wisdom (the *Bayt al-Hikma*) in A.D. 832 specifically as a center of translation and mastery of the secular foreign sciences. Al-Ma'mun sent emissaries to collect Greek scientific manuscripts from Byzantine sources for the House of Wisdom where families of scholar-translators, notably Ishāq ibn Hunayn and his relatives, undertook the Herculean task of rendering into Arabic the Greek philosophical and scientific tradition. As a result, virtually the entire corpus of Greek natural science, mathematics, and medicine was brought over into Arabic, and Arabic became *the* international language of civilization and science. Ptolemy's *Almagest,* for example—the very title, *al-Mageste,* is Arabic for "the greatest"—appeared in several translations in Baghdad early in the ninth century, as well as Euclid's *Elements,* several works of Archimedes, and many of Aristotle, beginning with his logical treatises. Aristotle became the intellectual godfather of Islamic theoretical science, spawning a succession of commentators and critical thinkers. A measure of the effort expended on translating Greek texts is that, even now, more Aristotelian writings— the works of Aristotle and his Greek commentators—supposedly are available in Arabic than in any European language.

Al Ma'mun supported his translators and the House of Wisdom, not merely out of the love of learning, but for practical utility deemed directly useful to the monarch, notably in such fields as medicine, applied mathematics, astronomy, astrology, alchemy, and logic. (Aristotle became assimilated initially for the practical value of his logic for law and government, and only later did the entire body of his scientific and philosophical works find its way into Arabic.) Medicine was the primary field naturalized by Islamic translators; Ishāq ibn Hunayn alone supposedly translated 150 works of Galen and Hippocrates. Thus, by A.D. 900, while Europe possessed perhaps three works of Galen, transcribed by solitary scholars, Islam had 129 produced under government patronage. The basis had been established for a great scientific civilization.

In the Islamic world the secular sciences were generally not valued for their own sakes, but rather for their utility; secular knowledge was normally not pursued by individualistic natural philosophers as an end in itself as in Hellenic Greece or later in Christian Europe. To this extent, the "marginality" thesis provides a degree of insight into the place of pure science in Islamic society. Nevertheless, such a view slights the ways in which science became patronized and institutionalized in a variety of social niches in Islamic culture. As social history, the "assimilationist" thesis more properly portrays the role and institutionalized character of science and natural knowledge in Islam.

Each local mosque, for example, was a center of literacy and learning, albeit largely religious. But mosques also had official timekeepers (the *muwaqqit*) who set times for prayer. This recondite and exact procedure could only be effected by competent astronomers or at least trained experts. Thus, for example, afternoon prayers occur when the shadow of an object equals the length of its shadow at noon plus the length of the object. Several esoteric geographical and seasonal factors determine these times, and the *muwaqqit* used elaborate timekeeping tables, some with upwards of 30,000 entries, supplemented by instruments such as astrolabes and elaborate sundials to ascertain when prayer should take place. (The astrolabe became a highly developed instrument capable of solving 300 types of problems in astronomy, geography, and trigonometry.) Similarly, the faithful prayed in the direction of Mecca, and therefore geographical knowledge also had to be applied locally to discover that direction. Astronomers determined the beginning of Ramadan, the month-long period of daily fasts, and the hour of dawn each day. Along these lines, each local Islamic community possessed a mathematically and legally trained specialist, the *faradi,* who superintended the division of inheritances.

The Islamic legal college, or *madrasa,* was an institution of higher learning wherein some "foreign sciences" were taught. Widespread throughout the Islamic world, the madrasa was primarily an advanced school for legal instruction in the "Islamic sciences"—law, not theology, being the preeminent science in Islam. The madrasa should not be equated with the later European university, in that the madrasa was not a self-governing corporation (prohibited in Islam). It did not maintain a standard curriculum, and it did not confer degrees. Technically a charitable endowment rigidly bound by its founding charter and prohibited from teaching anything contrary to the fundamental tenets of Islam, the madrasa operated more as an assemblage of independent scholars with whom students studied on an individual basis and where instruction emphasized memorization, recitation, and mastery of authoritative texts. Endowments supported instructors and paid the tuition, room, and board of students.

The secular sciences found a niche in these institutions of higher learning. Logic, for example, was taken over from Greek traditions,

Fig. 5.1. An astrolabe. This multifaceted device was invented in Islam to facilitate astronomical observation and to solve problems relating to time-keeping, geography, and astronomy.

and arithmetic was studied for the purposes of training the faradi for handling inheritances. Similarly, geometry, trigonometry, and astronomy, although tightly controlled, likewise came within the fold of Islamic studies because of the religious needs of determining proper times for prayer and the direction of Mecca. While not publicly professed, specialists also offered private instruction in the "foreign sciences" outside the formal setting of the madrasa. And secular scientific and philosophical books could be found in public libraries associated with madrasas and mosques. In a word, then, the student who wished to learn the natural sciences could do so at a high level of sophistication in and around the institution of the madrasa.

The library formed another major institution of Islamic civilization wherein the natural sciences were nurtured. Often attached to madrasas or mosques, usually staffed by librarians and open to the public, hundreds if not thousands of libraries arose throughout the Islamic world. Córdoba alone had seventy libraries, one containing between 400,000 and 500,000 volumes. Thirty madrasas existed in Baghdad in the thirteenth century, each with its own library, and 150 madrasas operated in Damascus in 1500 with as many libraries. The library attached to the observatory in Maraghah reportedly contained 400,000 volumes. Another House of Wisdom (the *Dār al-'ilm*) in tenth-century Cairo contained perhaps 2 million books, including some 18,000 scientific titles. One collector boasted that it would take 400 camels to transport his library; the estate of another included 600 boxes of books, manhandled by two men each. The tenth-century physician Ibn Sīnā (980–1037), known in the West as Avicenna, left an account of the impressive quality of the royal library in Muslim Bukhara on the Asian outskirts of Islam:

> I found there many rooms filled with books which were arranged in cases, row upon row. One room was allotted to works on Arabic philology and poetry, another to jurisprudence and so forth, the books on each particular science having a room to themselves. I inspected the catalogue of ancient Greek authors and looked for the books which I required; I saw in this collection books of which few people have heard even the names, and which I myself have never seen either before or since.

In sharp contrast, libraries in medieval Europe numbered only hundreds of items, and as late as the fourteenth century the library collection at the University of Paris contained only 2,000 titles, while a century later the Vatican library numbered only a few hundred more. But the love of learning alone could not have accounted for Islamic libraries. The formation of huge collections was clearly dependent on the willingness of caliphs and wealthy patrons to underwrite the costs. It was also dependent on paper-making, a new technology acquired from the Chinese in the eighth century which allowed the mass production of paper and much cheaper books. Paper factories appeared in Samarkand after 751, in Baghdad in 793, in Cairo around 900, in Morocco in 1100, and in Spain in 1150. In Baghdad alone 100 shops turned out paper books. Ironically, when the printing press appeared in the fifteenth century Islamic authorities banned it for fear of defiling the name of God and to prevent the proliferation of undesirable materials.

Although astronomers had previously observed the heavens, Islamic civilization created a new and distinctive scientific institution: the formal astronomical observatory. Underwritten by ruling caliphs and sultans, observatories, their equipment, and staffs of astronomers discharged several practical functions. Astronomers prepared increasingly accurate astronomical handbooks *(zij)* for calendrical and religious ends—to fix the times of prayer and of religious observances such as

Ramadan. The Islamic calendar was a lunar calendar, like that of ancient Babylonia, of 12 months of 29 or 30 days unfolding over a 30-year cycle, with trained observers determining when the new moon commenced. Geography was also closely connected to astronomy and, beginning with Ptolemy's *Geography*, Muslim astronomers developed navigational and geographical techniques serviceable to both sailors and desert travelers.

Islamic authorities formally distinguished between astronomy as the study of the heavens and astrology as investigating heavenly influence on human affairs. The distinction may have facilitated the social integration of astronomy, but the strongest single motive behind royal patronage of astronomy remained the putative divinatory power of astrology. Despite its occasional condemnation by religious authorities on the grounds that it misdirected piety toward the stars rather than God, astrology remained the most popular of the secular sciences, and it flourished especially in court settings, where regulations and exams fixed the qualifications, duties, and salaries of astrologers. Elsewhere, the local chief of police regulated astrology as a marketplace activity. Along with Ptolemy's *Almagest,* Muslim astronomer/astrologers had available his astrological treatise, the *Tetrabiblos,* and many used it and like volumes to cast horoscopes and gain patronage as court astrologers.

Observatories arose throughout the Muslim world. Al-Ma'mun founded the first around 828 in Baghdad. The best known, established in 1259, was the observatory at Maraghah in a fertile region near the Caspian Sea. It was formed in part to improve astrological prediction. A substantial library was attached to the observatory and actual instruction in the sciences was offered there with government support. Expert astronomers made up what can only be called the Maraghah school, and such men as al-Tūsī (d. 1274), al-Shīrāzī (d. 1311), and their successor, Ibn al-Shātir (d. 1375), far surpassed ancient astronomy and astronomical theory in perfecting non-Ptolemaic (although still geocentric) models of planetary motion and in testing these against highly accurate observation. But, the observatory at Maraghah, like many others, proved short-lived, lasting at most 60 years. Even though protected by non-Islamic Mongol rulers, the Maraghah observatory and several other Islamic observatories were closed by religious reaction against impious study of astrology.

Farther north and east, in fifteenth-century Samarkand, sustained by irrigated orchards, gardens, and cropland, the celebrated Muslim scholar-prince Ulugh Beg (1393–1449) founded a madrasa and a major observatory. The importance that Islamic astronomers attached to the precision of their observations necessitated the use of exceptionally large instruments, such as the three-story sextant at Samarkand with a radius of 40 meters (132 feet). These large instruments, along with the observatory structures, the staffs of astronomers and support personnel, and their affiliated libraries entailed costs so high that they could only be met through government support. Through its observatories medieval Islam established a tradition of observational and theoretical astronomy unequaled until the achievements of European science in the sixteenth and seventeenth centuries.

Islamic mathematics, while justly renowned, consistently displayed a practical trend in its emphasis on arithmetic and algebra rather than on the formal theoretical geometry of the Greeks. Medieval Islamic mathematicians also developed trigonometry, which greatly facilitated working with arcs and angles in astronomy. The adoption of easily manipulated "Arabic numerals" from Indian sources further reflects this practical orientation. While Islamic mathematicians solved what were, in effect, higher-order equations, many problems had roots in the practical world dealing with taxes, charity, and the division of inheritances. The ninth-century mathematician al-Khwarizmi, for example, who originally introduced "Arabic numerals" from India, wrote a manual of practical mathematics, the *al-Jabr* or what came to be known in the West as the *Algebra*. Not coincidentally, al-Khwarizmi worked at the court of al-Ma'mun.

Islamic medicine and its institutionalized character deserve special attention. The Arabs had their own medical customs, and the Quran (Koran) contains many sayings of the Prophet regarding diet, hygiene, and various diseases and their treatment. The Arabic translation movement made available to physicians all of the Hippocratic canon and the works of Galen, notably through the texts of ancient Greek medicine preserved at Alexandria. Islamic medicine also assimilated Persian and Indian traditions, in part from having taken over the medical school at Jundishapur and in part from direct contact with India through the drug and perfume trades. The resulting medical amalgam became thoroughly naturalized and integrated into the social fabric of Islam.

A handful of madrasas specialized in medical training, but the hospital became the primary institutional locus of Islamic medicine. Government-supported hospitals existed throughout the Islamic world, with especially notable medical centers in Baghdad, which eclipsed Jundishapur, Damascus, which saw the foundation of six hospitals between the thirteenth and fifteenth centuries, and Cairo. Many hospitals came to possess elaborate medical staffs, specialized medical wards, attached medical libraries, and lecture halls (*majlis*). Islamic hospitals thus evolved as centers of teaching and research, as well as dispensaries of medical treatment, including medical astrology. And, whereas guilds and corporate structures were never recognized in Islamic societies, governments licensed physicians through local police officials. Islamic doctors, such as al-Rāzī (Rhazes, 854–925), al-Majūsī (Haly Abbas, d. 995), Ibn Sīnā (Avicenna) and others developed unprecedentedly sophisticated and expert understanding of diseases and medical treatments.

The medical dimension may help explain a particular strength of Islamic science in optics. Especially in Egypt, where desert conditions contributed to eye ailments, a strong medical literature developed in ophthalmology, and Islamic physicians became expert in the treatment of the eye and the anatomy and physiology of vision. Although not a physician, the great Islamic physicist Ibn al-Haytham (Alhazen, 965–1040) worked in Egypt and wrote on eye diseases. His *Optics* is only the best known and most influential of a series of Islamic scientific works—many with an experimental approach—concerned with vision, refraction, the camera obscura, burning mirrors, lenses, the rainbow, and other optical phenomena.

Physicians enjoyed high public regard, and many Muslims who made scientific and philosophic contributions earned their living as court physicians or court-appointed administrators and legal officials. For example, Averroës (Ibn Rushd, 1126–98), known as "The Commentator" on Aristotle, worked as a court physician and religious jurist in Spain. The Islamic polymath Avicenna (Ibn Sīnā), renowned as the "Galen of Islam," accepted patronage as a physician in various courts in order to pursue philosophy and science. The noted Jewish philosopher and savant Moses Maimonides (Musa ibn Maymun, 1135–1204) acted as physician to the sultan at Cairo. In a word, court patronage provided institutionalized positions where physician-scientists could master and extend the secular sciences, and court positions afforded a degree of insulation from the dominant religious institutions and the supremacy of religious law in Islamic society at large.

Closely associated with courts and the patronage of rulers, a highly developed tradition of Islamic alchemy involved many scientists. Alchemy ranked among the sciences, being derived from Aristotle's matter theory. In the search for elixirs of immortality, Islamic alchemy also seems to have been influenced by Chinese alchemy, and it likewise subsumed work on mineralogy, which showed Indian and Iranian influences. Alchemy was a secret art, and adepts attributed some 3,000 alchemical texts to the founder of Islamic alchemy, the ninth-century figure Jābir ibn Hayyān, known as Geber in the Latin West. On one level, no doubt the one most appreciated by patrons, the transformation of base metals into gold and the creation of life-giving elixirs represented the goals of alchemy. To many practitioners, however, Islamic alchemy became a highly intellectual endeavor that primarily involved the spiritual refinement of the individual alchemist. In pursuing their science, Islamic alchemists invented new equipment and perfected new techniques, including distillation. Residues of Islamic alchemy remain in Arabic-derived terms, such as the word *alchemy* itself, *alcohol, alkali,* and *alembic.* Indeed, in such terms as *algebra, azimuth, algorithm,* and a host of others, the language of science to this day maintains the linguistic imprint of Arabic and the history of Islamic science.

The sheer institutional density of Islamic science accounts for some of its achievements and characteristics. Scholars and scientists staffed schools, libraries, mosques, hospitals, and especially observatories with their teams of astronomers and mathematicians. The opportunities and support that these institutions offered men of science produced a remarkable upsurge of scientific activity, as measured by the number of Islamic scientists which surpassed by an order of magnitude the handful of Europeans pursuing science before A.D. 1100. Another result was a characteristic research profile, like that of the ancient bureaucratic kingdoms, which exaggerated utility, public service, and the interests of the state.

Technology and industry in medieval Islam gave as little to and received as little from the realm of science as they had in the Greco-Roman world. Islamic science embraced much of ancient Greek learning, as we have seen, but Islamic technology remained more akin to that of Rome and the eastern kingdoms. In architecture the Muslims employed the Roman arch rather than the Greek post and lintel system of building. And agriculture depended heavily on hydraulic engineering as it had in the Roman provinces and in all civilizations in the Near East. Indeed, the Islamic conquest maps closely onto regions that lent themselves to hydraulic intensification; Greece and Italy, where artificial irrigation was less important, did not become Islamicized, while Spain saw a dramatic development of hydraulic technology under Muslim rule. The construction of large dams, waterwheels, and qanats (underground channels with earthenware pipes designed to tap ground water) all formed part of the Islamic engineering repertoire. In Iran qanats supplied fully half of the water used for irrigation and urban needs. Such were the feats of craftsmen and artisans divorced from the bookish worlds of theology and science.

Scholars disagree on when the vitality of scientific activity started to decline in the Islamic world. Some say that the decline began after the twelfth century, especially in the Western regions; others say that important new science continued to be done in the East until the fifteenth and sixteenth centuries. However, no one denies that Islamic science and medicine reached their historical golden age in the centuries surrounding the year 1000 and that decline in the creative level of original work eventually set in. It should be noted that such a consensus has tended to obscure the ways in which knowledge in mosques and madrasas continued to function in Islamic society for centuries, irrespective of any "decline" in the quality of original science. That point notwithstanding, several suggestions have been offered to account for the eventual decline of the Islamic scientific traditions, all of them external and sociological, for nothing in the internal logic of scientific ideas can account for the loss of vigor of Islamic science.

The main thesis has centered on the ultimate triumph of religious conservatives within Islam. As a religion, Islam emphasizes submission before the divine and unknowable nature of God/Allah. Thus, accord-

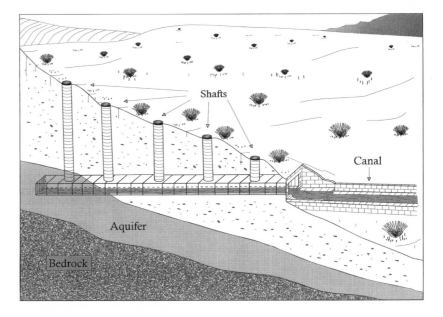

Fig. 5.2. Qanat technology. Artificial irrigation sustained Islamic agriculture and civilization. Islamic engineers developed sophisticated hydraulic techniques, including qanats, which tapped underground water sources.

Shafts

Canal

Aquifer

Bedrock

ing to the "marginality" thesis, the cultural values and legal tenets of Islam proved such that secular philosophy and learning were always suspect to varying degrees and remained peripheral to the mainstream of Islamic society. Individual jurists and religious leaders, for example, could and did sometimes issue religious rulings *(fatwas)* for impiety against those who became too expert in the secular sciences. Different factions within Islam contended over the value of human reason and rationality in pursuing understanding, but ultimately, so the argument goes, religious conservatives prevailed, and with increasing intolerance the creative spirit of Islamic science evaporated. Why it flourished and why it declined when it did lie beyond the reach of marginalist explanations.

A related suggestion notes that Islamic civilization was more pluralistic at its outset and that science declined as the Islamic world became culturally more homogeneous. In many conquered areas religious believers were initially in the minority. Islam began as a colonial power, and especially at the edges of the Islamic imperium multicultural societies flourished at the outset, mingling diverse cultures and religions— Persian, Indian, Arab, African, Greek, Chinese, Jewish, and Christian. As time went on, conversions increased, and Islam became religiously more rigid and culturally less heterogeneous. Not until the fourteenth century was Islamicization fully complete in many areas. Consequently, the cultural "space" for creative scientific thinkers narrowed and so, again, the scientific vitality of Islam weakened commensurately. However, this account flies in the face of the fact that in its heyday Islamic science often flourished in the most Islamicized centers, such as Baghdad.

War and sociocultural disruptions occasioned by war have likewise been invoked as factors in the decline of Islamic science. In Spain the Islamic world began to be pressured by Christian Europe in the eleventh century, with Toledo falling in 1085, Seville in 1248, and the *reconquista* completed in 1492. In the East, Mongol armies from the steppes of Asia attacked the Islamic caliphate, invading and capturing Baghdad in 1258. Mongol invaders led by Timur (Tamerlane) returned to the Middle East at the turn of the fifteenth century, destroying Damascus in 1402. Although Islamic culture and institutions in the East quickly rebounded from these invasions, the overall effect, or so it is claimed, reinforced religious conservatism and disrupted the conditions necessary for the pursuit of science.

Other experts have focused on the economic decline of Islamic civilization after 1492 as a contributing factor in the cultural decline of its science. That is, once European seafaring traders penetrated the Indian Ocean in 1497, the Islamic world lost its monopoly on the valuable East Asian spice and commodity markets. In such shrinking economic circumstances, the argument suggests that science could hardly have been expected to flourish, especially since it leaned heavily on government support.

Each of these interpretations doubtless possesses some truth, and further historical research will shed more light on understanding the decline of Islamic science. But commentators have also wanted to explain not the decline of Islamic science but the very different question of why modern science did not emerge within the context of Islamic civilization. The question often posed is why, given the advanced state of Islamic science, no Scientific Revolution developed within Islam—why did Islamic scientists not repudiate the earth-centered cosmology of antiquity, expound modern heliocentrism, and develop inertial, Newtonian physics to account for motion in the heavens and on Earth?

Much intellectual energy has been expended in dealing with the Islamic "failure" to make the leap to modern science. But to undertake to explain in retrospect the absolute myriad of things that *did not* happen in history confounds the enterprise of historians, who have a difficult enough time rendering plausible accounts for what *did* happen. As evident in this chapter, Islamic science flourished for several centuries, securely assimilated in observatories, libraries, madrasas, mosques, hospitals, and ruling courts. That was its positive achievement. Islamic scientists all labored within the pale of Islam, and they continued to do so for several centuries following the peak of Islamic scientific achievement. To suggest that science somehow "ought" to have developed as it did in the West misreads history and imposes chronologically and culturally alien standards on a vibrant medieval civilization.

CHAPTER 6

The Middle Kingdom

Although borders and political units fluctuated, Chinese emperors controlled a huge, densely populated territory about the size of Europe. Even China proper (excluding Manchuria, Mongolia, Tibet, and western regions) encompassed about half the area of Europe and was still seven times the size of France (see Map 6.1). From its first unification in 221 B.C. China was the world's most populous country, and except for being briefly eclipsed by the patchwork Roman Empire, the successive empires of China stood as the largest political entities in the world. The population of China proper reached 115 million in A.D. 1200, twice that of contemporary Europe and with nearly five times Europe's population density.

Geography isolated China from outside influences more than any other Old World civilization. Nomadic and pastoral peoples to the north and west had a large effect on Chinese history, to be sure, but mountains, deserts, and inhospitable steppe ringed China on the southwest, west, and north, and impeded contact with cultural and historical developments in West Asia and Europe. The earliest Chinese civilization arose in the valley of the Hwang-Ho (the Yellow River), and only later in historical periods did civilization spread to the valley and flood plain of the Yangtze River. China represents an archetypal hydraulic civilization whose cultural orientation faced eastward along these and related river and lake systems.

The technology of writing developed independently in China. A complex ideographic type of writing can be seen in the "oracle bone script" in the Shang dynasty (1520–1030 B.C.). It became highly developed with upwards of 5,000 characters by the ninth century B.C., and characters became standardized by the time of China's unification. Hundreds of basic signs could be combined into thousands (indeed, tens of thousands) of different characters. Because of this complexity and because each Chinese written word embodies phonetic and pictographic elements, Chinese writing was (and is) difficult to master. In adhering to

Map 6.1. China. Chinese civilization originated along the Yellow (Hwang-Ho) River in the second millennium B.C. Mountains, deserts, and steppe regions in the north and west effectively cut China off from the rest of Asia. The first unification of China occurred in the third century B.C., producing the largest and most populated political entity in the world. The map shows two of the great engineering works of Chinese civilization: the Great Wall and the Grand Canal.

the ideographic mode, Chinese writing did not simplify phonetically or syllabically as did ancient Egyptian, Sumerian, and Old Babylonian, but that "obstacle" did not impede the long, unbroken tradition of Chinese literacy and the impressive Chinese literary and scientific record from the second millennium B.C.

China embodies thousands of years of cultural continuity, and one cannot adequately trace here the intricate social and political changes observable in China's history, as various empires rose and fell (see Table 6.1). Nevertheless, the Sung dynasties (A.D. 960–1279) and the "renaissance" accompanying the Sung command attention. In many ways the Sung period represents the zenith of traditional China. The several centuries of Sung rule formed the golden age of Chinese science and technology, and they provide an effective point of contrast with contemporary developments elsewhere in the world.

The flowering of China under the Sung resulted from agricultural changes, notably the upsurge of rice cultivation in South China and in the Yangtze basin beginning in the eighth century. Rice paddies produce a higher yield per acre than any other cultivated crop, so the mere introduction of rice inevitably produced significant social and cultural consequences. After 1012 the government introduced and systematically distributed new varieties of early-ripening and winter-ripening rice from Indochina. Some varieties ripened in 60 days, allowing two and even three harvests a year in favored locales. Other varieties required less water, which meant that new lands could be brought under cultivation. The Sung made major efforts to extend rice production by reclaiming marshlands and lakesides, by terracing, and by improving irrigation, all under government direction. The new technique of planting out seedling rice plants eliminated the need for fallow, and the introduction of new tools for cultivating rice, such as the rice-field plow and paddle-chain water-lifting devices, likewise improved efficiency and productivity enough to provide increasingly large surpluses.

The consequences proved dramatic. The population of China more than doubled from 50 million in A.D. 800 to 115 million (one census reports 123 million) in 1200. The center of gravity of Chinese civilization shifted south, with more than twice as many Chinese living in the south than the north by 1080. Urbanization likewise skyrocketed. According to one report, Sung dynasty China contained five cities with populations of more than a million, and another estimate puts the urban population at 20 percent of the total, a remarkably high figure for an agrarian society, one not reached in Europe until the nineteenth century. A leisured middle class arose along with the commercialization of agricultural commodities, increased trade, and expanded manufacturing.

Centralization of power in the hands of the emperor and rule by a governing bureaucracy—the mandarinate—reached new heights under the Sung. The "mandate of heaven" dictated that the Chinese emperor rule all of China, and an improved civil service to support that mandate proved pervasive in Chinese life. The bureaucracy was huge and monolithic; a later report from Ming times puts the number of state civil servants at 100,000, exclusive of military officers. Such organization allowed direct control by the emperor down to the village level. No intermediary or independent bodies existed in China to challenge the authority of the emperor and the mandarinate. Different traditional provinces and linguistic regions acted as something of brakes to centralizing forces, but no other formal centers of power existed. Towns and cities were neither autonomous nor separate administrative units. Such an exclusive and centralized administration prevented the rise of independent institutional entities, notably colleges or guilds. The omnipresence of the Chinese mandarinate seems also to have restricted any neutral social or intellectual space for science or technology outside of official channels.

The teachings of the Chinese sage Confucius (551–479 B.C.) dra-

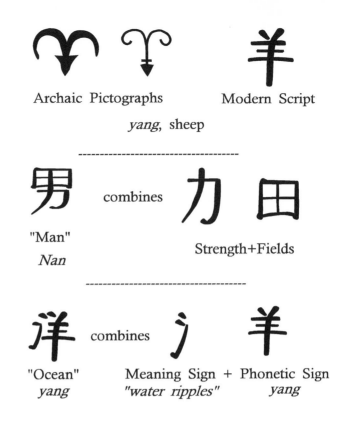

Fig. 6.1. Chinese pictographic writing. Written Chinese languages derive from pictographic antecedents. Various characters can be combined to form new word signs. In some cases parts of characters indicate the sound of the word and/or distinguish the general class of things to which the word in question belongs. Unlike other languages with pictographic origins, written Chinese never simplified into a wholly phonetic or sound-based script, and literacy in China still entails mastery of hundreds of separate word signs. Such difficulties did not prevent the Chinese language from developing sophisticated technical and scientific vocabularies.

matically shaped Chinese high culture, particularly through the official state ideology of Neo-Confucianism elaborated by commentators in Sung times. The Confucian outlook focused on the family, humanity, and society, not nature or the world outside of human affairs. Confucianism was a practical philosophy that emphasized the ethical and moral dimensions of behavior and statecraft and the maintenance of a just and harmonious society. Thus, custom, etiquette, virtuous behavior, filial piety, respect for one's elders, submission to authority, the moral example of the sage, and justice (but not law) became the watchwords of Confucianism in Sung times. In these ways Confucianism sustained the status quo and the paternalistic and patriarchal society of the day.

The power and appeal of the imperial bureaucracy drained talent that might have flowed into science. The bureaucracy skewed scholarship toward the humanities and the Confucian classics, and it helped enforce a deep divide between learned culture and the crafts. Under the Sung, the imperial bureaucracy operated as a true meritocracy open to talent. The state recruited functionaries not through political or hereditary connections, but rather based on ability and performance on exacting state civil-service exams, which provided virtually exclusive access to political power. Already in Han times (206 B.C.–A.D. 220) Chinese officials instituted the system of state examinations, one effect of which was to restrict the political power of the nobility. The Sung dynasties reformed the system, and it reached its high point under their rule and continued in effect in China until 1904.

An official board of examiners offered three levels of examination (local, regional, and national) every two to three years. Some unfortunate students devoted their whole lives to taking and retaking the rigorous exams. Passage of even the lowest level exam brought special privileges, such as exemption from the labor of the corvée. Passing at the high levels incurred obligations since laureates could not refuse employment in the bureaucracy. Based on a standardized subject matter, the exams focused on Confucian classics, on esoteric literary and humanistic studies, and, under the Sung, on administrative problems. Memorization took pride of place, along with recitation, poetry, and calligraphy. With the emphasis on moral learning and the goal of producing a scholar-gentry to rule the country, the civil service exams shaped the values and efforts of the best and brightest Chinese minds for nearly 2,000 years. Certain exceptions aside, science and technology did not figure in the exam system.

Outside the bureaucracy, other elements of society lacked the power and autonomy to be independent sources of any nascent scientific tradition. If the exam system effectively precluded rule by nobles, civilian authority also managed to subordinate military and merchant classes to its power. From the third century B.C. China possessed large armies— on the order of a million soldiers. (Sung armies numbered 1,259,000 men in 1045.) Yet, the military remained subject to civilian control. Military power was divided, units split up, and overlapping commands established. Merchant activity was likewise tightly controlled so that, unlike in Europe, merchants never rose to social or institutional positions of consequence. From the Confucian point of view, merchant activity, profit, and the accumulation of private wealth were disdained as antisocial vices. Merchants occasionally flourished and achieved great wealth, but periodic prosecutions and confiscations ensured the marginality and low status of merchants as a class in Chinese society. Likewise, the suppression of religious institutions in A.D. 842–845, following a period of Buddhist prominence, meant that no clergy could challenge the predominance of the bureaucracy.

The Flowering of Chinese Technology

Learned culture in traditional China was largely separate from technology and the crafts. Calendrical astronomy benefited the state and society, and mathematics played a role in the solution of practical problems, but economic, military, and medical activities were, on the whole, carried out on the strength of traditional techniques that owed nothing to theoretical knowledge or research. Craftsmen were generally illiterate and possessed low social status; they learned practical skills

Table 6.1.
A Simplified Chronology of Chinese Dynasties

First Unification	
Ch'in	221 B.C.–207 B.C.
Han	209 B.C.–A.D. 220
Third Unification	
T'ang	A.D. 613–903
Third Partition	
Five Dynasties Period	907–960
Fourth Unification	
Northern Sung (at Kaifeng)	960–1126
Falls to Chin	
Southern Sung (at Hangchow)	1127–1279
Falls to Mongols	
Chin (Jurchen Tartar)	1115–1234
Annexes North China, 1126	
Falls to Mongols, 1234	
Yüan (Mongol)	1260/1271–1368
Annexes Chin, 1234	
Conquers Southern Sung, 1276–79	
Ming	1368–1644
Chhing (Manchu)	1644–1911

through apprenticeship and experience, and they plied their trades without benefit of scientific theory. Scholars and "scientists," on the other hand, were literate, underwent years of schooling, enjoyed high social status, and remained socially apart from the world of artisans and engineers. The exam system and the bureaucracy, by institutionally segregating scholar-bureaucrats from artisans, craftsmen, and engineers, strengthened the separation of science and technology. The value system of traditional China, like that of Hellenic Greece, looked down upon crass technology. Scholars and literati repudiated working with their hands and preferred more refined concerns such as poetry, calligraphy, music, and belles-lettres.

In considering Chinese technology one must be wary of a tendency to record the priority of the Chinese over other civilizations for this or that invention: the wheelbarrow, the south-pointing chariot, lacquer, gunpowder, porcelain china, the umbrella, the fishing reel, suspension bridges, and so on. While such "firsts" are interesting, they are of limited analytical value in historical inquiry. Rather, the starting point for any investigation of Chinese technology must be the realization that the totality of its advanced technologies, regardless of their originality or priority, made China a world leader in technology through the Sung era and beyond.

Government control of industries was a characteristic feature of Chinese technology. The government nominally owned all resources in the country, and it monopolized production in key sectors by creating government workshops and state factories for such industries as mining, iron production, salt supply, silk, ceramics, paper-making, and alcoholic beverages. Through these monopolies run by bureaucrats, the Chinese state itself became a merchant producer, in large part to provide for its enormous military needs. The government commanded a vast array of specialized craftsmen, and anyone with technical skills was ostensibly subject to government service. The Yüan emperors, for example, enlisted 260,000 skilled artisans for their own service; the Ming commanded 27,000 master craftsmen, each with several assistants; and in 1342 17,000 state-controlled salt workers toiled along the lower Yangtze.

State management of technology and the economy reached a high point in the Sung period when more government income came from mercantile activity and commodity taxes than from agricultural levies. One result was the spread of a monied economy. Coinage issuing from government mints jumped from 270,000 strings (of a thousand coins) in 997 to 6 million strings in 1073. As a result of that increase, the Sung began issuing paper money in 1024, and paper money became the dominant currency in twelfth- and thirteenth-century China. The technology of paper money is significant not as a world historical "first," but because it facilitated the growth and functioning of Chinese civilization.

Hydraulic engineering represents another basic technology underpinning Chinese civilization. We earlier encountered the essential role of irrigation agriculture in discussing the initial rise of civilization along the Hwang-Ho river in the second millennium B.C. While many canals and embankments existed in China from an early date, the first elements of an empire-wide inland canal system appeared about A.D. 70. Engineers completed the nearly 400 miles of the Loyang to Beijing canal in A.D. 608 and by the twelfth century China possessed some 50,000 kilometers (31,250 miles) of navigable waterways and canals. Completed in 1327, the Grand Canal alone stretched 1100 miles and linked Hangchow in the south with Beijing in the north, the equivalent of a canal from New York to Florida. After the Ming took power they repaired 40,987 reservoirs and launched an incredible reforestation effort in planting a billion trees to prevent soil erosion and to supply naval timber. Of course, such impressive engineering was impossible without the central state to organize construction, to levy taxes, and to redistribute the agricultural surplus. Canals allowed rice to be shipped from agricultural heartlands in the south to the political center in the north. One report has 400,000 tons of grain transported annually in the eleventh century. In Ming times 11,770 ships manned by 120,000 sailors handled inland shipping. Considerable maintenance and dredging were obviously required, all of it carried out by corvée labor, and the neglect of hydraulic systems inevitably led to famine and political unrest.

Pottery was an ancient craft that reached unprecedented artistic heights after the eleventh century. The imperial government owned its own industrial-scale kilns and workshops which came to employ thousands of craftsmen mass-producing both commonplace and luxury items. The Chinese originated porcelain—a mixture of fine clays and minerals fired at a high temperature—at the end of Han times and perfected porcelain wares in the twelfth century. The enduring art and technology of Chinese porcelain represent one of the great cultural achievements of the Sung and Ming eras. They bespeak a wealthy and cultivated society, and, indeed, ceramics became a major item of internal and international commerce and of tax income for the state. Chinese pottery made its way through the Islamic world and to Africa. From the Middle Ages onward Europeans came to covet Chinese porcelains, and efforts to duplicate Chinese ceramic technology proved a spur to the pottery industry in Europe at the time of the Industrial Revolution in the eighteenth century.

Textiles constitute another major industry in traditional China. One twelfth-century Sung emperor, for example, purchased and received in taxes a total of 1.17 million bolts of silk cloth. The Chinese textile industry is especially notable because of its mechanized character. Sources document the presence of the spinning wheel in China from A.D. 1035, and Chinese technologists also created elaborate, water-powered reeling machines to unwind silkworm cocoons and wind silk thread onto bobbins for weaving into cloth. And paper manufacturing, possibly evolving out of the textile industry, provided a product that facilitated the administration of imperial China. Solid evidence exists for paper from late Han times early in the second century A.D., although the technology may have originated several centuries earlier.

Chinese bureaucracies depended on writing, literary traditions, and libraries, which already existed in the Shang dynasty in the second millennium B.C. Although paper entered Chinese society at an early date, the technology of taking rubbings from carved inscriptions may have delayed the advent of printing until the first decade of the seventh century. Printing—block printing—at first simply reproduced seals for religious charms. The first book printed by means of whole pages of carved woodblock appeared in A.D. 868, and the technology of printing soon recommended itself to government authorities who used it to print money, official decrees, and handbooks, particularly useful ones in medicine and pharmacy. An official printing office supplied printed copies of the classics to be studied for the civil-service exams, and overall the Chinese government produced an impressive output of printed material in the service of the bureaucracy. The first emperor of the Sung dynasty, for example, ordered a compilation of Buddhist scripture, and the work, consisting of 130,000 two-page woodblocks in 5,048 volumes, duly appeared. In 1403 an official Chinese encyclopedia numbered 937 volumes and another of 1609 comprised 22,000 volumes written by 2,000 authors.

The Chinese invented movable type around 1040, first using ceramic characters. The technology developed further in Korea where the Korean government had 100,000 Chinese characters cast in 1403. But movable type proved impractical compared to woodblock printing, given the style of Chinese writing with pictograms and the consequent need for thousands of different characters. Block printing thus proved not only cheaper and more efficient, but it allowed illustrations, often in multicolors. The ability to reproduce pictures put China well ahead of the West in printing technology even after Gutenberg developed movable type in Europe.

Chinese superiority in iron production likewise helps account for the vibrancy of its civilization. Possibly because of limited resources of copper and tin for bronze, Chinese metallurgists early turned to iron. By 117 B.C. iron production had become a state enterprise with forty-eight foundries, each employing thousands of industrial laborers. Production zoomed from 13,500 tons in A.D. 806 to 125,000 tons in 1078 in the Sung period, doubtless because of increased military demands. (By contrast England produced only 68,000 tons of iron in 1788 as the Industrial Revolution got under way in Europe.) Technologically innovative and advanced, the Chinese iron industry used water-powered bellows to provide a blast and smelted the ore with coke (partially combusted coal) by the eleventh century, some 700 years before like processes arose in Europe. By dint of such superior technology Sung military arsenals turned out 32,000 suits of armor and 16 million iron arrowheads a year, as well as iron implements for agricultural use.

The invention of gunpowder in mid-ninth-century China, and, more significantly, the application of gunpowder to military ends beginning in the twelfth century redirected the course of Chinese and world history. Gunpowder seems to have emerged from traditions of Chinese alchemical research, and its initial use in fireworks was intended not as a military device but as a means to ward off demons. Only as they became threatened by foreign invasion did Sung military engineers improve the formula for gunpowder and develop military applications in rockets, explosive grenades, bombs, mortars, and guns.

Unlike paper, the magnetic compass was a technology that Chinese civilization could get along without, but the case illuminates the few ties between science and technology in traditional China. The mysterious properties of the loadstone—the natural magnetism of the mineral magnetite—were known by 300 B.C. and first exploited for use as a fortuneteller's device. Knowledge attained by 100 B.C. that a magnetic needle orients itself along a north-south axis was then applied in geomancy or *fêng-shui*, the proper siting of houses, temples, tombs, roads, and other installations. An elaborate naturalistic theory later arose to explain the movement of the compass needle in response to energy currents putatively flowing in and around the earth, an example of how, contrary to conventional wisdom today, technology sometimes fosters speculations about nature rather than the reverse.

Sources fail to attest to the use of the compass as a navigational tool at sea until Sung times early in the twelfth century. China entered late as a major maritime power, but from the period of the Southern Sung through the early Ming dynasties, that is, from the twelfth through the early fifteenth centuries, China developed the largest navy and became the greatest maritime power in the world. Hundreds of ships and thousands of sailors composed the Sung navy. Kublai Khan, Mongol founder of the Yüan dynasty, attempted an invasion of Japan in 1281 with a navy of 4,400 ships. The Ming navy in 1420 counted 3,800 ships, of which 1,300 sailed as combat vessels. The Ming launched an official shipbuilding program and constructed 2,100 vessels in government shipyards between 1403 and 1419. With compasses, watertight compartments, up to four decks, four to six masts, and the recent invention of a sternpost rudder, these were the grandest, most seaworthy, technologically sophisticated vessels in the world. The largest approached 300 feet in length and 1,500 tons, or five times the displacement of contemporary European ships. Armed with cannon and carrying up to 1,000 sailors, they were also the most formidable.

The Ming used their powerful navy to assert a Chinese presence in the waters of South Asia and the Indian Ocean. From 1405 to 1433 they launched a series of seven great maritime expeditions led by the Chinese admiral Cheng Ho (also known as Zheng He). With several dozen ships and more than 20,000 men on each voyage, Cheng Ho sailed to Vietnam, Thailand, Java, and Sumatra in southeast Asia, to Sri Lanka and India, into the Persian Gulf and the Red Sea (reaching Jedda and Mecca), and down the coast of East Africa, possibly as far as Mozambique. The purpose of these impressive official expeditions seems to have been political, that is, to establish the authority and power of the Ming dynasty, and on at least one occasion Cheng Ho used force to assert his authority. With these initiatives, the Ming acquired a number of vassal states, and at least two Egyptian diplomatic missions wound their way to China.

Then, abruptly, the extraordinary maritime thrust of the Ming came to an end. Official shipbuilding ceased in 1419, and a decree of 1433 put an end to further Chinese overseas expeditions. No one can say whether the course of world history would have been radically different had the Chinese maintained a presence in the Indian Ocean and rebuffed the Portuguese when they arrived with their puny ships at the end of the same century. Several explanations have been offered to account for the stunning reversal of Chinese policy. One notion suggests that the Chinese repudiated overseas adventures because Cheng Ho

太保相宅圖

太保

Fig. 6.2. Chinese geomancer. Prior to laying out a new city, an expert geomancer or *fêng-shui* master consults a compass-like device to ascertain the flow of energy *(chi)* in the locale. He will then use his readings to situate artificial structures in harmony with their natural surroundings.

was a Muslim and a eunuch, qualities reminiscent of the oppressive Mongol/Yüan years and greatly in disfavor among the nationalistic Ming. Another envisions the expeditions as merely the somewhat idiosyncratic initiative of two Ming emperors, and not as growing organically out of contemporary Chinese society and economy. A strong technical argument has also been advanced. Restoration of the Grand Canal began in 1411–15, and in 1417 the construction of deepwater ("pound") locks on the Grand Canal allowed a year-round link between the Yangtze and Yellow Rivers. Accordingly, the Ming trans-

Fig. 6.3. European and Chinese ships. Chinese civilization abandoned maritime exploration and left the Indian Ocean early in the fifteenth century, just decades before European sailors entered the region. The ships of the Chinese admiral Cheng Ho were much larger than European vessels.

ferred their capital from Nanking in the south to Beijing in the north, and as a result, the need for a strong navy or foreign adventures supposedly disappeared.

One way or another, Ming China turned inward, and a degree of technological stagnation set in. China remained a great and powerful civilization, but the dynamism and innovative qualities of the Sung era no longer obtained. Only with its encounter with the West beginning in the seventeenth century would technological innovation once again move China.

The World as Organism

In approaching the subject of the natural sciences in traditional China, one must avoid the tendency, similar to that already observed with regard to Chinese technology, to place undue emphasis on a search for "first" honors in scientific discovery: first recognizing the nature of fossils, first using Mercator projections in maps and star charts, discovering Pascal's triangle and the mathematics of binomials, foreshadowing the even-tempered musical scale, or, particularly far-fetched, crediting alternations of yin and yang as anticipations of the "wave theory" of today's quantum physics. Such claims reflect a perverse judgmentalism and a desire, in the name of multicultural relativism, to inflate the accomplishments of Chinese science while devaluing those of the West. Instead, the present section emphasizes the social history of Chinese science rather than a chronology of discovery, and it strives to show that the relationship between science and society in traditional China parallels the other primary civilizations of the Old World: useful knowledge patronized by the state and developed in support of statecraft and civilization generally.

Any historical evaluation of Chinese science must overcome several further obstacles. The Western concept of science or natural philosophy remained foreign to intellectual thought in traditional China. As one author put it, "China had sciences, but no science." That is, learned experts pursued various scientific activities—in astronomy, astrology, mathematics, meteorology, cartography, seismology, alchemy, medicine, and related studies—but nothing united these separate endeavors into a distinct enterprise of critical inquiry into nature. Indeed, the Chinese language possessed no single word for "science." China, like Egypt and the other bureaucratic civilizations, lacked natural philosophy in the Hellenic sense, and one gathers that Chinese thinkers would have been perplexed by the notion of pure science pursued for its own sake. Chinese society provided no social role for the research scientist, and no separate occupation or distinct profession of science existed. Instead, elite amateurs and polymaths pursued scientific interests, often, perhaps furtively, when employed to gather and apply useful knowledge in a bureaucratic setting.

The traditional Chinese outlook conceived of nature in more holistic and organismic terms than did the West. Already in Han times, a conception emerged that envisaged the universe as a vast, single organism in which the world of nature and the social world of humankind merge in a complete unity. Heaven and Earth along with man and nature harmoniously coexisted, the celestial and the human linked through the person of the emperor. From the Chinese philosophical perspective, the two great complementary forces of yin and yang governed change in nature and in human affairs. In addition, the constituent five "phases" of metal, wood, water, fire, and earth played dynamic roles in making up the world. The outlook was qualitative, and it emphasized recurring cycles, as yin, yang, and one or another of the elemental "phases" assumed predominance over the others. In considering Chinese scientific thought, then, one must acknowledge that Chinese intellectuals lived in a world separated from the West by more than geography.

On a more mundane level, although schools abounded in China, Chinese educational institutions did not incorporate or provide instruction in the sciences. Founded in the eighth century A.D., an Imperial Academy in the capital topped a complex educational structure, with a central Educational Directorate superintending a standardized Confucian curriculum for the empire. A host of private academies following the standard curriculum also blossomed. Unlike European universities, none of these schools possessed a legal charter granting them a permanent, independent existence. All existed by tradition and the will of the emperor. They could be and were closed simply by decree. Furthermore, the entire focus of these schools—public and private alike—was careerist and directed to preparing students to take the state civil-service exams. None granted degrees. Even the Imperial Academy was

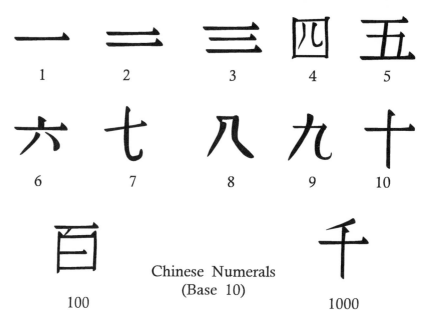

Fig. 6.4. Chinese numerals (base 10). Chinese civilization developed a decimal, place-value number system early in its history. Supplemented with calculating devices such as the abacus, the Chinese number system proved a flexible tool for reckoning the complex accounts of Chinese civilization.

一	二	三	四	五
1	2	3	4	5
六	七	八	九	十
6	7	8	9	10

百
100

Chinese Numerals
(Base 10)

千
1000

merely another bureau at which scholarly functionaries taught for limited periods of time, and only one such academy existed in the whole of China, compared to Europe in the following centuries with its scores of autonomous colleges and universities. Although authorities established separate schools of law, medicine, and mathematics around the year A.D. 1100, none survived very long. The sciences simply did not figure in Chinese education or educational institutions.

These cultural and institutional impediments notwithstanding, the necessities of imperial administration dictated that from its inception the Chinese state had to develop bureaucratically useful knowledge and recruit technical experts for its service. In a typical fashion, like writing, applied mathematics became a part of the workings of Chinese civilization. By the fourth century B.C. the Chinese developed a decimal place-value number system. Early Chinese mathematics used counting rods and, from the second century B.C., the abacus to facilitate arithmetical calculations. By the third century B.C. Chinese mathematicians knew the Pythagorean theorem; they dealt with large numbers using powers of 10; they had mastered arithmetic operations, squares, and cubes, and, like the Babylonians, they handled problems we solve today with quadratic equations. By the thirteenth century A.D. the Chinese had become the greatest algebraists in the world.

While the record occasionally indicates Chinese mathematicians engaged in the seemingly playful exploration of numbers, as in the case of the calculation of π to 7 decimal places by Zu Chougzhi (A.D. 429–500), the overwhelming thrust of Chinese mathematics went toward the practical and utilitarian. The first-century A.D. text, *Nine Chapters on the Mathematical Art (Jiu Zhang Suan Shu)*, for example, took up 246 problem-solutions dealing with measurements of agricultural fields, cereal exchange rates, and construction and distribution problems. To solve them, Chinese mathematicians used arithmetic and algebraic techniques, including simultaneous "equations" and square and cube roots. Indian influences made themselves felt in Chinese mathematics in the eighth century, as did Islamic mathematics later. Characteristically, Chinese mathematicians never developed a formal geometry, logical proofs, or deductive mathematical systems such as those found in Euclid. The social history of Chinese mathematics reveals no reward system for mathematicians within the context of the bureaucracy. Mathematicians worked mostly as scattered minor officials, their individual expertise squirreled away in separate bureaus. Alternatively, experts wandered about without any institutional affiliation. The three greatest contemporary Sung mathematicians (Ts'in Kieou-Chao, Li Ye, and Yang Housi), for example, had their works published, but did not know each other, had different teachers, and used different methods. In considering the character and social role of Chinese mathematics, one must also factor in a strong element of numerology and traditions of mathematical secrets, all of which tended to fragment communities and disrupt intellectual continuity.

A pattern of state support for useful knowledge, characteristic of centralized societies, is nowhere more evident than in Chinese astronomy. Issuing the official calendar was the emperor's exclusive prerogative, a power apparently exercised already in the Hsia dynasty (2000–1520 B.C.). Like their Mesopotamian counterparts, Chinese calendar-keepers maintained lunar and solar calendars—both highly accurate—and they solved the problem of intercalating an extra lunar month to keep the two in sync like the Babylonians by using the so-called Metonic cycle of 19 years and 235 lunations, that is, twelve years of twelve lunar months and seven years of thirteen lunar months.

Because disharmony in the heavens supposedly indicated disharmony in the emperor's rule, astronomy became a matter of state at an early period and the recipient of official patronage. Professional personnel superintended astronomical observations and the calendar even before the unification of China in 221 B.C., and soon an Imperial Board or Bureau of Astronomy assumed jurisdiction. Astronomical reports to the emperor were state secrets, and because they dealt with omens, portents, and related politico-religious matters, official astronomers occupied a special place in the overall bureaucracy with offices close to the emperor's quarters. Chinese astronomers played so delicate a role that they sometimes altered astronomical observations for political reasons. In an attempt to prevent political tampering, astronomical procedures became so inflexible that no new instruments or innovations in technique were permitted without the express consent of the emperor, and edicts forbade private persons from possessing astronomical instruments or consulting astronomical or divinatory texts.

Marco Polo (1254–1324), the Italian adventurer who served for 17 years as an administrator for the Mongol/Yüan dynasty, reported that the state patronized 5,000 astrologers and soothsayers. Special state exams—given irregularly outside the standard exam system—recruited mathematicians and astronomers for technical positions within the bureaucracy. Unlike the rest of the bureaucracy families tended to monopolize positions requiring mathematical and astronomical expertise, with jobs handed down from one generation to another. The rules prohibited children of astronomers from pursuing other careers and, once appointed to the Astronomical Bureau, individuals could not transfer to other agencies of government.

The Chinese developed several theories of the cosmos, including one wherein the celestial bodies float in infinite empty space blown by a "hard wind." From the sixth century A.D. the official cosmology consisted of a stationary earth at the center of a great celestial sphere. Divided into twenty-eight "lunar mansions" corresponding to the daily progress of the moon in its monthly passage through the heavens, this sphere turned on a grand axis through the poles and linked heaven and earth. The emperor, the "son of heaven," stood as the linchpin of this cosmology, while China itself rested as the "middle kingdom" among the four cardinal points of the compass.

Although weak in astronomical theory, given the charge to search for heavenly omens, Chinese astronomers became acute observers. With reliable reports dating from the eighth century B.C. and possibly from the Shang dynasty, the range of Chinese observational accomplishments is impressive. The richness of documentary material reveals that, already in the fourth century B.C., Chinese astronomers measured the length of the solar year as 365¼ days. The north star and the circumpolar stars that were always visible in the night sky received special attention from Chinese astronomers who produced systematic star charts and catalogues. Chinese astronomers recorded 1,600 observations of solar and lunar eclipses from 720 B.C., and developed a limited ability to predict eclipses. They registered seventy-five novas and supernovas (or "guest" stars) between 352 B.C. and A.D. 1604, including the exploding star of 1054 (now the Crab Nebula), visible even in the daytime but apparently not noticed by Islamic or European astronomers. With comets a portent of disaster, Chinese astronomers carefully logged twenty-two centuries of cometary observations from 613 B.C. to A.D. 1621, including the viewing of Halley's comet every 76 years from 240 B.C. Observations of sunspots (observed through dust storms) date from 28 B.C. Chinese astronomers knew the 26,000-year cycle of the precession of the equinoxes. Like the astronomers of the other Eastern civilizations, but unlike the Greeks, they did not develop explanatory models for planetary motion. They mastered planetary periods without speculating about orbits.

Government officials also systematically collected weather data, the earliest records dating from 1216 B.C.; to anticipate repairs on hydraulic installations, they gathered meteorological data on rain, wind, snowfalls, the aurora borealis ("northern lights"), and meteor showers. They also studied the composition of meteorites and compiled tide tables beginning in the ninth century A.D. The social utility of this research is self-evident.

Three waves of foreign influences impacted on Chinese science. The first wave broke in the years A.D. 600–750, coincident with Buddhist and Indian influences in T'ang times. Chinese Buddhists undertook pilgrimages to India from the early fifth century A.D. in search of holy texts. A significant translation movement developed, wherein over time nearly 200 teams of translators rendered some 1,700 Sanskrit texts into Chinese. As part of this movement, the secular sciences of India, including works in mathematics, astrology, astronomy, and medicine, made their way to China.

A second wave of foreign influence (this time Islamic) had a strong impact beginning with the Mongol conquest of China by Kublai Khan in the thirteenth century. Although not Muslims themselves, Mongol rulers employed Islamic astronomers in the Astronomical Bureau in Beijing and even created a parallel Muslim Bureau of Astronomy alongside the one for traditional Chinese astronomy; and later Ming emperors continued the tradition of a parallel Muslim Astronomical Bureau. Muslim astronomers deployed improved astronomical instruments, including a 40-foot-high gnomon, sighting tubes, and armillary spheres and rings adjusted for the Chinese (and not Western) orientation to the north celestial pole. Across the greater Mongol imperium, reciprocal Chinese-Persian contact developed in Yüan (Mongol) times (1264–1368) that included Chinese contact with Islamic astronomers at the Maraghah observatory. This tie put Chinese astronomers in touch with the works of Euclid and Ptolemy, but, consistent with their indifference to abstract science, the Chinese did not translate or assimilate these pillars of Western science before the third wave and the arrival of Europeans in the seventeenth century.

Before and after the Mongols, the Chinese used complex astronomical clocks and planetaria known as orreries. About A.D. 725 a Chinese artisan-engineer, Liang Ling-Tsan, invented the mechanical escapement, the key regulating device in all mechanical clocks. Using the escapement, a small tradition of clock and planetarium construction thereafter unfolded in China. This tradition reached its height at the end of the eleventh century when Su Sung (1020–1101), a Sung dynasty diplomat and civil servant, received a government commission to build a machine that would replicate celestial movements and correct embarrassing shortcomings in the official calendar then in use. The Jurchen Tartars moved Su Sung's tower in 1129 after they captured Kaifeng from the Sung. Finally, lightning struck it in 1195, and some years later, for want of skilled mechanics, Su Sung's great machine fell into complete disrepair. With it, Chinese expertise in mechanical horology declined, to the point where officials expressed amazement at Western clocks when they came to China in the seventeenth century. Su Sung's clock and like instruments did not seriously affect traditional practices within the Chinese Astronomical Bureau, but the case represents another historical example, not of technology as applied science but, quite the converse, of technology applied in the service of science and scientific research.

Earthquakes seriously affected China—800,000 people are reported to have died in a catastrophic earthquake in 1303, for example. Because it fell to the government to provide relief to outlying areas, the study of earthquakes became a practical matter of state. Earthquake records date from 780 B.C., and from Han times state astronomers of the Astronomical Bureau had the duty of recording them. Pursuant to that charge, in the second century A.D. Chang Heng created the remarkable "earthquake weathercock," an ingenious seismograph or earthquake detector. Many such machines existed in traditional China, and later in Mongol times they passed to Islam and the Maraghah observatory.

Cartography or map-making became yet another notable dimension of Chinese scientific expertise developed and deployed in the service of state administration. Chinese map-makers created highly accurate maps of the Chinese empire using various grid systems including what became known in the West as Mercator projections with unequal spacing of latitudes. They also produced relief maps, and in 1027 under the Northern Sung they designed a wagon for measuring distances overland, and Ming cartographers produced several atlases after Cheng Ho's maritime expeditions into the Indian Ocean.

As befitted a highly centralized society, medicine was strictly regulated by the state and the practice of medicine was considered a form of public service. An Imperial Medical College came into existence in T'ang times (seventh to the tenth century A.D.), and physicians had to pass strict examinations. Court physicians occupied well-paid positions, and medical expertise, like astronomical, ran in families. Hospitals, or at least hospice-like organizations, arose in China out of Buddhist and Taoist philanthropic initiative, but these became state institutions after the suppression of religious foundations in A.D. 845. To guide physicians, the central government issued many official textbooks dealing with general medicine, pharmacy, pediatrics, legal medicine, gynecology, and like subjects. One Sung pharmaceutical document dating from around A.D. 990 contained 16,835 different medical recipes. The numerous botanical and zoological encyclopedias also deserve note, in part for their medicinal advice; a government official, Li Shih-Chen, compiled the *Pen Tsao Kang Mu,* or *Classification of Roots and Herbs,* which listed 1,892 medicaments in fifty-two volumes. Illustrations graced many of these books. The fact that works of natural history seem to take a special interest in insects, notably the silkworm, or that artificial breeding programs for the silkworm began early in Chinese

Fig. 6.5. Su Sung's astronomical clock. Built in 1090 Su Sung's clock was an impressive feat of mechanical engineering and the most complex piece of clockwork to that point in history. Housed within a 40-foot-high tower, powered by a waterwheel, and controlled by complicated gearing, Su Sung's machine counted out the hours and turned a bronze armillary sphere and a celestial globe in synchrony with the heavens.

history make plain once more that the state exploited useful knowledge across a wide range of applications.

Finally along these lines, one must not overlook magic, alchemy, and the occult sciences in traditional China. An element of the magical and the divinatory ran through Chinese medicine, astronomy, geography, and mathematics, the latter especially concerned with propitious numbers. Chinese alchemy became the most developed branch of esoteric knowledge, closely associated with Taoist religious philosophy. Popular from Han times, alchemy in the East, as in the West, was a practical science concerned with making elixirs of immortality and transmuting base metals into silver and gold, but Chinese adepts engaged in these efforts less for crass monetary benefit than from contemplative, spiritual motivations and the goal of spiritual transcendence. In some instances at least, alchemy attracted official patronage, as in the case of the Northern Wei emperor who supported an alchemical laboratory from A.D. 389 to 404. Alchemists sought to duplicate natural processes carried on within the earth. They built elaborate furnaces and followed intricate alchemical procedures, and, as we saw, gunpowder emerged as an inadvertent by-product of alchemical experimentation.

As in so much else in Chinese history, a certain rigidity and decline began to affect Chinese science, medicine, and technology during the

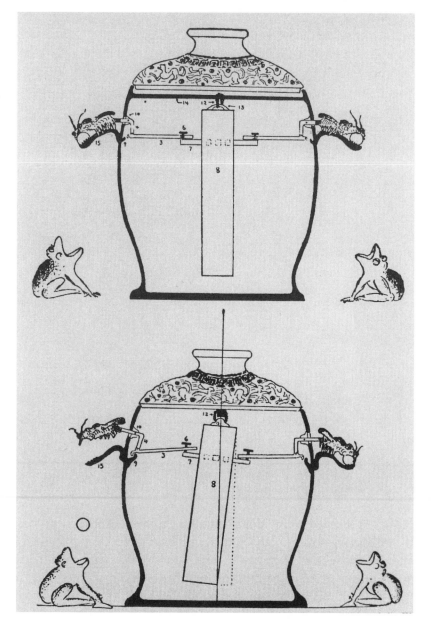

Fig. 6.6. Chinese seismograph. Earthquakes regularly affected China, and the centralized state was responsible for providing relief for earthquake damage. As early as the second century B.C., Chinese experts developed the device depicted here. An earthquake would jostle a suspended weight inside a large bronze jar, releasing one of a number of balls and indicating the direction of the quake.

Ming dynasty in the fourteenth and fifteenth centuries A.D. The reasons may well have been political. Unlike the expansive and innovative Sung or the internationally open Mongols, Ming China turned inward and developed isolationist, conservative policies. Two centuries after the apogee of Chinese algebra under the Sung, for example, Chinese mathematicians could no longer fathom earlier texts. A century after the great clockmaker Su Sung died, to repeat an example, no one could repair, much less duplicate, his handiwork. By the time Europeans arrived in China at the turn of the seventeenth century, the stagnation from the glory days of the Sung had taken its toll for several centuries.

The third wave of foreign influence impacting on Chinese science emanated from Western Europe. The Jesuit scientist and missionary Matteo Ricci (1552–1610) arrived in Macao on the Chinese coast in 1582 and finally gained admission to Beijing in 1601. The Ming emperor, the court, and Chinese society generally remained hostile to Ricci's religion and his efforts to win converts, but they took special interest in what he could communicate of Western mathematics, astronomy, the calendar, hydraulics, painting, maps, clocks, and artillery, and the ability he brought to translate Western technical treatises into Chinese. Indeed, Ricci himself became a court astronomer and mathematician and the titular deity of Chinese clockmakers. With Ricci leading the way, the Jesuits succeeded in their mission in China primarily because of their greater calendrical and astronomical expertise. In fact, the emperor handed over operational control of the Astronomical Bureau to the Jesuits. Ironically, Ricci brought with him not the new heliocentric astronomy of Copernicus, Kepler, and Galileo but, instead, perfected forms of Ptolemaic astronomy that Europeans had derived from Islamic sources and antiquity. In other words, the European science Ricci brought to China cannot be retrospectively praised because it was more "correct" than contemporary Chinese science. Rather, his Chinese hosts and employers valued it by the only measure that counted, its superior accuracy and utility in a Chinese context.

With the arrival of Ricci in China the subsequent history of Chinese science largely becomes its integration into ecumenical, world science.

Illicit Questions

As the diversity and sophistication of Chinese scientific traditions have become more evident to scholars over the last decades, a fundamental explanatory question has emerged: why the Scientific Revolution did not occur in China. As detailed in part 3, the umbrella term "Scientific Revolution" refers to the historical elaboration of modern science and the modern scientific worldview in Europe in the sixteenth and seventeenth centuries: the shift to a sun-centered planetary system, the articulation of a universal principle to explain celestial and terrestrial motion, the development of new approaches to the creation of scientific knowledge, and the institutionalization of science in distinct institutions. Since medieval China was scientifically and technologically more developed than Europe in many fields, it does indeed seem surprising that the Scientific Revolution unfolded in Europe and not in China. Over and over again, therefore, the question arises of what "went wrong" in China, what "handicapped" Chinese science, or what "prevented" the Scientific Revolution from happening there.

Historians to date have introduced several different explanations of why the Scientific Revolution failed to occur in China. The complexities of written and spoken Chinese may have made it less than an ideal medium for expressing or communicating science. That is, because mandarin Chinese and related languages are monosyllabic and written in pictographs, they are ambiguous and ill-suited as precise technical languages for science. But other experts dispute this suggestion, pointing to exact technical vocabularies in Chinese.

Chinese "modes of thought" may have proved inimical to logical, objective scientific reasoning of the sort that developed in the West. Historians have identified a persistent cultural pattern in China variously labeled as analogical reasoning or as correlative or "associative" thinking. This style of thinking, it is said, strove to interpret the world in terms of analogies and metaphorical systems of paired correspondences between diverse things (such as virtues, colors, directions, musical tones, numbers, organs, and planets) based on the fundamental forces of yin and yang and the five "phases" of metal, wood, water, fire, and earth. Yin and yang thus parallel female and male, day and night, wet and dry, the emperor and the heavens; "wood" becomes associated with "spring" and the cardinal direction "east," and so on. In a related way, the famous divinatory work, the "Book of Changes," the *I Ching*, purportedly exercised a negative influence on Chinese thought in that it rigidly defined analytical categories and unduly dominated the attention of Chinese intellectuals by promoting analogical reasoning.

Commentators have also blamed the related lack of a scientific method in China for the stagnant quality of Chinese science. They point to the suppression of two early schools of thought in China, the Mohists and the Legalists, whose tenets resembled Western scientific approaches and whose methods conceivably could have engendered Western-style science and a Scientific Revolution in China. The Mohist school, derived from the thought of Mo Ti (fifth century B.C.), primarily dealt with political matters, but its followers, together with a related group known as the Logicians, emphasized logic, empiricism, and deduction and induction as means of knowing, and thus conceivably could have given rise to a scientific tradition akin to what developed in the West. Gaining prominence in the fourth and third centuries B.C., the other school of thought, the Legalists, sought to develop a universal law code. Their efforts at classification and quantification, had they succeeded politically, might also have established a basis for the rise of modern science in China. The harsh approach of the Legalists won them little favor, however, and with the advent of the Han dynasty in 202 B.C. both they and the Mohist school found themselves repudiated and replaced by the more mainstream but less strictly scientific philosophies of Taoism and Confucianism.

Traditional Chinese thought also lacked a concept of "laws of nature." Unlike Islam or the Christian West, Chinese civilization did not entertain notions of a divine, omnipotent lawgiver who issued fixed commandments for humans and for nature. Especially after the failure of the Legalists, Chinese society by and large was not subject to strictly defined positive law and law codes; the more flexible concepts of justice and custom generally governed Chinese legal proceedings. As a result, it made no sense for Chinese intellectuals to inquire into laws of nature or to find motivation for scientific efforts to discover order in God's handiwork.

Another notion advanced to explain the "failure" of Chinese science concerns the felt cultural superiority of the Chinese. That is, China was a great and ancient civilization, culturally homogeneous, inward-looking, with a long written tradition and with a strong emphasis on traditional wisdom. China thus had no reason to overturn its traditional view of the world or to investigate or assimilate scientific knowledge of "barbarians" outside of China.

The dominant philosophies of Confucianism and Taoism likewise have been censured for stultifying scientific inquiries in traditional China. Several features of the Confucian outlook did indeed prove antithetical to the pursuit of science in the Western manner: the focus on society and human relations (and not a separate "nature"), the disdain of the practical arts, and the repudiation of "artificial" acts (i.e., experiment). Based on the Tao—"the way"—and the idea of universal harmony through cooperation, the Taoist outlook dictated that followers take no action in conflict with or contrary to nature. The very idea of any special inquiry into an "objective" nature, much less a prying, experimental prodding of nature, was foreign to Taoism. From these points of view, the Western conception of nature and scientific inquiry remained alien to the Chinese experience.

A final proposal suggests that because the merchant classes remained largely peripheral to Chinese civilization, modern science could not emerge in traditional China. Had entrepreneurs and free-market capitalism been encouraged in China and not subordinated to monolithic bureaucratic control, then, this argument suggests, perhaps more of a free market of ideas might have evolved, independent institutions akin to the university might have developed, and modern science conceivably resulted.

Each of the preceding explanations of why the Scientific Revolution did not unfold in China doubtless reflects some insight into circumstances in China before the coming of Europeans. However, akin to the previously encountered case of Islamic science, it is crucial to repeat that the negative question of why the Scientific Revolution did *not* occur in China is foreign to the historical enterprise and not one subject to historical analysis. The number of such negative questions is, of course, infinite. This particular question retrospectively and fallaciously presupposes that somehow China *should* have produced the Scientific Revolution and was only *prevented* from doing so because of obstacles or because China lacked some elusive necessary condition. It is a gross mistake to judge Chinese science by European standards, and only a retrospective projection of later European history onto the history of Chinese science would demand that China necessarily could and should have taken the same path as Europe. Quite the contrary, despite its comparative limitations, science in traditional China functioned perfectly well within its own bureaucratic and state context. Such is not a moral judgment of the high and ancient civilization of China, just good history.

The question thus remains why the Scientific Revolution unfolded in Europe rather than why it did not happen elsewhere. Perhaps it is not too early to suggest that in an ecological context where government support but also government control was less pervasive, individual thinkers had more space and freedom to apply critical faculties to abstract questions.

CHAPTER 7

Indus, Ganges, and Beyond

Dharma and Karma

Urban civilization flourished continuously on the Indian subcontinent for at least 1,500 years before the first university appeared in Europe. As we might expect, Indian experts pursued professional and highly exact work in mathematics, astronomy, medicine, and several other sciences.

In recent decades the scholarly study of science and civilization in China has influenced historians concerned with the history of science and technology in India. But, alas, no comprehensive synthesis has yet appeared to match the studies of China. Historians have examined the *texts* of Indian astronomers, mathematicians, and doctors, sometimes with a now-familiar attitude that attributes all sorts of "firsts" to early Indian scientists. Although circumstances are changing, much more research remains to be done to fathom the Indian case. Here we can only suggest that the earmarks of a typical bureaucratic civilization again present themselves in India: irrigation agriculture, political centralization, social stratification, urban civilization, monumental architecture, and higher learning skewed toward utility.

Compared to China or the Islamic world, traditions of research in the natural sciences were less vigorous in India. In part at least, the otherworldly, transcendental character of Indian religions militated against the direct study of nature. In various ways, the major religions of Hinduism, Buddhism, and Jainism envision the everyday world as a grand illusion with a transcendental theological reality underlying an ephemeral world of appearances. In these philosophies, unlike Platonic or later Christian traditions, no correspondence unites the world we see with the abstract realm of a greater reality. Truth, then, remains wholly metaphysical and otherworldly, and the goal of knowledge becomes not the understanding of the ordinary world around us, but rather to transcend this world, to escape its debilitating karma, and to ascend to a higher plane. Spiritually very rich, such views did not focus traditional Indian thinkers on the natural world itself or on any underlying regularities in nature or nature's laws.

Civilization arose along the Indus River Valley in the third millennium B.C. (see chapter 3), but declined after 1800 B.C. for reasons that remain unclear, but that probably resulted from shifting ecological patterns. The society that followed was not an urban civilization but rather consisted of decentralized agricultural communities, tribally organized and each headed by a king and chief priest. In time, settlements spread from the Indus to the Ganges basin in eastern India. Four orders or estates constituted early Indian society: priests, warrior-nobles, peasants or tradesmen, and servants, a social division out of which later emerged the full complexities of the Indian caste system. This fourfold division tended to break down local or regional identities in favor of "class" identities. The priestly class (the Brahmin or Brahman) guarded lore and ritual expertise, without which the universe would supposedly collapse. The Brahmin monopolized education, enacted ceremonies, advised kings, participated in statecraft, and drank of the hallucinogenic beverage *soma*.

The historical picture of India down to the sixth century B.C. remains fuzzy, depending wholly on literary evidence from religious texts known as the Vedas of the period 1500–1000 B.C. and auxiliary Brahmanic commentaries compiled in the succeeding 500 years. Originally oral works, they became codified only with the advent of writing in India in the sixth century B.C. Certain obscurities aside, these early texts reveal the existence of scientific knowledge directed at the maintenance of the social and cosmic orders.

Given the centrality of the sacred Sanskrit texts and the "magical" power of their oral recitation, linguistics and grammatical studies became the first "sciences" to develop in India. The Sanskrit language and the Vedas formed the basis of all study, and many grammatical and linguistic guides were produced to lead novices and experts through their intricacies. The fifth-century B.C. Sanskrit grammar of Panini, for example, set out 3,873 aphoristic rules concerning grammar, phonetics, meter, and etymology. The importance of oral recitation of the Vedic scriptures likewise led to traditional studies of acoustics and analyses of musical tones.

A smaller, subsidiary group of Vedic and Brahmanic texts concerned astronomy and mathematics. They make plain that a high-status professional class of priests, astrologers, stargazers, and calculators functioned within Vedic society. Experts created and maintained a calendar in order to regulate Brahmanic ceremonies and sacrifices that had to take place on specific days and in specific months and years. They developed multiple solutions for dividing the solar year into months and for intercalating an extra month to keep the religious calendar in synchrony with the solar year. The moon possessed astrological significance, and, like the Chinese, early Indian astrologers divided its monthly course across the heavens into twenty-seven (sometimes twenty-eight) constellations or "lunar mansions" *(naksatras)*. The Vedic-Brahmanic calendar thus united lunar and solar cycles. The construction and orientation of altars was a related affair of high importance, for which mathematical competence proved essential. At early stages of Indian history, Indian mathematicians also explored very large numbers in keeping with Hindu and Buddhist notions of great cosmic cycles, giving names for numbers up to 10^{140}.

The Indian subcontinent was more open to outside influences than was China, and so, too, were Indian scientific and technical traditions. The invasion of India by the Persians in the sixth century B.C. and their subsequent 200-year occupation of the Indus Valley opened the door to Persian and Babylonian influences on Indian astronomy. Similarly, the invasion by Alexander the Great in 327–326 B.C. allowed Greek science to begin to diffuse into India. Conversely, Indian scientific and technical accomplishments influenced developments in the Islamic world, China, and Europe.

At least one relatively powerful kingdom (the Magadha) arose in India by the fourth century B.C. Until then no single polity united India but, triggered by the invasion of Alexander the Great, the Indian adventurer Chandragupta Maurya forged the first unified empire on the subcontinent, reigning as the first Mauryan king from 321 to 297 B.C. [Map 7.1.] His grandson Aśoka expanded the realm during his reign from 272 to 232 B.C. One study claims that the contemporary Mauryan empire, centered on the Ganges, was then the largest empire in the world.

With the advent of the Mauryan empire greater clarity emerges in the historical record. The Mauryan empire was first and foremost a great hydraulic civilization. A Greek traveler, Megasthenes, spent time at Chandragupta's court around 300 B.C. and relates that more than half of the arable land was irrigated and that, as a result, Indian agriculture produced two harvests a year. A special department of state supervised the construction and maintenance of a well-developed irrigation system with extensive canals and sluices, and the same bureau planned and directed the settlement of uncultivated land. Land and water were regarded as the property of the king, and the Mauryans levied charges on water taken for irrigation. With no intervening agencies between peasants and state tax collectors, peasants held lands in a kind of tenancy, and henceforth in Indian history the state received its main revenues in the form of ground rents. Irrigation thus proved essential to both food production and state revenue, and it also fortified political centralization. Archaeological evidence of ancient irrigation systems remains elusive, largely because rivers have so changed their courses since the onset of historical times in India. Documentary evidence, however, reflects the importance of the hydraulic infrastruc-

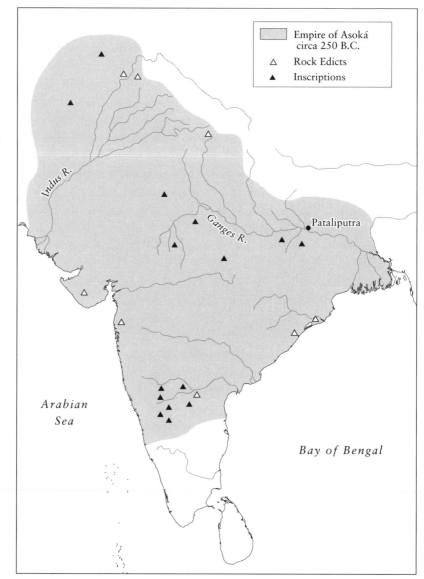

Map 7.1. India. One of the major civilizations, India spread from its initial origin in the Indus River Valley eastward to the Ganges River and to the south of the Indian subcontinent. In the third century B.C. India became united under Chandragupta Maurya. The map shows the extent of the Mauryan empire under his grandson Aśoka.

Legend:
- Empire of Asoká circa 250 B.C.
- △ Rock Edicts
- ▲ Inscriptions

Labels on map: Indus R., Ganges R., Pataliputra, Arabian Sea, Bay of Bengal

ture—it tells that under the Mauryans breaching a dam or tank became a capital offense, punishable by drowning.

The Mauryan empire did not lack the other hallmarks associated with hydraulic civilizations. An elaborate bureaucratic structure administered the empire. In addition to the department concerned with rivers, "digging," and irrigation, a number of regional and urban superintendents—all salaried officials of the king—dealt with commerce, weights and measures, excise, the mint, registration of births and deaths, supervision of foreigners, and the overseeing of such state industries as weaving, salt provision, mining, and iron-making. State control of the economy was a characteristic feature of Mauryan society and, indeed, artisans owed some royal service. Mauryan political success originated with and depended upon its military strength, and a complex war office with six subsidiary departments administered and provisioned a paid standing army of nearly 700,000 men and thousands of elephants. The existence of an elaborate bureaucracy of royal spies bolstered the autocratic nature of Mauryan government.

The growth and increasing wealth of cities under the Mauryans are additional earmarks of a developing civilization. Sixty-four gates, 570 towers, and a 25-mile defensive wall guarded the capital city at Pataliputra (present-day Patna) at the confluence of the Ganges and Son Rivers. Within the city, amid two- and three-story houses, the Mauryans erected a monumental wooden palace, replete with gilded pillars and an ornamental park with lakes and an arboretum. The Mauryans undertook other public works, including a communication system linking the empire with tree-lined roads, public wells, rest houses, and a mail service.

Although the details remain sketchy, it seems evident that expert knowledge continued to be deployed under the Mauryans. The social position of the Brahmin with their priestly expertise was not seriously undermined during the period, despite Aśoka's conversion to Buddhism. Mauryan cities became centers of arts, crafts, literature, and education; the administration of the empire clearly required literacy and numeracy. We know that the superintendent of agriculture, for example, compiled meteorological statistics and used a rain gauge. One of Aśoka's rock-edicts—carved inscriptions erected across his empire— also refers to his having established infirmaries for people and animals. And Babylonian and Hellenistic influences came to be felt in India at this point, especially in astrology. For example, the Greco-Babylonian zodiac of twelve houses or signs of 30 degrees each entered Indian astronomy and helped establish its astrological nature. Doubtless, further research will reveal more of Mauryan astronomers and astrologers and their attachment to powerful patrons.

The Mauryan empire declined after Aśoka's death, and India splintered into a host of smaller kingdoms and principalities. More than 500 years passed before India once again regained a unified status, this time under the reign of the Guptas in the fourth century A.D. The founder of this dynasty, Chandragupta (not to be confused with Chandragupta Maurya), ruled from 320 to 330, and his better-known grandson Chandragupta II (Chandragupta Vikramditya) held power from 375 to 415. The period of the Guptas continued until roughly 650 with some discontinuities and represents the golden age of classical Indian civilization. The Gupta empire resembled that of the Mauryans in its strong central power, public works, regulation of trade, and revenues from ground rent. The Gupta period is noted for the flourishing of Hindu art and literature, for traditions of liberal royal patronage, and for systematic scholarship in astronomy, mathematics, medicine, and linguistics. It formed the high-water mark of classical Indian science.

No less than earlier, Indian astronomy under the Guptas remained a practical activity. Trained professionals created calendars, set times for religious exercises, cast horoscopes, and made astrological predictions, with "lucky days" for agriculture as well as personal fortune. Indian astronomy was not especially observational or theoretical, and it did not delve into the physics of celestial movements. The emphasis remained entirely on astrological prediction and computational expertise. Furthermore, because of its putative roots in the ancient Vedas, Indian astronomy remained a conservative, backward-looking enterprise that placed no premium on theoretical innovation. Isolated from the rest of Indian intellectual life, astronomers acted more like specialized priests, with technical expertise passing down in families from one generation to another. Unlike astronomy in China, the Islamic world, or Europe, where consensus generally united scientific traditions, some six regional schools of Indian astronomy-astrology competed for intellectual allegiance and material patronage.

Despite its limitations and divisions, Indian astronomy became highly technical and mathematical in the period of the Guptas. From the fourth through the seventh centuries various Indian astronomers produced a series of high-level textbooks (*siddhānta* or "solutions") covering the basics of astronomy: the solar year, equinoxes, solstices, lunar periods, the Metonic cycle, eclipses, planetary movements (using Greek planetary theory), seasonal star charts, and the precession of the equinoxes. Aryabhata I (b. 476 A.D.) lived in Pataliputra, composed a *siddhānta*, trained students, and held the unorthodox view that the earth rotates daily on its axis (despite his knowledge of Ptolemy's *Almagest*). In his *siddhānta* in the following century the astronomer Brahmagupta (b. 598 A.D.) repudiated Aryabhata's notion of a moving earth on the grounds that it violated common sense and that, were it true, birds would not be able to fly freely in every direction. Brahmagupta's estimate of the circumference of the earth was one of the most accurate of any ancient astronomer.

Indian astronomy depended on precise arithmetical calculations, and Aryabhata and Brahmagupta obtained renown as mathematicians no less than as astronomers. Algebraic and numerical in character, Indian mathematics by and large reflected practical concerns and eschewed general solutions in favor of "recipes." Aryabhata employed a place-value system and decimal notion using the nine "Arabic" numerals and zero in his work. (The appearance of zero within the context of Indian mathematics may possibly be due to specifically Indian religiophilosophical notions of "nothingness.") He calculated the value of π to four decimal places, a value later Indian mathematicians extended to nine decimal places. In his *siddhānta* Brahmagupta extended earlier work on measurement, algebra, trigonometry, negative numbers, and irrational numbers such as π. Indian mathematical work became known to the West primarily through reports by the eleventh-century Islamic scientist Al-Bīrūnī in his *History of India*.

As in civilizations elsewhere, the world of doctors and medicine became solidly institutionalized and developed. Wealthy and aristocratic families patronized physicians, and court physicians possessed especially high status, in part because of their expertise regarding poisons and snakebites. Top-level physicians seemingly differentiated themselves from empirics through training and licensing. A traditional medical text, the *Charaka Samhitā,* for example, speaks of a process of apprenticing with a master physician-teacher and getting royal permission before practicing medicine. The religious center at Nalanda flourished as a medical school from the fifth through the twelfth centuries A.D. Thousands of students (reports vary from 4,000 to 10,000) and hundreds of teachers studied and taught at this vast complex, more than a mile square with 300 lecture rooms and a large library. Tuition was free, supported by the king and by rich families. Other teaching centers existed at Taxila and Benares. As mentioned, the Mauryan king Aśoka established medical infirmaries, and charitable dispensaries also existed in the Gupta period. Not surprisingly, veterinary medicine for war horses and elephants reached a high level of competence in India from the fourth century B.C.

Medical theory and practice became quite developed early in Indian history. The Vedic oral tradition reported anatomical information, particularly of the sacrificial horse, based on dissection, as well as botanical information and descriptions of diseases. The tradition known as the *Ayurveda*—or the "science of life"—began to be codified in the sixth century B.C., and it came to include sophisticated medical and physiological theories and treatments that involved maintaining equilibrium balances between and among various bodily humors. Ayurvedic medicine is famous for its rational approaches to diseases and their cures and, indeed, it possessed a self-conscious epistemological dimension in assessing the processes of medical reasoning and judgment. The standard medical compendium by the physician Charaka (the *Charaka Samhita*) appeared around the first century A.D. Reflecting the Hindu penchant for naming and listing, the *Charaka Samhita* identifies 300 different bones, 500 muscles, 210 joints, and 70 "canals" or vessels; its associated nosology of diseases was no less elaborate. A related "collection" by the physician Susruta (the *Susruta Samhita*) became a bible for Indian surgery. At their heights Indian medicine and surgery were probably the most developed and advanced of any contemporary civilization.

Alchemy, another science deemed to be useful, also flourished in India, perhaps having arrived from China. Closely associated with medicine and the sect of Tantric Buddhism, Indian alchemical treatises focused on various forms of mercury, on preserving health, and on the creation of an undecayable body. Practitioners came to master a sophisticated corpus of chemical knowledge that found applications in medicine through elixirs, aphrodisiacs, and poisons.

Quite apart from these scientific developments, traditional India became a highly evolved technological civilization. Indeed, although not heavily mechanized, India has been labeled "an industrial society" for the period before European colonialism and the Industrial Revolution established themselves on the Indian subcontinent. The major industry in India was textiles, and India was then the world's leading textile producer. The caste of weavers, for example, stood second in numbers only to the agricultural caste, and textile production gave rise to subsidiary industries in chemicals and the dyeing and finishing of cloth. Shipbuilding, which supplied ocean-going vessels for the substantial Indian Ocean trade, was likewise a major industry of traditional India. Indian shipwrights developed construction techniques especially suited to monsoon conditions of the Indian Ocean, and the importance of the Indian shipbuilding trade actually increased after Europeans entered those waters. Although iron smelting in India dates from 1000 B.C., it was practiced on a comparatively small scale until the Islamic Mughal empire and the advent of gun manufacture in the sixteenth century. The competence of Indian foundrymen is no better illustrated than by the commemorative iron pillar 24 feet high made at Delhi under Chandragupta II in the fourth century A.D. (It reportedly shows no sign of rust even to this day.) Indian artisans also engaged in pottery-making, glass-making, and a myriad of other practical crafts befitting a great civilization. Given its technological complexity, India actually underwent an astonishing process of deindustrialization with the coming of formal British rule in the nineteenth century.

The caste system became more rigid in the Gupta period, with the definition of some 3,000 different hereditary castes. While the significance of the caste system for the history of technology in India was probably less than previously thought, the system remains noteworthy in that different technical crafts and craft traditions became socially separated into distinct castes and guild-like bodies, it being nominally forbidden to ply a trade outside of one's caste. Although caste barriers were sometimes breached, the separation of technology from scientific traditions is as evident in India as in China or ancient Greece.

A Hun invasion of Gupta territory in A.D. 455 proved disruptive, and a partial breakup of the empire ensued in the decade 480–90. Subsequent sixth-century Indian kings reestablished the empire, but the unity of classical Indian civilization collapsed completely in 647 after the death of the heirless king Harsha. A succession of minor Hindu states followed, and Islamic influences and incursions began to be felt in India after A.D. 1000. Islam exercised wide appeal, in part because it repudiated caste divisions. An independent Delhi sultanate ruled over the Indus and Ganges in northern India from 1206 to 1526, and, by dint of superior cannon technology, the Islamic Moghal empire governed North India from 1526 nominally to 1857. Muslim rule brought improved irrigation and hydraulic technology to North Indian agriculture, including the use of artificial lakes. The great Moghul emperor Akbar (1556–1605) established a special government canal department, and at the height of the Moghul empire, fully one-third of the water used for irrigation flowed in manmade canals. As part of the Islamic imperium, India also fully assimilated Islamic science, most visibly in the spread of Islamic astronomical observatories. The cultural and institutional success of Islam spelled the end of traditional Hindu science and learning in those areas touched by the teachings of the Prophet.

Traditional Indian culture continued in the non-Islamic south within the borders of states and wealthy cities dependent on intensified agriculture. The Chola kingdom, for example, flourished from 800 to 1300. Chola engineers built irrigation works on a huge scale, including the damming of rivers and the creation of an artificial lake 16 miles long. Bureaucratic supervision is evident in the special committees in charge of irrigation tanks. In Mysore in South India 38,000 tanks remained in the eighteenth century, and 50,000 tanks in Madras in the nineteenth. This tantalizing evidence notwithstanding, the major manifestation of centralized agriculture, science, and state patronage occurred not on the Indian subcontinent itself, but rather in the spread of Indian civilization offshore to Sri Lanka and to Southeast Asia.

Greater India

The correlations between science and hydraulic civilization are evident in the case of Buddhist Sri Lanka (ancient Ceylon). Founded by legendary "water kings," a quintessential hydraulic civilization arose on the island after invasions from the Indian mainland in the sixth century B.C., and a distinctive Sinhalese civilization maintained itself there for 1,500 years. Using thousands of tanks and catchments to collect irregular rainfall, irrigation agriculture and grain production spread in the dry zone in the north of the island. The hallmarks of hydraulic civilization likewise appeared: centralized authority, a government irrigation department, corvée labor, agricultural surpluses, and monumental building, including shrines, temples, and palaces built with tens of millions of cubic feet of brickwork on a scale equaling that of the Egyptian pyramids. Large urban population concentrations inevitably followed. Indeed, the main city of Polonnaruwa reportedly ranked as the most populous city in the world in the twelfth century A.D.

The details remain sketchy, but records indicate royal patronage of expert knowledge in ancient Sri Lanka for work in astronomy, astrology, arithmetic, medicine, alchemy, geology, and acoustics. A bureaucratic caste, centered on temple scholars, also seems to have existed, with the chief royal physician a major government figure. Following the pattern established in India by Aśoka, the state diverted considerable resources to public health and medical institutions such as hospi-

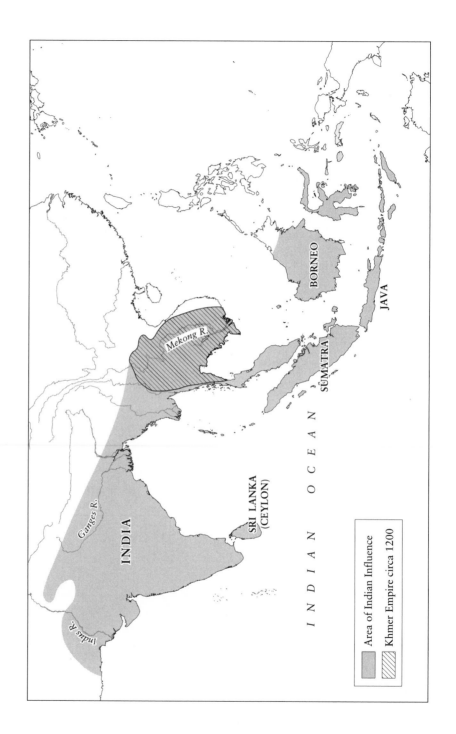

INDIA

Indus R.

Ganges R.

INDIAN OCEAN

SRI LANKA
(CEYLON)

Mekong R.

SUMATRA

BORNEO

JAVA

Area of Indian Influence

Khmer Empire circa 1200

tals, lying-in homes, dispensaries, kitchens, and medicine-halls. In all, Sri Lanka reveals a typical pattern of useful, patronized science.

From an early date in the first millennium A.D. Indian merchants voyaged eastward across the Indian Ocean. By dint of extensive trade contact and sea links to Sumatra, Java, and Bali in Indonesia and through cultural contact with Buddhist missionaries from Sri Lanka, a pan-Indian civilization arose in Malaysia and Southeast Asia. A third-century account by a Chinese traveler, for example, reported the existence of an Indian-based script, libraries, and archives in the Funan kingdom in what is modern Vietnam. Indian influence in the region increased in the fourth and fifth centuries. Brahmin from India were welcomed as local rulers bringing with them Indian law and administrative procedures. Sanskrit became the language of government and learned religious commentaries, while Hinduism and Buddhism coexisted as the dominant faiths.

The great Cambodian or Khmer empire provides the most remarkable and revealing example of this extension of Indian cultural influence. A prosperous and independent kingdom for over six centuries from 802 to 1431, the Khmer empire at its height under King Jayavarman VII (reigned 1181–1215) was the largest political entity ever in Southeast Asia, covering parts of modern Cambodia, Thailand, Laos, Burma, Vietnam, and the Malay Peninsula.

The Khmer empire arose along the alluvial plains of the lower Mekong River, and the great wealth of Khmer society derived from the most substantial irrigation infrastructure in Southeast Asian history. The annual monsoon flooded the Mekong and its tributaries, with the Cambodian Great Lake (Tônlé Sap) turning into a natural reservoir. Using impounded water techniques, Khmer engineers built an enormous system of artificial lakes, canals, channels, and shallow reservoirs with long embankments (called *barays*) to control the river system and to hold water for distribution in the dry season. By 1150 A.D. over 400,000 acres (167,000 hectares) were subject to artificial irrigation. The East Baray at Angkor Wat alone stretched 3.75 miles long and 1.25 miles wide. Hydrologic conditions along the Mekong were ideal for cultivating rice, that exceptionally bountiful crop that produced dramatic effects whenever it was introduced, as we saw in the case of China. Such a productive capability supported dense population concentrations, an immense labor force, and a wealthy ruling class.

Yet again social and scientific patterns associated with hydraulic civilization repeated themselves in the Khmer empire. Khmer kings, living deities like Egyptian pharaohs, exercised a strong centralized authority. A complex bureaucracy, headed by an oligarchy of learned Brahmins and military officers, ran the day-to-day affairs of the empire. One source labels this bureaucracy a welfare state, perhaps because Jayavarman VII supposedly built 100 public hospitals. Various libraries and

Map 7.2. Greater India. An Indian-derived civilization arose on the island of Sri Lanka (Ceylon), and Indian-inspired cultures also developed in Southeast Asia, notably in the great Khmer Empire that appeared in the ninth century B.C. along the Mekong River. *(opposite)*

Map 7.3. The Khmer empire. Based on rice production and the hydrologic resources of the Mekong River and related tributaries, this Indian-inspired empire flourished magnificently in the twelfth and thirteenth centuries. Based on substantial irrigation and impounded water technologies, the Khmer empire constituted the largest political entity in Southeast Asian history. It exemplified the typical trappings of high civilization, including monumental building, literacy, numeracy, astronomical knowledge, and state support for useful science. With the demise of its irrigation infrastructure Khmer civilization disappeared in the early fifteenth century.

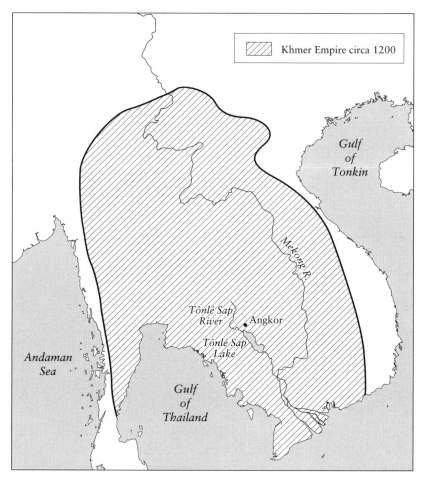

archives also testify to the bureaucratic nature of the state and of higher learning. In addition to irrigation projects and a highway system (with rest houses) linking various parts of the empire, Khmer royal power directed prodigious construction projects, notably in the capital district of Angkor, built up over a 300-year period. As an urban center Angkor covered 60 square miles and consisted of a whole series of towns along coordinate axes running 19 miles east-west and 12 miles north-south. Among the 200 temples in the region, each with its own system of reservoirs and canals of practical and symbolic significance, the complex at Angkor Wat is the largest temple in the world. Surrounded with a moat almost 660 feet wide, the temple is made of as much stone as the great pyramid at Cheops, and with virtually every square inch of surface area carved in bas-relief. The complex itself was completed in A.D. 1150, after fewer than 40 years of construction. Nearly a mile square, Angkor Wat itself contained twelve major temples, and its central spire soared to nearly 200 feet. Even more formidable, the magnificent administrative and temple complex at Angkor Thom, finished in 1187, enclosed a walled city of almost four square miles. Among their other uses, these great temples functioned as mausoleums for Khmer kings.

The Khmer court patronized science and useful knowledge. The court attracted Indian scholars, artists, and gurus, and with them Indian astronomy and alchemy came to Cambodia and Southeast Asia. Alongside ruling Brahmins and military leaders, a separate caste of teachers and priests plied their trades, teaching Sanskrit texts and training new generations of astrologers and court ceremonialists. The existence of Khmer "hospitals" suggests the organization of medical training and practice at a high level within the empire. The unity of astronomy, calendrical reckoning, astrology, numerology, and architecture is evident in the structure of Angkor Wat, meticulously laid out along lines dictated by Indian cosmology, with special moats and an architectural sacred mountain. Several of the thousands of bas-relief carved into the buildings indicate concern with elixirs of immortality. The complex also has built-in astronomical sight lines recording solar and lunar motion along the horizon. The spring equinox, which evidently marked the onset of the calendar year, receives special emphasis. Given these

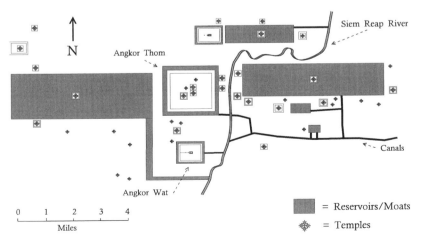

Fig. 7.1a–b. Angkor Wat. Among the 200 temples in the region, each with its own system of reservoirs and canals, Angkor Wat is the largest temple complex in the world. Surrounded by a moat almost 660 feet wide, the temple is made of as much stone as the great pyramid at Cheops, and with virtually every square inch of surface area carved in bas-relief. The complex was completed in A.D. 1150, after fewer than 40 years of construction.

sight lines built into the monument, eclipse prediction is possible at Angkor Wat, but whether Khmer astronomers actually predicted eclipses is a matter of speculation.

Overbuilding may have exhausted the state and sapped the vitality of Khmer civilization. Beginning in the fourteenth century, the Khmer empire suffered repeated invasions from neighboring Thai and Cham (Vietnamese) peoples. These attacks destroyed the irrigation infrastructure on which Khmer civilization depended: maintenance activities ceased, war created damage, and demands for soldiers reduced the corvée. As a result, populations collapsed, and so did the Khmer empire itself. The Thais conquered; Sanskrit ceased to be the learned language of Southeast Asia; a new, less ornate style of Buddhism prevailed; and Angkor itself was abandoned in 1444, to be swallowed by the encroaching jungle. The French "discovered" and brought to the world's attention the ruins of Angkor and of Khmer civilization only in 1861. Although lost for four centuries, Khmer civilization testifies to the now-familiar pattern of agricultural intensification, bureaucratic centralization, and patronage of useful sciences.

CHAPTER 8

New Worlds

Roughly coincident with Old World scientific civilizations in Islam, China, and India, a series of cultures arose in the New World: the Maya, the Aztec, and the Inca. The pattern of convergence with Old World civilizations is especially striking because developments in America un-folded along a separate technological trajectory, without plows, draft animals, or bronze or iron metallurgy. But like their ancient counter-parts on the other side of the globe, these American civilizations came to deploy scientific experts and expertise in the running of their states. And they did so without the distinctive element of disinterested natural philosophy that was the fruit of the Hellenic tradition.

Lands of the Jaguar

Based on antecedent American cultures and an intensified agriculture that exploited the productive capabilities of lowland swamps, Mayan civilization arose after 100 B.C. and flourished for a thousand years in the area radiating outward from modern Belize in Central America (see Map 8.1). In a characteristic fashion and more than in any other American civilization, the Maya came to deploy institutionalized knowl-edge for the maintenance of their societies.

As a result of the vandalism by Spanish conquerors in the New World, what we know of Mesoamerican "glyph" writing remains lim-ited. The Spanish destroyed thousands of codices—fanfold bark- and deerskin-paper books, of which only four Mayan examples survive. The primary source of our knowledge consists of the 5,000 Mayan texts engraved on stone stelae and architectural elements, some with hun-dreds of glyphs or carved signs. As a result of significant recent ad-vances in deciphering ancient Mayan, passages are being translated at an accelerated rate, and about 85 percent are currently decoded. Based on Olmec roots, Mayan writing is now known to embody phonetic and pictographic components of a distinct Mayan language. The 287 hiero-

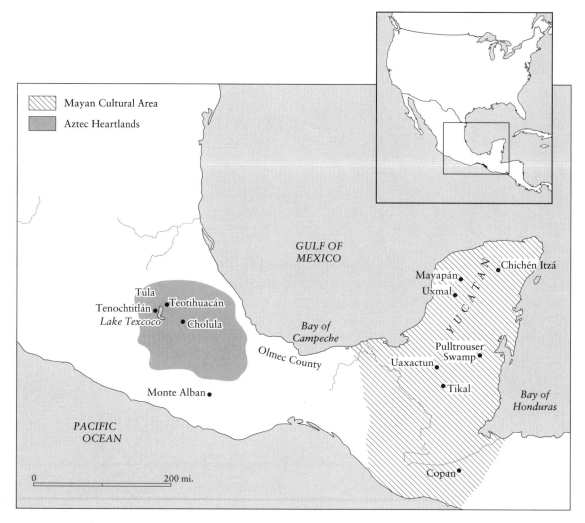

Map 8.1. Civilization in Mesoamerica. High civilization arose independently in Central America. Mayan civilization centered itself in the humid lowlands surrounding the Bay of Honduras, and, later, the Aztec empire arose in a more desert-like setting around a lake where modern Mexico City stands.

glyphic signs now deciphered, of which 140 carry phonetic meaning, presently give a clear picture that ritualistic sculpture-writing primarily recorded historical events—the reigns and deeds of kings, dynasties, and ruling families. Undoubtedly, records of dynastic legends are overrepresented in the medium of public sculpture. If the written records produced over hundreds of years had not been destroyed, a more mundane picture of Mayan society would emerge.

The inherent difficulty of mastering Mayan writing would suggest the existence of a specialized caste of Mayan scribes and the necessity of extensive training to join such a cadre. Other evidence shows that the scribal profession formed an exclusive, specialized occupation, one that enjoyed great status and prestige at the top of a highly stratified society. Scribes were drawn from the upper ranks of Mayan nobles and were often the princely second sons of kings. Their positions were probably hereditary; they functioned as high courtiers and royal confidants, closely associated with ruling kings, and apparently they sometimes vied for political power. At least in some instances, such as the late Mayan center of Mayapán, a separate "academy" existed for training priests and scribes. The scribal caste had its own patron deity, the high-god Itzamná, the inventor of writing and the patron of learning. Scribes wore a special headdress as a symbol of their profession and used special utensils, and they were sometimes buried with codices and with great pomp. Given what is known of Mayan scribes and their accomplishment, one may well speak of a class of Mayan intellectuals.

The Maya used a vigesimal, or base-20 system of arithmetic in which dots represent 1s and bars represent 5s. The suggestion has been made that the choice of the five-unit and the vigesimal system stemmed from a correspondence with the five digits of the hand and the total of twenty fingers and toes. In any event, the Maya developed a place-value system with a sign for zero, and they used it to reckon extraordinarily

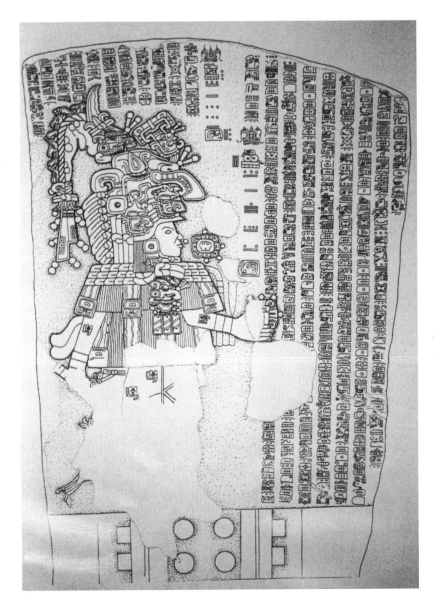

Fig. 8.1. Pre-Mayan stela. The complex writing system used in Mayan civilization has now been largely deciphered. Based on earlier antecedents like the second-century A.D. stone carving shown here, the Mayan form of writing used glyphs that possessed phonetic or sound values. Mayan inscriptions had a public character, commonly memorializing political ascendancy, warfare, and important anniversaries.

large numbers. They did not develop fractions, but like their Babylonian counterparts several millennia earlier, Mayan mathematicians created tables of multiples to aid in calculation. Mayan mathematical expertise functioned primarily in connection with numerology, ritual astronomy, and an elaborate calendrical system.

Based on Oaxacan antecedents, the Mayan calendar and dating systems became quite complex, the most complex in the Americas and probably ever in the world. The Maya juggled four or five different timekeeping systems simultaneously, and an earlier view that the Maya were obsessed with cycles of time has not been seriously undermined by the more recent understandings of the historical content of Mayan writing.

The most important of the Mayan calendars was the so-called *tzolkin*, a 260-day sacred cycle, itself generated by thirteen 20-day periods. The 260-day cycle is evident in Mayan history as early as 200 B.C. and may be related to the period of human gestation; later Maya developed related, more elaborate cycles based on multiples of 260. Along with the *tzolkin* the Maya inherited a "Vague Year" calendar consisting of eighteen 20-day months and a period of five unlucky days for a separate

"ix"
A Day Name

"mol"
A Month Name

"cu - tz(u)"
Turkey

"tzu - f(u)"
Dog

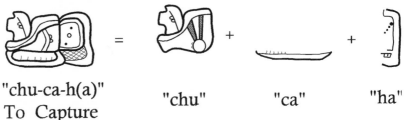

"chu-ca-h(a)"
To Capture

"chu"   "ca"   "ha"

Fig. 8.2. Mayan word signs. The originally pictographic character of pre-Mayan and Mayan languages came to assume phonetic values. These signs could be combined to represent other words and concepts.

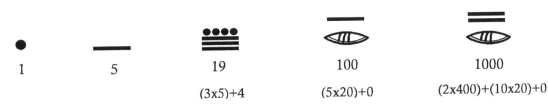

1	5	19	100	1000
		(3x5)+4	(5x20)+0	(2x400)+(10x20)+0

Fig. 8.3. The Mayan vigesimal number system. The Mayan number system was a base-20 place-value system with separate signs for 0, 1, and 5. Note that the "places" are stacked on top of one another.

365-day year. The Maya did not rectify the quarter-day discrepancy with the actual solar year, and, as a result, their "Vague Year" calendar, like that in ancient Egypt, gradually lapped the seasons over a period of 1,460 years. Meshing the *tzolkin* and "Vague Year" calendars, the Maya produced the so-called Calendar Round, a fantastic combination of the 365-day and 260-day cycles that turned on each other like gears in a great clock repeating every 52 years. The Calendar Round operated as an elaborate fortune-telling machine. Each day in the long cycle possessed its own name and became associated with various omens, and specialized priests used the Calendar Round for astrological divination, research, and prophecy.

Mayan calendar keepers and astronomers also maintained a separate lunar calendar and a fourth timekeeping system, a sequential count of days, the so-called Long Count that enumerated a linear succession of days in six units of time ranging from one day to almost 400 years. The starting point for the Maya Long Count has been calculated to have been August 13, 1314 B.C. (alternatively, sources give 1313 B.C.), and the end of the world was foreseen for A.D. December 23, 2012. In other mythological constructs Mayan time-reckoners employed time in lengths that ran into millions of years.

Closely linked to the calendar, Mayan astronomy formed part of a single astrological enterprise devoted to sacred purposes. Anonymous cadres of Mayan court astronomers observed the heavens in state institutions such as the well-known "observatory," the Caracol at Chichén Itzá. The Caracol was aligned on the horizon points marking the extremes of Venus's rising and setting in A.D. 1000, and it incorporated built-in astronomical sight lines for the equinoxes, the summer solstice, lunar settings, and cardinal points of true south and true west. Other Mayan centers at Copán, Uaxactún, and Uxmal embodied related sight lines. Indeed astronomical orientations seem to be a primary element in all Mayan public architecture and city planning, as evident in the seemingly skewed axes of Mayan buildings and towns which mirror the risings and settings of the planet Venus as well as the solstices and equinoctial points. No less important are zenith passage markers, which track the sun as it seasonally passes highest in the sky at noon, the event most likely used to monitor the agricultural cycle.

A handful of surviving codices make plain that Mayan astronomy became a highly sophisticated research endeavor. Mayan astronomers

Fig. 8.4. Mayan Calendar Round. Like their counterparts in all other civilizations, Mayan astronomers developed a complex and reliable calendar. Theirs involved elaborate cycles of days, months, and years. It took 52 years for the Mayan Calendar Round to repeat itself.

computed the length of the solar year to an accuracy greater than 365¼ days, although they used a 365-day year for calendar purposes. Judging from a stele at Palenque, Mayan astronomers in the seventh century A.D. determined the length of the lunar month to the equivalent of three decimal places or 29.530 days. (At Copán in the eighth century, another virtually identical calculation of the lunar month exists.) Given their mastery of the periodic motion of the sun and the moon, the Maya possessed the knowledge to solve the Babylonian "new moon" problem discussed earlier. Like the Babylonians, the Maya could predict

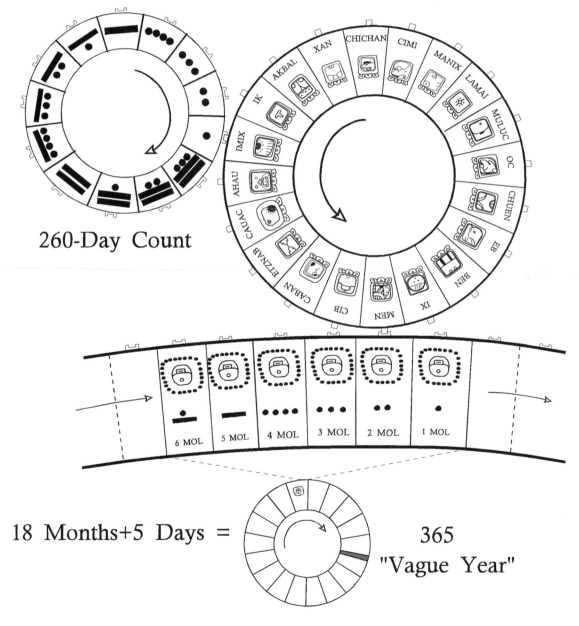

13 Day Cycle   20 Named Days

260-Day Count

18 Months+5 Days =   365 "Vague Year"

Fig. 8.5. The Mayan "observatory" at Chichén Itzá. From the observing platform of the Caracol ancient Mayan astronomers had a clear line of sight above the treetops. The various windows in the structure align with the risings and the settings of Venus and other heavenly bodies along the horizon.

eclipses accurately, and they created eclipse tables that calculated probabilities of eclipses.

The planet Venus, the object of special veneration in Mayan culture, was carefully monitored. Mayan astronomers kept a separate Venus calendar, and actively undertook research to improve the accuracy of their Venus tables to within two hours in 481 years. Like their earlier Babylonian counterparts, Mayan astronomers harmonized cycles of Venusian years and solar years, and they elaborated even more complex cycles, including one that integrated 104 solar years, 146 sacred *tzolkin* cycles, and 65 Venus years. Experts may well have produced tables for Mars and for Mercury. Other inscriptions indicate that Jupiter possessed astrological significance; certain stars had special meaning as well. Clearly, Mayan astronomers engaged in specialized research at a high level of precision and expertise.

A range of utilitarian motivations prompted this mass of esoteric astronomical work. Calendrical mastery at the most simple level gave Mayan rulers an understanding of the seasons and agricultural cycles. In its more complex formulations the Mayan calendar governed elaborate religious and ritual activities, and the full complexities of Mayan astrology produced magic numbers, prognosticated the fates of individuals, and predicted propitious and unfavorable times for a host of activities. That the Venus cycle was used to time military campaigns, for example, gives an indication of the political uses of learning and the ways in which Mayan astronomy and astrology were integrated into the power structure and ruling ideology of Mayan society.

The high culture of the Maya came under tremendous stress around A.D. 800, and the major centers in the central regions of Mayan civilization collapsed entirely around 900. A resurgence of Mayan power developed on the Yucatan peninsula—notably in Chichén Itzá—in the following period to 1200, but from the eleventh century the Long Count fell into disuse, and the record indicates a decline in the rigor of training for scribes and priests. Thereafter the high civilization of the Maya passed into history. An array of causes have been put forward and debated to explain the protracted death of Mayan civilization. Endemic warfare among confederations of city-states may have played a role, and the inevitable pressures of population measured against a fragile system of food production may have produced radical demographic fluctuations. Compounding such possibilities, two centuries of drought—the worst in 8,000 years—affected the Mayan lowlands in 800–1000 and no doubt took a heavy toll. Researchers have recently highlighted the problems of deforestation stemming from Mayan techniques of using large amounts of wood to make lime for stucco with which Mayan monumental architecture was plastered. Deforestation probably disrupted rainfall patterns and, at least in places, led to soil erosion and thereby the ruination of agriculture. Mayan civilization gradually waned, and with it the exquisite system of understanding nature that the Maya had achieved.

Cactus and Eagle

Central America also saw the rise of Toltec and Aztec civilizations. Based on irrigation agriculture, between A.D. 900 and 1100 the Toltec city of Tula had 35,000–60,000 inhabitants, and the Toltecs built what is technically the largest pyramid in the world, a manmade mountain of 133 million cubic feet, 1,000 feet on a side, and 170 feet high covering 45 acres at Cholula.

The Aztecs began as a seminomadic tribe, and in the fourteenth and fifteenth centuries they established the most powerful empire in Central America. They built their city of Tenochtitlán in 1325 on a lake where Mexico City presently stands; according to legend, an omen—an eagle perched on a cactus—drew them to the lake. The Aztecs proved to be master hydraulic engineers. The lake, Lake Texcoco, was saltwater, and Aztec engineers built a huge dike across it to separate a fresh-water portion (fed by springs) from the brine; they also installed floodgates to regulate lake levels and aqueducts to bring additional fresh water to the lake. Each year millions of fish and ducks were taken from lakes, which also provided a nutritious algae paste. The Aztecs developed an intensive style of lake-marsh (or lacustrine) agriculture that involved dikes, dams, drainage canals, and land reclamation, all produced as public works under state management. Agricultural production was literally based on floating paddies known as *chinampas*. Measuring 100 meters by 5–10 meters (328 feet by 16–33 feet) and fertilized by human waste and bat guano, these plots proved extraordinarily productive and produced up to seven crops a year and the surplus which supported urbanized Aztec civilization. Prior to the Conquest, Aztec farmers cultivated over 30,000 acres of *chinampas*.

The chief city of Tenochtitlán covered five square miles and became filled with monumental pyramids, palaces, ceremonial centers, ball courts, markets, and roads built with corvée labor. Aqueducts brought fresh water into the city, and 1,000 public employees swept and watered the streets daily. On the eve of the European conquest Tenochtitlán had a population estimated at 200,000 to 300,000, the largest ever in the Americas.

Predictably, Aztec kings, a full-time priest caste, a bureaucracy, and the military ruled this most powerful state in pre-Columbian Mesoamerica. They controlled an empire of 5 million people, possibly twice the population of ancient Egypt. The bureaucracy ran local and imperial administrations, collected taxes and tribute, and performed judicial duties; positions in the bureaucracy were hereditary in aristocratic lineages, and separate schools existed for commoners and for the nobility and priests. The Aztecs formed a militaristic society heavily involved in ritual sacrifice and probably cannibalism—Aztec priests offered up tens of thousands of human sacrifices a year. But the Aztecs also developed extensive trading and mercantile communities, which required mathematics and recordkeeping like other civilizations. The chocolate bean served as a unit of currency.

The Aztecs shared common writing and number systems, astronomy, and theologies with earlier Mesoamerican societies. The Aztec system of writing was less developed than the Mayan, being more pictographic, but it possessed some phonetic elements. The subjects of surviving Aztec books and codices concern religion, history, genealogy, geography, and administrative records, including tribute lists, censuses, and land surveys; some Aztec books were specialized instruction manuals for priests. Aztec numeration was a simple dot system. They inherited the Mayan 52-year Calendar Round, with its 260-day and 365-day cycles. Architect-astronomers aligned the *Templo Major* in Tenochtitlán to the setting sun, and other buildings were sited along equinoctial lines. The calendar governed an elaborate train of seasonal festivals and feasts, wherein offerings to various manifestations of the Aztec sun god Tezcatlipoca played a central role. Indeed, theology required the ritual sacrifice of blood in order to keep the sun in its course and the earth bountiful.

The Aztecs also possessed sophisticated botanical and medical knowledge. Priests functioned as medical specialists, and medical knowledge passed from fathers to sons. Based on empirical researches, Aztec doctors developed an elaborate and apparently effective medical pharmacopeia that was at least the equal of that of their Spanish conquerors. (Aztec life expectancy exceeded that of Europeans by 10 years or more.) Aztec medicine and astronomy were linked through the belief that the heavens and the human body were mutually linked. In a pattern of patronage that is now familiar, in 1467 the Aztec emperor Montezuma I established a botanical and zoological garden wherein expert knowledge was cultivated and transmitted. In this connection one might mention the domestication of the red cochineal insect used as a dye for textiles throughout Mexico, Central America, and in Europe after the Conquest.

The Spanish adventurer Hernán Cortés landed on the coasts of Mexico in 1519 with 500 fellow conquistadors. They subdued the great civilization of the Aztecs in two years.

Heads in the Clouds

A similar story of cultural and scientific development repeated itself in South America. There a series of independent civilizations arose which mirrored the pattern of their sister cultures around the globe. Cultural development in South America reached its peak in the great civilization of the Inca (or Inka) that spread for more than 2,000 miles up and down the west coast of South America between the Andes and the Pacific in the thirteenth and fourteenth centuries A.D. In an earlier chapter we had occasion to mention the hydraulic basis of the Inca empire, with its highly stratified and bureaucratized society and monumental building.

Although less is known about science and civilization in pre-Columbian South America than most of the other pristine civilizations, all evidence conforms to the now-familiar emergence of science and expert learning in the service of the state. The ancient Inca did not develop writing or formal mathematical systems, but they originated their acclaimed *quipu* or the method of recording information by means of complex knotting of strings. To this extent the Inca were not an exception to the common pattern of mathematical and recordkeeping systems in the first civilizations. The Inca empire was organized on the basis of a decimal system, with units of 10 to 10,000 people, and the Inca possessed a set of standard weights and measures. Quipu recorded information that included tax and census records and imperial history, and as part of the enormous imperial bureaucracy a hereditary class of accountants memorized information contained in quipu.

Inca astronomer-priests divided the heavens into quarters according to seasonal changes in the inclination of the Milky Way across the southern sky. The mountains of the Andes created natural horizon

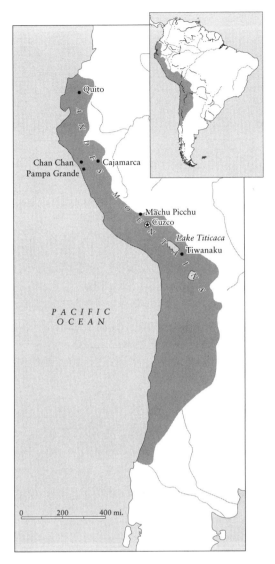

Map 8.2. The Inca empire. On the west coast of South America the Inca built on cultural developments of the previous 1,000 years and created a high civilization between the Andes Mountains and the Pacific Ocean where engineers terraced mountains and built irrigation systems tapping numerous short rivers to intensify agriculture.

markers against which the Inca could track the periodic motion of the sun, moon, planets, constellations, and even apparently void spaces referred to as "dark clouds." The Inca also built artificial stone pillars along the horizon to mark solstices. Forty-one sight lines *(ceques)* radiated from the grand ceremonial center of the Coricancha temple in Cuzco, along which other markers indicated lunar positions, water sources, and political subdivisions of the Inca empire. In other words, the Inca incorporated a calendar and map of their empire into the design of the city of Cuzco—it became an architectural quipu. Other Inca sites likewise had astronomical orientations, such as the summer solstice indicator built at Machu Picchu.

The Incas reportedly employed lunar and solar calendar systems, but because calendrical knowledge was not written down, local lords instituted a variety of local timekeeping systems. In the Inca capital of

Fig. 8.6. Inca recordkeeping. All high civilizations developed systems of recordkeeping, usually in the form of writing. The Inca used knotted strings called quipu for recording numerical and literary information.

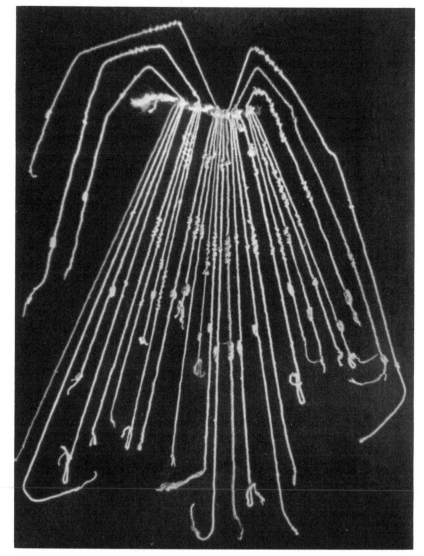

Cuzco authorities maintained a seasonal, 365-day solar calendar of twelve 30-day months (each of three 10-day weeks) and five festival days; they corrected the quarter-day discrepancy with the solar year by resetting the calendar at the summer solstice (in December). The Inca also kept track of the zenith motion of the sun, and sources speak of Inca observatories and an office of state astrologer. They also maintained a 12-month lunar calendar of forty-one eight-day weeks comprising a year of 328 days. (Other sources speak of an Inca lunar calendar of 12- and 13-month years.) Observations of the seasonal first rise of the constellation of the Pleiades in the night sky corrected the inherent discrepancies in a 328-day lunar year. The annual appearance of these stars regulated other ritual and seasonal events, including the deployment of corvée labor, probably because it coincided with the coming of rains in August.

Like that of the Aztecs, Incan medical and botanical knowledge was quite sophisticated. Specialized classes of "doctors" and surgeons existed along with state-appointed collectors of medicinal herbs. Inca medical personnel performed amputations and in urgent cases trepanned patients (i.e., cut holes in their skulls), presumably to forestall the fatal effects of brain swelling. Like the ancient Egyptians, the Incas mastered the arts of mummification.

Inca civilization fell in 1532 to an invading Spanish force led by the conquistador Francisco Pizarro. Along with the collapse of the Aztecs a decade earlier the history of the Americas and the history of science and technology in the Americas thereby became inextricably linked to developments in Europe.

Sun Daggers

In contrast to Central and South America, the full panoply of an indigenous high civilization did not arise in pre-Columbian North America. The continent did indeed have great rivers, but in its eastern two-thirds these flowed through vast, unrestricted expanses of temperate forest and plain. Population densities never crossed critical thresholds, and bureaucratically centralized societies did not arise. At first, a Paleolithic economy universally governed the lives of North American hunters and gatherers. In certain areas an intensified form of the Paleolithic way of life came to involve systematic exploitation of deer, fowl, wild grain, and nuts. Then, from roughly 500 B.C., as groups began to cultivate beans and as the practice of growing corn and squash spread outward from Central America, characteristically Neolithic societies appeared in North America. As one might anticipate, the increased wealth of the Neolithic mode of production resulted in somewhat more stratified societies; permanent settlements, towns, and ceremonial centers; trading networks; and larger-scale building. Commonly known as Mound Builders, these cultures, such as the Hopewell and Adena, maintained themselves over centuries and are characterized by the large earthen works they left behind. The Great Serpent Mound in modern Ohio is a well-known example. It and many related constructions served as mortuary sites and may also have been redistribution centers. Such works and the cultures that gave rise to them are strongly reminiscent of Neolithic Stonehenge in terms of social complexity. The so-called Mississippian culture that flourished in the area of the American midwest from 750 into the 1600s represents a high point of social evolution in pre-Columbian North America. Based on more complex agricultural systems for producing corn, Mississippian groups created a true city, Cahokia, in modern Illinois; it covered six square miles with hundreds of mounds and temples and a population of 30,000–40,000 in A.D. 1200.

Patterns of culture in North America thus paralleled developments

Fig. 8.7. Great Serpent Mound. Like Neolithic societies elsewhere around the world, native American groups regularly built substantial structures, often with astronomical orientations. Here an early illustration depicts the archaeological site known as Great Serpent Mound (Adams County, Ohio), probably built and used by the Adena culture between 100 B.C. and A.D. 700. Perched on a bluff overlooking Brush Creek, the four-foot-high mound structure uncoils to a length of a quarter of a mile. The effigy represents a snake ingesting an egg. The site was probably a regional center that several separate Neolithic groups came to for trading and ceremonial purposes.

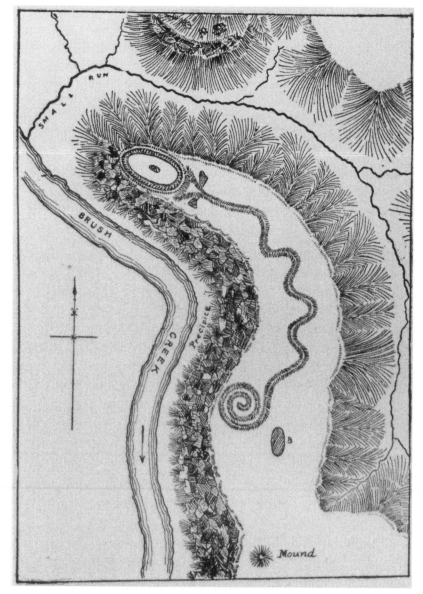

seen elsewhere in particular locales around the world with successive Paleolithic, intensified Paleolithic, Neolithic, and intensified Neolithic stages of production. Each stage evolved a set of appropriate technologies, knowledge systems, and cultural forms necessary to maintain the populations involved, often for extended periods. Not unexpectedly, Amerindian groups developed their own practical astronomies. The solstice orientation of many of the works of the Mound Builders, the extensive petroglyph astronomical record that includes a possible observation of the Crab Nebula explosion of 1054, and the stone circles known as "medicine wheels" of the Plains Indians with their solstice markers—all these evidence a diverse but familiar response of hunter-gatherers, pastoralists, and Neolithic farmers to problems posed by the natural world around them.

In the desert regions of the American Southwest a further social development occurred that displays a partial, but revealing interplay of ecology, technology, and science similar to other early civilizations. As Amerindian groups such as the Hohokam and the Anasazi utilized intermediate levels of irrigation in their agriculture, they consequently achieved intermediate levels of development between Neolithic and high civilization. What might be termed "incipient hydraulic society" produced intermediate political centralization, intermediate social stratification, intermediate population densities, intermediate monumental building, and intermediate levels of scientific development. This American case is revealing of the forces and factors at play in the initial appearance and character of pristine and other high civilizations around the world.

The Hohokam peoples migrated from Mexico to Arizona around 300 B.C., bringing irrigation technology with them. Hohokam engineers tapped the Gila River, and by A.D. 800 they completed a substantial irrigation system that involved 50 kilometers (some sources say "several hundred miles") of canals, including main feeder canals several meters wide that flowed for 16 kilometers. Coordinated effort was obviously required to build and maintain these irrigation works. Hohokam irrigation technology watered 30,000 acres and produced two crops a year. Increases in agricultural productivity on this scale produced predictable social repercussions for the Hohokam: population growth, political centralization, public building, and territorial expansion.

After centuries of interaction with the Hohokam, the Anasazi cultural group to the North coalesced around A.D. 700 and flourished over two centuries from 950 to 1150 in the Four Corners area of the American Southwest. The area is barren desert that receives an average of only nine inches of rain a year and suffers extreme conditions of freezing winters and sweltering summers. Yet here the Anasazi settled and produced a vibrant and expansive society. At its peak Anasazi culture consisted of 75 relatively small communities more or less evenly spread out over an area of 25,000 square miles. The total population numbered on the order of 10,000 people who resided in distinctive cliff "cities." Anasazi architects also fashioned an 800-unit stone and masonry structure of four and five stories at the main Anasazi center at Chaco Canyon. (They brought in timber for beams and construction from a distance of 80 kilometers.) With several large ceremonial buildings, the Chaco Canyon complex could harbor 7,000 temporary or permanent residents. A system of regional roads hundreds of miles in length linked Anasazi settlements. These roads were wide—up to nine meters (30 feet) across—and deliberately engineered with roadbeds. Agricultural production was possible only by means of irrigation, and

Map 8.3. Civilization in the American Southwest. Around 1,500 years ago civilizations began to form north of Mexico. The Anasazi Indians flourished in the American Southwest for 200 years around A.D. 1050. A regional intensification of agriculture led to dispersed population centers and an extensive network of roads.

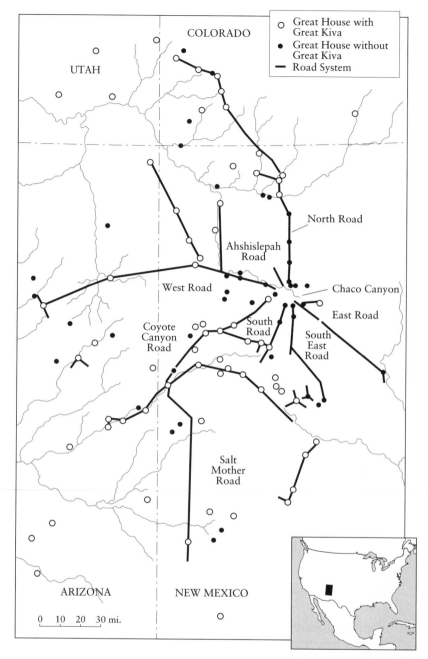

with extensive works that tapped the San Juan River Basin and seasonal streams and that involved canals, check dams, and terraced hillsides, Anasazi agriculture yielded corn production at modern levels. Because of the persistent uncertainty of the weather and the harvest, it would seem that Anasazi opted to spread population and production over a wide area, so that if the crop failed in one place, success in another could maintain the whole group. The creation of such sizable residences and ceremonial complexes and irrigation and transportation-communication systems could not have been the work of individuals or even local segments of the Anasazi population.

What about Anasazi "science"? Would this small group of Amerindians dependent on irrigation agriculture in a marginal ecological zone display the same earmarks of scientific knowledge and expertise characteristic of more affluent civilizations elsewhere? A major archaeological find in 1977 proved such to be the case. Archaeologists discovered an exquisite Anasazi seasonal observatory that accurately indicated the summer and winter solstices and the spring and fall equinoxes. As the sun passes overhead at these seasonal junctures, beams of light created by the manmade rock formation move across a spiral

Fig. 8.8. The Great Kiva at Pueblo Bonito, Chaco Canyon. The Anasazi peoples settled in the American Southwest in the eighth century A.D. Given the ecological uncertainties of their desert setting, their society flourished based on water management techniques. They built cliff dwellings to house much of the population and large ceremonial centers, known as kivas, to store surplus corn and to conduct ritual activities. Not surprisingly, these and other Anasazi constructions embody astronomical knowledge.

design inscribed on the facing stone wall. Perched on Fajada Butte 450 feet above the valley floor in Chaco Canyon, the Anasazi "Sun Dagger" is distinctive among archaeoastronomical artifacts in marking these seasonal passages at the sun's zenith and not along the horizon. The Fajada Butte construction also records the maxima and minima points of the moon's 18.6-year cycle. Other research elsewhere in Chaco Canyon has shown that the Anasazi ceremonial kivas likewise had an astronomical orientation and embodied astronomical knowledge. Built as a circle to mirror the heavens, the main door of the Great Kiva at Chaco aligned with the North Star, and at the summer solstice the sun's rays would pass through a window and strike a special niche. Clearly, the Anasazi, like so many groups before them faced with the necessity of gaining control of the calendar, developed the requisite expertise, and some among them became masters of observational astronomy.

Because of the limits imposed by the ecological setting, the levels of intensified agricultural production, population density, monumental building, political centralization, and institutionalized scientific expertise among the Anasazi never rose to match those of fully developed civilizations elsewhere. Nevertheless, that the Anasazi displayed elements of a comparable pattern speaks volumes about the associations between cultural development and scientific expertise. Unfortunately for the Anasazi, a severe drought affected their region in the years 1276–99, and, given the marginal viability of their mode of existence in the first place, this drought spelled the end of their remarkable cultural achievements.

Intermission

Let us step back and briefly consider the state of science and systems of natural knowledge on a world scale at roughly the year A.D. 1000. Plainly, no cultural group was without some understanding of the natural world. The point applies to the not-insignificant number of peoples who continued to live by Paleolithic rules and to forage for their food, as well as to tribes of nomadic pastoralists great and small, to villages of simple farmers who followed in the footsteps of their Neolithic ancestors, and to the centers of urban civilization in the Islamic world, classical India, Sung China, Mesoamerica, and Peru. What distinguishes the science and scientific cultures of these latter civilizations is that they *institutionalized* knowledge and patronized the development of science and scientific expertise in order to administer the comparatively huge social, political, and economic entities that constituted their respective complex civilizations.

In the year A.D. 1000 none of the worldviews held by any of the world's peoples envisioned the earth other than at the center of their

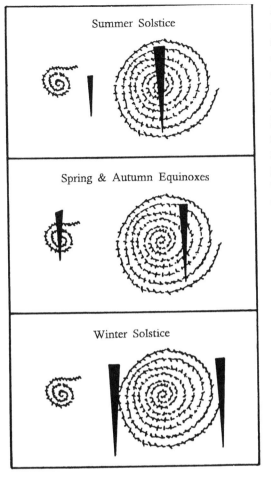

Summer Solstice

Spring & Autumn Equinoxes

Winter Solstice

Fig. 8.9. Anasazi astronomy. Like all agricultural civilizations the Anasazi developed reliable systems of calendrical reckoning. At Fayada Butte in New Mexico archaeologists have found a unique marker whereupon the noonday sun sends "daggers" of light to indicate the solstices and equinoxes.

respective universes. Similarly, nowhere except in the Islamic world—and there its status proved increasingly precarious—did savants pursue the intellectual game of disinterested theoretical inquiry into nature uniquely initiated by the ancient Hellenic Greeks.

A related but separate pattern pertains to the nature and distribution of technology on a world level at the outset of the second millennium A.D. No society lacked technology—the very notion is absurd. "Paleolithic" groups lived by means of appropriate "Paleolithic" technologies. "Neolithic" groups lived by means of appropriate "Neolithic" technologies. And the universally more diverse and complex character of urban civilization depended on a myriad of specialized crafts and trades that kept the machinery of cities and civilization going.

Only in those handful of subject areas where societies required and patronized specialized knowledge—astrology/astronomy, literacy, numeracy, aspects of engineering, and medicine, to name the major ones—is it at all meaningful to speak of a limited existence of applied science. Otherwise, the worlds of technology and learned science remained sociologically and institutionally poles apart. The vast bulk of technology was not applied science and had developed according to sociologically distinct craft traditions.

To measure the full historical force of the pattern of hydraulic civilization seen in the Islamic world, in China, in India, pan-India, and in contemporary American civilizations and to fully evaluate their attendant scientific cultures, one must compare these great civilizations with the rise of a secondary civilization in rain-watered Europe where the ecological conditions did not call for any government management or control of the basic agrarian economy.

PART III

Europe

Compared with the East and with medieval Islam, Christian Europe at the turn of the first millennium A.D. was truly an "empty quarter." Latin Christendom numbered 22 million souls in 1000, versus 60 million in the heartlands of China, 79 million on the Indian subcontinent, and perhaps 40 million under Islamic governance. By the year 1000 the population of Rome had declined to 35,000 from its peak of 450,000 in antiquity; only 20,000 people lived in Paris, and only 15,000 inhabited London. In contrast, the population of Córdoba in Islamic Spain topped 450,000 (some estimates range as high as 1 million), Constantinople held 300,000, Kaifeng in China reached 400,000, and the population of Baghdad—the largest city in the world—was nearly 1 million. Contemporary Europe was a cultural, intellectual, economic, technological, and demographic backwater that paled before the technological and scientific vitality of contemporary centers of civilization in Islam, Byzantium, India, China, Mesoamerica, and South America.

Through the first millennium of the Christian era Europe was dotted with rural settlements and displayed only a thin veneer of literate culture. A patchwork of tribal societies existing under an essentially Neolithic economy knitted together western Europe in the early Middle Ages, and repeated Viking incursions after the ninth century put at risk what fragile social and institutional bonds there were. Prior to the "Renaissance of the twelfth century," the state of learning in Europe north of the Alps also remained meager. Lacking large urban centers, significant ports, wealthy royal or aristocratic patrons, and highly developed cultural institutions, Europe stood in sharp contrast to both contemporary civilizations and those that had come before.

Civilization arrived in Europe by a unique route and developed under unique physical and social conditions. In the East the first civilizations formed in constricted semiarid river valleys where centralized governments managed the basic agricultural economies. In Europe, where rain fell in the spring and summer, no such governmental intervention estab-

lished itself, nor indeed did an urbanized civilization arise at all until around the tenth century, when a unique form of agricultural intensification began to change Europe into an urbanized civilization. Once it found a way to intensify its agriculture Europe achieved a very different character. Demographically, its population swelled to match India and China, and technologically, economically, and politically Europe became a major player on the world scene. Beginning in the fifteenth century, based on its mastery of the technologies of firearms and the oceangoing ship, European power began to spread beyond its own borders and to establish world-transforming overseas empires. Western Europe also became the world center of scientific learning and research. Indeed, the origin of modern science unfolded in the Scientific Revolution in Europe in the sixteenth and seventeenth centuries and nowhere else.

These remarkable historical developments raise several questions. First, how could the destitute "empty quarter" of Europe transform itself materially and intellectually in such profound and historically far-reaching ways? How did European scientists come to accept heliocentrism and the earth as a planet spinning through space? And how did European society eventually harness this new knowledge to the quest for solutions of practical problems?

CHAPTER 9

Plows, Stirrups, Guns, and Plagues

A series of interlocked technical innovations—an agricultural revolution, new military technologies, and a dependence on wind and water for the generation of power—shaped the history of medieval Europe. This technical perspective enables us to address the questions of why and how Europe transformed itself from a cultural backwater based on an economy scarcely more advanced than that of traditional Neolithic societies to a vibrant and unique, albeit aggressive, civilization that came to lead the world in the development of science and industry.

"Oats, Peas, Beans and Barley Grow"

The Agricultural Revolution of the Middle Ages represents a response to problems that resulted from a rising population in combination with a land shortage. The population of Europe as a whole rose 38 percent between 600 and 1000. The increase in France stood closer to 45 percent, and geographical pockets undoubtedly experienced larger increases. In medieval Europe land was put to many uses, not only to produce food and fiber on cropland but also to pasture dairy animals for food, cattle and horses for traction, and sheep for wool, while expanding cities reduced the acreage available for agricultural production. Moreover, tracts of forest land provided timber for construction, shipbuilding, and both heating and industrial fuel. In particular, wood was used as the fuel in making iron, an industry that consumed vast quantities of timber and placed a heavy strain on land use. By the ninth century the people of Europe began to face the kind of ecological crises that thousands of years earlier had induced the Neolithic settlers of the river valleys of the East to intensify their agriculture and make the transition to civilization.

In Europe, agricultural intensification did not and could not follow the same pattern that it had in the ancient East where artificial irrigation provided a technological solution. Europe was already irrigated naturally by sufficient rainfall which fell during the spring and summer. Instead, the European farmer could increase his production only by plowing the heavy soils of districts that had resisted the encroachments of the light Mediterranean scratch plow. The unique constellation of technological innovations adapted to the unique ecological conditions of Northern Europe produced the European Agricultural Revolution.

The first innovation involved the introduction of the heavy plow. This behemoth of wood and iron, mounted on wheels and armed with an iron cutter, tore up the soil at the root line and turned it over, forming a furrow and eliminating the need for cross-plowing. The heavy plow was resisted by enormous friction and therefore had to be pulled by as many as eight oxen. By contrast, the Mediterranean scratch plow, adapted to light soils, was essentially a hoe dragged through the ground by one or two oxen, with fields plowed twice. The heavy plow, which the Romans had invented but rarely used, increased agricultural production by allowing the farmer to cultivate the wet lowlands of Europe.

A second innovation that contributed to an increase in agricultural production involved the substitution of the horse, with its greater speed and endurance, for the ox as a draft animal. The traditional neck harness, which the ox with its short neck could tolerate, was unsuitable for the horse. Europeans, it seems, adapted the horse collar from the Chinese who employed it several centuries before. The device transferred the pressure points from the windpipe to the shoulders and thereby increased the horse's traction four- or fivefold. In combination with the iron horseshoe, another European innovation, it resulted in a shift to the horse from the ox as the principal draft animal.

Still another component of the Agricultural Revolution of the Middle Ages was the development of the three-field rotation system. The classic two-field farming system of the Mediterranean regions of antiquity typically involved farming one field while leaving another fallow. In the new three-field pattern that arose on the European plain, arable land was divided into three fields with plantings rotated over a three-year cycle: two seasonal plantings employed two of the fields, a winter wheat crop and a spring crop of oats, peas, beans, barley, and lentils, with the third field left fallow.

These new technologies produced a variety of social consequences, both progressive and problematical. The deep plow made it possible to farm new lands, particularly the rich, alluvial soils on the European plain, and this ability helps account for the northward shift of European agriculture in the Middle Ages. Then, because the heavy plow and its team of oxen was an expensive tool, beyond the capacities of individual peasant farmers to own, it brought collective ownership and patterns of communal agriculture and communal animal husbandry, thus solidifying the medieval village and the manorial system as the bedrock of European society at least through the French Revolution.

Similarly, the shift to horses allowed for larger villages because of the larger "radius" horses could work, presumably leading to socially more diverse and satisfying village life. Horses also decreased the cost of transporting goods, so that a greater number of villages could participate in regional, national, and international economies.

The three-field rotation system produced important advantages. The spring crop of vegetables and oats significantly improved the diets of the common people of Europe. The three-field system also increased the productive capabilities of European agriculture from 33 to 50 percent, an extraordinary surplus of food production that fed the rise of Europe and European cities in the High Middle Ages.

A richer, more productive, more urbanized Europe resulted from the Agricultural Revolution of the Middle Ages, a Europe destined to create modern science and lead the world in technological progress, but also a Europe that contained the seeds of many future problems—land shortage, timber famine, population pressure, imperial ferocity, devastating epidemics, world war, and finally, as a result of its technological success, ecological disruption on a global scale.

By 1300 the population of Europe to the Ural Mountains trebled to 79 million from a low of 26 million in 600. Paris increased in population by more than 10 times to 228,000 in 1300, and then to 280,000 in 1400. Coincident with urbanization and this growth in population came a wave of cathedral building (Canterbury begun in 1175, Chartres in 1194), university founding (Bologna in 1088, Paris in 1160) and the full complement of high medieval culture including the ethos of chivalry and poetic tales of damsels in distress and knights in search of the Holy Grail.

Agriculture was not the only activity in which technology contributed to the rise of medieval Europe. In military affairs technological innovations produced some of the unique developments that characterize European feudalism and that begin to account for Europe's eventual global dominance. One of the definitive figures of European feudalism, the armored knight mounted on an armored charger, was created by a key piece of technology—the stirrup. In Europe prior to the eighth century the mounted warrior remained mounted only until he reached the field of battle, where he dismounted and fought on foot. Without stirrups to provide stability only the most skilled horsemen could fight as true cavalry and could swing a sword or stretch a bow without losing his mount. The Chinese invented the stirrup in the fifth century A.D., and it thereafter diffused westward. With no moving parts, the stirrup is a deceptively simple piece of technology, but in stabilizing a warrior on his horse it allows for fighting on horseback without dismounting. With stirrups, a rider with a lance hooked to his armor became a formidable unit where momentum replaced muscle in a new style of warfare—"mounted shock combat." The European knight evolved into the medieval equivalent of the tank, with ever more heavily armored knights and horses forming the most powerful arm on the battlefield.

The new European technology of mounted shock combat meshed easily with the manorial system brought about by the Agricultural Revolution. The knight replaced the peasant-soldier common in the early Middle Ages, and being a knight became a full-time job. The cost of equipping the traditional knight in shining armor, while substantial, lay within the means of a local lord. The system resulted in truly feudal relations, wherein vassal knights pledged their loyalty and their arms to a higher feudal lord in exchange for part of the lord's domain to be governed and taxed in the lord's name. Such local relations were especially apt for the decentralized character of European societies in the Middle Ages. No strong central government, comparable to those of bureaucratic civilization, was required to manage an agricultural economy that needed no hydraulic infrastructure. The manorial system was well adapted to the European ecology. The advent of the knight and European feudalism further forged appropriately local relations between villages and the knights and lords who governed them. The knight-village relation became characteristic of European feudalism and the manorial system, wherein the village owed "dues" to the church and the knightly manor house. Transformed through the Agricultural Revolution, the village now produced the surplus needed to support cadres of knights, and those knights policed, taxed, and enforced justice on a local level.

Given primogeniture, the custom of feudal lands passing to the firstborn son, the number of landless knights rose, and ultimately more knights populated Europe than could be easily accommodated. As a result, a first wave of European expansion erupted with the Crusades. Pope Urban II launched the first Crusade in 1096. These unruly expeditions continued to pour forth from Europe for nearly 200 years; the seventh and last Crusade began in 1270. Since the European invaders encountered civilizations that were technologically their equal and culturally their superiors—Byzantine and Islamic civilization—there was little chance that they would prevail.

Coincident with these changes European engineers developed a fascination for new machines and new sources of power, and they adopted and developed novel methods of generating and harnessing it. Indeed, medieval Europe became the first great civilization not to be run primarily by human muscle power. The most outstanding example concerns the development of water-powered machines and their incorporation into the fabric of village life and European society generally. The waterwheel became widely used to wring energy from the profusion of streams that run through many parts of Europe, and it powered a variety of other machines including sawmills, flour mills, and hammer mills. In some districts windmills increased cropland by reclaiming land from the sea. The need for water-driven mills may be attributed to a lack of surplus labor and to the increased production levels generated by the Agricultural Revolution. That is, with more grain to grind, a widespread shift to water- or wind-powered milling machines was only to be expected. The mill existed in antiquity, but saw limited use perhaps because of the availability of slaves and human muscle power to do the work required to grind grain. It is no coincidence, therefore, that slavery withered away in western Europe coincident with the advent of labor-saving machines.

Anonymous medieval engineers also used wind to turn windmills and tidal flow to drive tidal mills. In so doing they mastered older kinds of mechanical gearing and linkage and invented new ones. Europeans perfected water- and wind-driven mills, the spring catapult (or trebuchet), and a host of other devices, and in so doing they drew on new sources of nonhuman motive power. Their civilization was literally driven by comparatively more powerful "engines" of wind and water which tapped more energy of one sort or another than anywhere else in the world. Medieval Europeans have been described as "power-conscious to the point of fantasy," and by dint of the medieval fascination with machines, more than other cultures European civilization came to envision nature as a wellspring of power to be exploited technologically for the benefit of humankind. This distinctive attitude toward nature has had powerful and increasingly dire consequences.

The impressive array of technological innovations that led to the transformation of European society and culture owed nothing to theoretical science, in large measure because science had little to offer. The science ultimately inherited from the ancient world and from medieval Islam had even less applicability in medieval Europe than it had in those earlier civilizations. Some knowledge of geometry (but not proofs of theorems) and calendrical reckoning proved useful, but none of it had any application in the development of the machines and techniques for which medieval Europe became justly famous.

But the development of European civilization created new external conditions for science and natural philosophy and set the stage for a vital new culture of learning that emerged in Europe. In what is known as the "Renaissance of the twelfth century," European scholars came into contact with, and themselves began to build on, the philosophical and scientific traditions of antiquity and Islam. And just as Europe was singular in its method of intensifying agriculture and in its use of machinery, it was also singular in founding an institution to harbor higher learning—the university.

Books and Gowns

Europe north of the Alps had never been the scene of much higher learning prior to the twelfth century, and, hence, it is misleading to speak of Europe as having fallen into a "Dark Age." Since Roman times a veneer of literate culture, manifested in monastery schools and the occasional scholar, covered a core of essentially Neolithic village life in northern

Fig. 9.1. Water power. Europeans began to use wind and water power on unprecedented scales. Undershot waterwheels were constructed at many sites along the profuse streams with which Europe is favored.

Europe. Monasteries spread across Europe after 500 A.D., and with their scriptoria (or rooms for copying texts) and libraries they maintained themselves as minor centers of learning. Catholic priests had to be minimally literate, and in 789 the Frankish king and later Holy Roman emperor Charlemagne issued an ordinance establishing "cathedral schools" in each bishopric in order to guarantee a supply of literate priests for an otherwise illiterate society. Of necessity, the level of learning and instruction in the cathedral schools and monasteries of the early Middle Ages remained quite low, centered essentially on rudiments of the "Seven Liberal Arts" (grammar, rhetoric, logic, arithmetic, geometry, music, and astronomy) inherited from classical antiquity. Some knowledge of astronomy was needed for astrological and calendrical purposes, especially for setting the date for Easter. But beyond these elementary approaches to higher learning in the early Middle Ages, the intellectual emphasis remained on theology and religious affairs rather than on science. Almost no original scientific research took place.

Paradoxically, at the very edge of Europe, monasteries in Ireland achieved a reputation for their theological sophistication and general learning, including knowledge of Greek, which was otherwise essentially lost to Europeans. And every now and again a truly learned individual appeared on the scene, such as Gerbert of Aurillac (945–1003), who became Pope Sylvester II in 999. In addition to mastering the Bible, the writings of the Church Fathers, and what little of classical pagan knowledge had passed to Europe, Gerbert studied the mathematical sciences in monasteries in northern Spain, where Islamic learning had filtered through and from where he brought back to France knowledge of the abacus and the astrolabe, that convenient device for sighting on stars and performing simple astronomical calculations. Gerbert's idiosyncratic and still narrow mastery of the mathematical sciences notwithstanding, intellectual life in early medieval Europe remained comparatively elementary and socially insignificant.

Against the background of weakly organized learning in the early Middle Ages, the appearance of the European university in the twelfth century and its rapid spread across Europe mark an institutional watershed in the history of science and learning. Instruction in medicine arose in the independent principality of Salerno in Italy in the ninth century, but the union of students and faculty that developed at Bologna usually ranks as the first university in Europe. The University of Paris followed by 1200, Oxford by 1220, and perhaps eighty additional universities appeared by 1500. The rise of the European university coincided with burgeoning cities and growing wealth made possible by the Agricultural Revolution, for universities were decidedly urban institutions, not rural like the monasteries, and they depended (and depend) on an idle student body with the means to pay for and the job prospects to justify attending universities.

Despite occasional claims to the contrary, the European university was a unique institution. Modeled after the craft guilds of medieval Europe, universities evolved as nominally secular communities of students and master teachers, either as guilds of students (as in Bologna) who employed professors or as guilds of master teachers (as in Paris) who took fees from students. Moreover, universities did not depend on state or individual patronage like the scribal schools of antiquity or the Islamic *madrasa*. They were not state organs, but rather remained independent, typically feudal institutions—chartered corporations with distinct legal privileges under only the loose authority of the church and state. Privileges included the institutional right to grant degrees and freedom from town control. As essentially autonomous, self-governing institutions, universities thus fell in a middle ground between the total state control typical of the bureaucracies of the great empires and the wholly individualistic character of Hellenic science.

The early universities were vibrant institutions serving the invigorated societies of late medieval Europe. They functioned mainly to train the clergy, doctors, lawyers, administrators, and teachers increasingly required to run the affairs of the state, the church, and the private sector as Europe in the Middle Ages continued to flourish. The graduate faculties of theology, law, and medicine provided instruction and training in those advanced subjects to select students, while the undergraduate liberal arts faculty taught all students at the first stages of their university careers. The natural sciences found a secure home within the arts faculty in the core curriculum concerned with logic and the last four of the seven liberal arts (the quadrivium), devoted to arithmetic, geometry, astronomy, and music. Baccalaureate arts graduates who went on to pursue advanced degrees in theology, law, or medicine typically took a master's degree in arts, where natural philosophy was a prime component of the course of study. Master's-level students often taught in the undergraduate arts faculty while undertaking graduate studies. In this way the natural sciences became subjects for more intense study by some scholars within the university context as a preliminary stage in a larger program of professional training. Unlike universities today, the medieval university was not primarily a research institution, nor was science pursued primarily as an end in itself.

Before Greek and Islamic science could provide the basis of a scientific curriculum for the university, that corpus of knowledge had to be made available through extensive programs of translation into Latin. The great Muslim city of Toledo fell to the Christians in 1085 (another indication of the new power of an expansive European civilization), and Toledo became the center of translation activity where teams of translators rendered classic scientific and philosophical texts from Arabic into Latin. Jewish intellectuals in Spain played an important role in this activity by translating from Arabic into Hebrew for themselves and into Spanish for their Christian collaborators and patrons, who would further render the work from Spanish into Latin. Translations also took place in southern Italy and in Sicily ("liberated" by Norman knights in the second half of the eleventh century), where scholars rendered Latin versions from Greek as well as Arabic originals. Note that the motivation for these remarkable efforts was not entirely an abstract love of knowledge, for the sought-after documents primarily concerned the putatively useful sciences of medicine, astronomy, astrology, and alchemy.

As a result, by 1200 Europeans recovered much of ancient science along with the several centuries of scientific and philosophical accomplishment produced within the Islamic world. Adelard of Bath (fl. 1116–42) translated Euclid's *Elements* (from Arabic) and other Arabic mathematical texts in the 1120s. The most noted translator, Gerard of Cremona (1114–87), traveled to Spain around 1140 to locate a copy of Ptolemy's *Almagest* and stayed for 40 years to translate not only the *Almagest* in 1175, but also—while heading a team of translators—a total of seventy to eighty other books from Arabic originals, including many major Islamic treatises along with works of Archimedes, Galen, Hippocrates, Aristotle, and Islamic commentaries on Aristotle. Previously, Europe knew Aristotle only from a few minor works in logic. After 1200 his paramount importance as "the Philosopher" emerged clearly. Later in the Renaissance better translations would be made from older, more authentic Greek originals, but by 1200 the "Western" tradition finally made it to western Europe.

If the twelfth century was a period of translation, the thirteenth represents a period of assimilation wherein the learned of Europe began to absorb the scientific and philosophical traditions of antiquity and medieval Islam. Much of the process of assimilation amounted to attempts to reconcile a traditional Christian worldview with Aristotle and other pagan Greek traditions. The great intellectual synthesis of Thomas Aquinas (1224–74) in large measure completed this process of assimilation. Whether Aquinas Christianized Aristotle or aristotelianized Christianity, or both, matters little, for one way or another Aristotle came to provide a complete intellectual system upon which medieval scholastics raised the edifice of rational thought about God, man, and nature. Aristotle's logic and analytical categories became virtually the exclusive conceptual means of investigating any subject. The elaboration and defense of Aristotle's works became a mission of the universities, and the resulting intellectual amalgam of Christian theology and Aristotelian science produced a coherent and unified vision of the world and humanity's place in it.

Consider, for example, the vision of the cosmos revealed in *The Divine Comedy*, the famous poem by the medieval Italian poet Dante Alighieri (1265–1321). The earth remains motionless and fixed at the center of the world. The sublunary realm of flux—the earth with its four elements and their natural and violent motions—provides the setting for the human drama. In the heavens celestial spheres carry the planets and the stars in their courses. Hell sits at the center, purgatory in the middle, and paradise beyond. A great hierarchical chain of being unites all creatures in ranked order, from the lowest worm to the noblest king or pope and then, through a series of angels and archangels, to God. The physical laws are Aristotelian. The divine laws are God's. The whole arrangement is temporary, created at a particular moment in the past and awaiting the end of time. Such a powerful and unified vision must have provided considerable intellectual and spiritual satisfaction to common people and intellectuals alike.

Medieval scholars interpreted the world primarily from a theological point of view, but they believed that reason can play a role in human understanding of the divine and that we can learn of the existence and nature of God from his work as well as from his word—that is, from the study of nature as well as from the study of the Bible. But in the overall context of the medieval worldview secular natural science took second place whenever Aristotelian natural philosophy clashed with traditional Christian theology.

Since the Enlightenment of the eighteenth century it has become customary in Western democracies to emphasize the importance of religious freedom and the separation of church and school. In medieval Europe, unity—not separation—was required. As Aristotelian teachings began to infiltrate Europe and the new European universities in the thirteenth century, faith and reason had to be harmonized. Certain points in Aristotle clearly conflicted with traditional Catholic teachings—his views that the world was eternal, that there was no creation, that the human soul was not necessarily immortal, that limits existed on Divine potency.

Institutional conflicts compounded the intellectual problems of assimilating Aristotle into standard theology, as teachers in the arts faculty promoted philosophy and reason as independent and equally valid

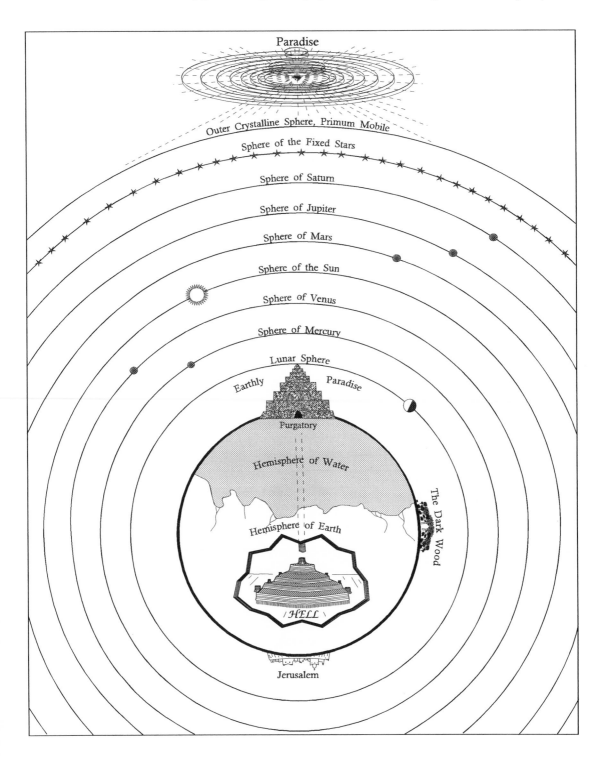

routes for arriving at truth, as opposed to the theology faculty, which naturally resisted the threat of secular philosophy and the natural sciences as competing systems of knowledge. A series of intellectual skirmishes between theologians and philosophers unfolded across the thirteenth century, culminating in the condemnation of 1277, wherein the bishop of Paris with the backing of the pope condemned the teaching of 219 "execrable errors" in Aristotle and subjected anyone who held or taught them to excommunication.

On the surface, the condemnation seems to represent a decisive victory for conservative theology, the suppression of liberal and independent philosophy and science, and the institutional subordination of the arts faculty to the theology faculty within the university. Some commentators, however, see the condemnation of 1277 as only a minor outbreak of hostilities which in the end produced harmony rather than discord, especially given that the condemnation held only for a few decades at the University of Paris, while at Oxford less restrictive measures operated, and elsewhere none at all. Still other investigators go further and argue that by freeing medieval thinkers from the yoke of strict obedience to Aristotle, the condemnation of 1277, in effect, liberated them to conceive new alternatives in solving long-standing problems in Aristotelian science and natural philosophy. From this point of view, the Scientific Revolution did not begin with Copernicus in the sixteenth century, as is usually held, but 250 years earlier with Catholic scientific intellectuals and their response to the condemnation of 1277.

Questions concerning continuities versus discontinuities in late medieval and early modern history and the history of science have long been debated and remain far from settled among scholars even today. The thesis regarding the essentially modern character of late medieval science remains a notable focus of those debates. For the present we can refer to a middle-ground interpretation of the condemnation of 1277 which sees its effect neither as wholly squelching scientific inquiry nor as launching the Scientific Revolution outright. The condemnation produced a paradoxical effect, in that by subordinating philosophy to theology and by forcing the admission that God could have fashioned the world in any number of ways, given his omnipotence, a path opened for friars in the arts faculty to consider any and all scientific possibilities, as long as they stayed out of theology and did not claim that their intellectual games had any necessary relation to the world as God's artifact. An extraordinary flourish of what might be termed "hot-house" science resulted, theologically inoffensive work wherein scholastic natural philosophers entertained all variety of scientific possibilities, but only hypothetically, on the basis of "suppositions" or thought experiments, or as products of their ingenious imaginations. Thus, for example, Jean Buridan (1297–1358) and Nicole Oresme (1320–82), among other medieval scientists, examined the possibility

Fig. 9.2. Dante's scheme of the universe. In *The Divine Comedy* the early Italian poet Dante Alighieri developed the medieval portrait of the cosmos as a Christianized version of Aristotle's worldview. *(opposite)*

of the earth's daily rotation on its axis, and each offered what seem like compelling rational arguments for admitting that such motion occurs in nature. Each believed that science on its own might well lead to that conclusion. Yet both men ended up rejecting the possibility of a moving earth. Oresme reached his conclusion not on scientific grounds, but on the basis of apparent conflicts between that hypothesis and passages in the Bible and because of his view of the inherent superiority of theology as a means of arriving at truth.

By the fourteenth century the major question facing scientific intellectuals of the European Middle Ages no longer concerned simply uncovering new texts or assimilating Aristotle's natural philosophy to scripture or even of purging Aristotle of his anti-Christian elements, but rather building on the Aristotelian paradigm and breaking new ground. Under the general conceptual framework provided by Aristotle, scholastic natural philosophers actively and creatively pursued a wide range of scientific investigations. With two translations of Ptolemy's *Almagest* appearing by 1175, for example, an indigenous tradition of observational and mathematical astronomy arose in western Europe. The Alfonsine Tables (ca. 1275), calculated in an extra-university setting by the astronomers of the king of Castile, was one result, as were pathbreaking though ineffectual calls for calendar reform. Geoffrey Chaucer's *Treatise on the Astrolabe* figures among the astronomical works of the fourteenth century, Chaucer being better known, of course, for his poetical writings. Translation in the 1130s of Ptolemy's great work in astrology, the *Tetrabiblos*, actually preceded his purely astronomical text by half a century. Serious and sustained research in astrology arose alongside medieval astronomy, often associated with medicine and medical practice. Moreover, building on a strong Islamic tradition and propelled by the religious connotations associated with "light," medieval investigators carried on research in optics, including improved understanding of vision and the rainbow. In the realm of mathematics, Leonard of Pisa (ca. 1170–1240), known as Fibonacci, introduced Europeans to "Arabic" (actually Indian) numerals and sophisticated algebraic problems in his *Liber abaci* of 1228. A number of works attributed to Jordanus de Nemore (ca. 1220) took up mechanical questions concerning statics and a "science of weight" in the thirteenth century. The translation and assimilation of the medical treatises of Galen, the great Roman physician of late antiquity, revitalized medical theory and practice after 1200, as did the creation of medical faculties, which became separate centers of science within the medieval university alongside the arts faculty. Closely associated with developments in medieval medicine and Aristotelian traditions in biology and natural history, a number of more narrowly scientific texts appeared which touched on the life sciences, notably in works by Albert the Great (Albertus Magnus, 1200–1280), *On Vegetables* and *On Animals*. And since little distinction was drawn between rational and occult knowledge, one cannot dismiss significant efforts directed toward alchemy and natural magic by medieval alchemists, philosophical magicians, and doctors. Needless to say, women were not part of the medieval university scene, but some, like Hildegard of Bingen (1098–1179), served in positions of authority as abbesses and collected unto themselves bodies of useful knowledge of nature.

The men who undertook inquiries into nature in the Middle Ages were hardly monolithic in approach or slavishly Aristotelian in outlook. Rather, the record reveals a diversity of often conflicting points of view and approaches to studying nature. For example, the University of Paris, dominated by the religious order of the Dominicans, tended to be more purely Aristotelian and naturalistic in outlook, while at Oxford the Franciscan fathers shaded more toward the Platonic and abstract, with consequent differences in how they viewed the role of mathematics in interpreting nature. Scholastics also divided over the role of experiment and hands-on approaches to discovering new knowledge. In opposition to the traditional "scholastic" preoccupation with book learning, Robert Grosseteste (1168–1253), the first chancellor of the university at Oxford, argued for active investigations of nature and on that account is sometimes hailed as the father of the experimental method in science. Greatly influenced by Grosseteste, Roger Bacon (ca. 1215–92) (not to be confused with the later seventeenth-century propagandist for science Francis Bacon, 1561–1626) proposed that human ingenuity ought to be applied to the creation of useful mechanical devices such as self-propelled ships and carts. Similarly, in his *Letter on the Magnet* of 1269 Petrus Peregrinus emphasized the value of experiment in discovering new facts about nature. Retrospectively, the views of these men are suggestive to us today because they seem to anticipate a later experimental style in science. However, in their own context they represent minority approaches among savants of the Middle Ages. Franciscan authorities, for example, restricted the circulation of Roger Bacon's experimental researches, possibly because of their associations with magic. And none of these radical experimentalists of the Middle Ages questioned the general theological outlook of their era or suggested that the natural sciences were anything but the subordinate handmaiden to theology.

Two historically significant examples can illustrate the character and accomplishment of medieval scientific thought. The first concerns the explanation of projectile motion defended by Jean Buridan, the celebrated fourteenth-century master of arts at the University of Paris. The reader will recall that Aristotle's physics required a mover to be in contact with the object moved for each and every case of forced (i.e., nonnatural) motion. The question of identifying a mover for the apparently anomalous case of projectiles (such as arrows, javelins, chalk thrown at students) stimulated a small research tradition within the broad sweep of the Aristotelian paradigm. Building on earlier medieval commentators and the work of the sixth-century Byzantine natural philosopher John Philoponus, Buridan proposed that an internal motive agent he called "impetus," implanted in a projectile by its mover, provided a motive quality that drove projectiles after they lost contact with any evident mover. Buridan envisioned his impetus as applying not only to projectiles, but also to the free fall of bodies and the perpetual revolution of the heavenly spheres. As a kind of self-propelling quality, Buridan's impetus at first glance resembles Newton's principle of inertia, which postulates that bodies remain in motion (or at rest) unless otherwise acted upon by an outside force. However, such a superficial resemblance to the later idea of inertia belies a vast conceptual gulf separating Buridan's impetus and, indeed, medieval physics, from Newton and the Scientific Revolution. That is, for early modern physics, as we will see, projectiles move of their own accord and require no cause of any description. From Newton's point of view, what needs to be explained is not motion itself but rather *change* in motion, why a projectile starts or stops moving or changes speed or direction. But Buridan approached projectile motion from a diametrically opposite point of view, in quite conventionally seeking to identify outside Aristotelian *movers* and an effective *cause* of motion for projectiles while in flight. In other words, rather than a radical break that equating impetus with inertia retrospectively and misleadingly suggests, Buridan placed his inventive explanation squarely within the tradition of Aristotelian science and Aristotelian research puzzles.

A second example of medieval scientific accomplishment concerns the great Parisian philosopher and church doctor Nicole Oresme—the outstanding man of science of the European Middle Ages. In his *On the Configuration of Qualities,* written around 1350, Oresme created visual constructs—graphs—to represent qualities and qualitative change geometrically. The somewhat modernized diagram shows how Oresme depicted uniformly accelerated motion, motion that he referred to as "uniformly difform" and that we see exemplified by falling bodies (see fig. 9.3). The horizontal axis represents time; the vertical axis represents the degrees of speed possessed by a uniformly accelerating body; the area under *AB* (the triangle *ABC*) represents the total distance covered by the body. Inherent in the diagram lie several mathematical laws pertaining to motion that Oresme and his compeers understood perfectly: for example, that the distance covered by a uniformly accelerated body equals that of another body moving with constant speed equal to one-half of the final speed of the accelerated body. (The constant motion may be represented by the horizontal line *DE;* because the triangle *ADF* equals triangle *FBE,* the area under *DE* equals the area under *AB,* and hence the distances covered by uniform motion and uniformly accelerated motion are equal.) Also lurking in the graph is the notion that the distance a body covers in uniformly accelerated

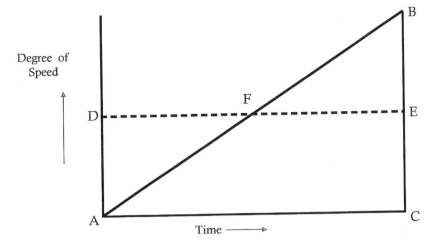

Fig. 9.3. Uniformly difform motion. In classifying different types of motion, the medieval scholastic Nicole Oresme (1320–82) depicted a motion that increases uniformly over time. He called it uniformly difform motion; we call it constant acceleration. Implicit in the diagram is the notion that the distance an accelerated body travels is proportional to the square of the time elapsed. Galileo later transformed Oresme's abstract rules into the laws of motion that apply to falling bodies.

motion is proportional to the square of the time a body is accelerated $(s \propto t^2)$.

The formula $s \propto t^2$ expresses Galileo's law of falling bodies, a law Galileo formulated 250 years after Oresme. The obvious question arises why Galileo and not Oresme gets credit for uncovering this fundamental law of nature. The answer is revealing, for after 1277 Oresme in no way associated his abstract inquiries into the character of accelerated motion with any motion in the real world. His interest lay in understanding accelerated and other motions and qualities on their own terms, on a wholly abstract basis, and purely as a theoretical intellectual exercise. In other words, that Oresme's "uniformly difform" motion characterizes the fall of bodies in the real world apparently never occurred to him or his contemporaries. Oresme's was a phenomenal intellectual achievement of a prodigious scientific imagination, but it remained a precocious one that by itself did not lead to Galileo or the Scientific Revolution.

Considering its rise out of the impoverished institutional and intellectual circumstances of the early Middle Ages in Europe, late medieval science seems remarkably productive in the rational study of nature and in exploring the limits of Aristotelian natural philosophy. The European Middle Ages created a new institutional foundation in the European university and an intellectual foundation in bringing a critical review of Aristotelian science to Europe. But the historical significance of the medieval scientific accomplishment may be less in what it meant for the Middle Ages than for having laid the institutional and intellectual foundations for further developments during the Scientific Revolution of the sixteenth and seventeenth centuries.

A series of disruptions ravaged large parts of Europe in the fourteenth century and in ecological and demographic terms interrupted several centuries of European prosperity that was characteristic of the later Middle Ages. The climate in Europe became cooler and wetter, disastrously affecting harvests and agricultural productivity. Unprecedented famines erupted across Europe in the years 1315–17, and the resulting economic depression, seriously exacerbated by an international banking crisis in 1345, lasted well into the next century. The epidemic of bubonic and pneumonic plagues—the Black Death—sweeping Europe in 1347–48 wiped out a quarter to a third of the population. Tens of thousands of European villages simply disappeared, and devastating outbreaks of plague recurred until the eighteenth century. Some experts put the death toll as high as 40 percent, with demographic recovery not occurring until 1600. Less dramatic in its import, but wrenching nonetheless, the removal of the papacy from Rome to Avignon for most of the fourteenth century broke the unity of Christendom and divided the allegiances of Catholics along several papal lines. And the Hundred Years' War between England and France, breaking out in 1338, periodically devastated the heartland of France through the 1450s. Peasant revolts and threats of social unrest were no less characteristic of the decades on either side of 1400. These doleful developments fell hardest on the lower orders and affected science directly through the mortality among scientists and indirectly through institutional closures and educational disruptions. Taken together, they mark a watershed in the development of European material culture and contribute to a sense of discontinuity in reviewing the history of medieval and early modern science.

After the disruptions of the later Middle Ages the essentials of European agriculture and feudal society based upon that hardy system of production revived. Universities resumed their growth and expansion, and if the number of scholars pursuing scientific inquiries had dipped after 1350 their numbers not only recovered but, as figure 9.4 indicates, ultimately rebounded as if the demographic horrors of the fourteenth century never occurred. Perhaps the alluring artistic accomplishments of the European Renaissance have seduced us into seeing more of a historical break at the end of the Middle Ages than is warranted.

Cannons and Sailing Ships

By the fourteenth century Europe had recapitulated some but not all of the earmarks of the earlier civilizations. Agriculture had been intensified, the population grew, urbanization took hold, building (in the form of the soaring cathedrals) became ever more monumental, and higher learning was institutionalized. But in a rainfall environment that forestalled the need for public works to build and maintain a system of hydraulic agriculture, neither a centralized authority nor a universal corvée came into being. Only later, beginning in the sixteenth century, did these components of civilization arrive on the European scene. The historical dynamic that produced those consequential innovations was

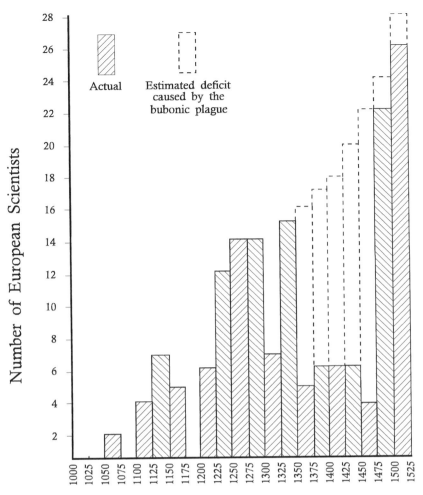

Fig. 9.4. The Plague. Also known as the Black Death, the plague struck Europe in 1347, depressing the amount of scientific activity. After more than 100 years Europe and European science began to recover.

a sweeping military revolution that was, like Europe's agricultural system, cathedrals, and universities, a unique development.

Gunpowder technologies originated in Asia. The Chinese invented gunpowder in the ninth century A.D. and developed fireworks and rockets before 1150. By the mid-1200s Chinese armies employed Roman-candle-style "fire lances" and explosive bombs thrown by catapults, and by 1288 the Chinese created metal-barreled guns. In an early instance of technology transfer the Mongols acquired gunpowder technology from the Chinese, whence it probably passed into Europe across the steppes of central Asia. Possibly through contact with Chinese engineers and technicians, the technology also passed into the Islamic world, as gunpowder was used against European Crusaders in 1249. Europeans may also have learned of the technology from travelers in the East like Marco Polo who worked for the Mongol emperor in China from 1275 to 1292.

While gunpowder and early firearms technology originated in China, large cannon seem to have originated in Europe in the decade 1310–

20. The technology then spread quickly back to the Middle East and Asia, cannon finding their way to Islam by the 1330s and to China by 1356. By 1500 the manufacture of guns had become a universal technology in the Old World with centers in China, the Moghal empire in India, the Ottoman empire, and Europe, and these powers further spread cannon technology to client-states throughout the Old World.

Early cannon and bombards were huge. Turkish guns deployed in the siege of Constantinople in 1453, for example, were so large they were cast on the site and could not be moved. "Mons Meg," a cannon cast for the duke of Burgundy in 1449, measured almost 10 feet long, weighed 17,000 pounds, and threw a stone ball of nearly two feet in diameter. Probably because they became engaged in a more intense cycle of competition, European military engineers and foundrymen actively developed the technologies of cannon-making, and with a superiority of gun design, Europeans soon surpassed their Asian counterparts from whom they initially learned the technology. The great size of guns—suitable for battering walls of established fortifications—gave way to smaller, relatively portable bronze and then to much less costly cast-iron cannon, especially after 1541 when the English mastered techniques of casting iron cannon under the initiative of King Henry VIII. Smaller cannon not only proved more mobile on land, they produced powerfully changed conditions when brought aboard ships.

Already in the fifteenth century, gunpowder and firearms began to play a decisive role on the battlefields of Europe, and by the end of the century they had transformed the politics, sociology, and economics of war. The "gunpowder revolution" undermined the military roles of the feudal knight and the feudal lord and replaced them with enormously expensive gunpowder armies and navies financed by central governments. The knight was not made useless by the new weaponry. In fact, knights continued to proffer their services and to maintain retinues of archers and pikemen. But neither knights nor noblemen could master the economic dimension of the revolution in warfare, for the new artillery lay beyond the means of any individual captain or lord and could be financed only by royal treasuries. At the outset of the Hundred Years' War (1337–1453), the primary means of combat remained the longbow, the crossbow, the pike, and the armored knight mounted on an armored charger. At the end, gunpowder artillery won out.

The military career of Joan of Arc (1412–31) nicely illustrates this transition in military and technological history. Joan of Arc, an illiterate peasant girl of 17, could defeat experienced English commanders partly because artillery was so new that previous military experience carried little advantage, as is the case with any new technology not derived from accumulated knowledge and tradition. Indeed, her fellow commanders praised Joan especially for her keen ability to place artillery in the field. (In what has been labeled the "Joan of Arc syndrome," it would seem that whenever a new technology appears on the scene, e.g., with computers, the young often surpass the old and may make significant contributions.)

The new weaponry appearing in Europe in the fifteenth century required large increases in the budgets of European governments. During the second half of the fifteenth century, for example, as the Military Revolution took hold, tax revenues in western Europe apparently doubled in real terms. From the 1440s to the 1550s, as a further example, the French artillery increased its annual consumption of gunpowder from 20,000 pounds to 500,000 pounds, and the number of French gunners rose from 40 to 275. Spanish military expenses went from less than 2 million ducats in 1556 to 13 million in the 1590s. To cope with these increasing expenditures Philip II of Spain tripled the tax burden on Castile and repeatedly repudiated the state debt, without ever meeting his military payroll on time.

The musket was introduced in the 1550s, and in their reforms Maurice and William Louis of Nassau instituted volley fire by extended rows of musketeers using standardized muskets and coordinated routines of loading and firing. Those reforms and standardized field artillery made for potent new armies from 1600. In the face of muskets and artillery, longbows, crossbows, broadswords, cavalry, and pikemen exercised diminished roles or vanished entirely from the scene of battle. Infantry, now bristling with handguns, once again became a dominant arm on the field of battle. As a result, over the next two centuries the size of standing armies of several European states jumped dramatically from the range of 10,000 to 100,000 soldiers. During the last 70 years of the seventeenth century alone the French army grew from 150,000 to perhaps 400,000 under the Sun King, Louis XIV.

Because cannon could reduce medieval castles and old-style city walls with relative ease, they mandated new and even more expensive defensive countermeasures that took the form of earthen ramparts studded with star-shaped masonry bastions known as the *trace italienne*. Fortified with guns of their own, these installations allowed a defender to rake the attacking formations. European governments poured money into the development of these new and expensive systems of fortifications, but they strained the resources of even the richest European states. Offense and defense alternated in an escalating pattern of challenge and response. Relentlessly, costs mounted and warfare became the province of centralized states.

Only larger political entities, notably centralized nation-states with taxing power or other mercantile wealth, could afford the new weaponry and its attendant fortifications. The Military Revolution, therefore, shifted power from local feudal authorities to centralized kingdoms and nation-states. The kingdom of France, for example—the most powerful of the early modern states in Europe—only emerged as an entity after the Hundred Years' War in the fifteenth century. The related development of the musket and the standing armies that resulted

Fig. 9.5. Military drill. Although the musket had a much lower rate of fire than the bow and arrow, it struck with much greater force. The complex firing procedure led to the development of the military drill. Organizing, supplying, and training large numbers of soldiers armed with guns reinforced political trends toward centralization.

after 1550 reinforced this trend, insofar as central governments ended up as the only agencies that could afford standing armies and that possessed the bureaucratic capabilities to organize, equip, and maintain them.

The substantial government assistance and intervention required by this historically unique military technology led European societies toward centralized authority. The Military Revolution introduced competition between states and dynamic social mechanisms that relentlessly favored technical development. The centralizing effects—social, political, and economic—were akin to those called forth by irrigation agriculture in the great hydraulic civilizations we have previously examined. Thus, Europe acquired, like Egypt and China thousands of years before, the full panoply of civilized institutions. From the fifteenth century onward, the creation of national armies and royal navies resulted in political centralization as inevitably as had the hydraulic projects of the ancient and medieval East. Arsenals, shipyards, and fortresses were maintained as state-owned and state-controlled public works, comparable to the dams and canals of hydraulic societies. And, beginning in the seventeenth century with the Swedish army of Gustavus Adolphus and later established as a national principle by the French Revolution, universal military conscription became the modern equivalent of the ancient corvée.

While Europe became increasingly centralized as a result of the Military Revolution, its ecology and geography precluded the rise of a cohesive European empire of the sort found in China, India, or the Islamic world. In comparison with the large-scale irrigation works of the East, which encompassed whole regions—the Nile Valley and the Euphrates-Tigris flood plain—a typical European military-political system, based on rain-watered agriculture, remained a more local affair in which a variety of ethnic, linguistic, and geographical factors combined to define a nation-state. As the primary outcome of the Military Revolution, then, a *group* of relatively centralized, competing nation-states emerged

Fig. 9.6. Fortification. As the castle fell to the cannon it was replaced by a complex system of fortification with projections known as the *trace italienne*. While effective as a defensive system it proved to be extraordinarily expensive.

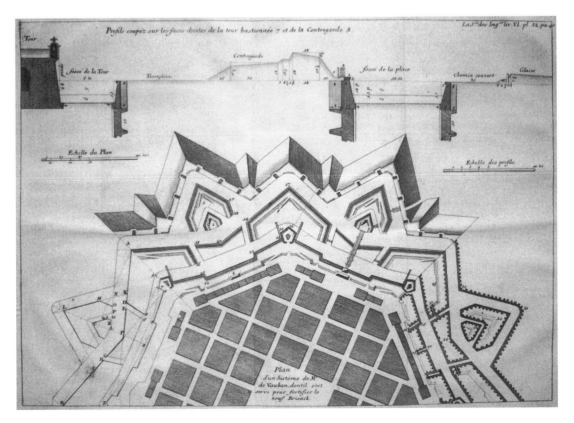

in Europe. They became locked in political, military, and economic competition, none sufficiently powerful to wholly dominate the others. More than anything else, then, the jostling that ensued between and among Spain, Portugal, France, England, the Low Countries, Prussia, Sweden, and later Russia created unique conditions that made Europe a hotbed of conflict and, at the same time, technologically poised for a world-historical role.

Other evidence supports the argument for limited centralization in Europe. The decentralized character of feudal Europe prior to the Military Revolution provides one point of contrast. For another, the new military technology was inherently innovative and, in a complete reversal of the traditions that had characterized hydraulic civilizations, the risks among European states were highest for those that failed to change. Lesser political units or nations (such as Poland) that did not or could not adapt to the Military Revolution simply disappeared as political entities, swept up by larger, more powerful neighbors. Most telling in this regard is the general absence of pan-European institutions that could conceivably have united a European empire. The papacy and the Holy Roman Empire—that nominal remnant of the West Roman Empire that persisted in Europe from its creation in 800 until its dissolution in 1806—represent the most potent of such institutions. Had the ecology of Europe demanded centralized authority, either the papacy or the Holy Roman Empire might have supplied a supranational structure of governance. In the event, however, neither brought about a unified Europe, and both remained weak institutions compared to the nation-states that emerged. Even when hegemony became artificially imposed on Europe, it was inevitably short-lived, as in the case of the Napoleonic Empire, which barely lasted a decade before collapsing in 1812.

A second major outcome of the Military Revolution, alongside political centralization within Europe, was a wave of European colonialism and the beginnings of European global conquest. A revolution in naval warfare which accompanied military changes on land formed the technological basis of the process. In part, this revolution in naval warfare entailed the creation of a new type of ship and new techniques of naval engagement. The comparatively lightly manned, wind-powered Portuguese caravel and its successor, the galleon, came to replace the human-powered oared galley, which required large crews and was typical of warships plying the Mediterranean. Similarly, whereas naval tactics for galleys such as that deployed by Turks and Christians in the battle of Lepanto in 1571 involved ramming and boarding, the emerging new style involved heavily armed gunned ships, artillery fire from broadsides and at a distance, and tactics to prevent boarding. Experts credit the defeat of the Spanish Armada sent against England in 1588 in part to the English having adopted broadside artillery fire "along the line" while the Spanish clung to the ramming and boarding techniques that had served them so well in the Mediterranean.

The development of the caravel/galleon shows how complicated technological change is in general. We must keep in mind that the players, the "actors," do not know the outcomes beforehand. Some shipbuilder did not explicitly set out to build an oceangoing gunned ship. Rather, technical, social/cultural, and geophysical factors interacted to produce the new ship. For example, sails and rigging had to be improved, gun ports cut, and gun carriages developed and installed. Captains had to master the compass (another Chinese invention) and determine their latitude at sea (i.e., how far north or south they were), techniques not wholly mastered for sailing south of the equator until the 1480s. It was not enough, of course, to sail to Africa, the Indian Ocean, or the Americas, the trick was to get back to Europe, and for Vasco da Gama, as for Columbus, that involved mastery of the technique known as the *volta,* whereby ships sailing northward along the African west coast headed *westward* back out into the Atlantic until they picked up winds to blow them eastward back to Iberia. Technological change embodies complex social processes, wherein strictly technical issues (in shipbuilding, say) interact with social factors of all sorts to produce technical and social outcomes that cannot be foreseen in advance. The case of the gunned ship likewise makes clear that we cannot isolate an autonomous "technology" in the world and then investigate its separate impact on society.

The global results of this new technological capability were stunning. The Portuguese made their first contacts along the sub-Saharan coast of Africa in 1443 and reached the Cape of Good Hope in 1488. Vasco da Gama's first voyage to the Indian Ocean by way of the Cape in 1497–98 involved four small ships, 170 men, and 20 cannon; he returned to Portugal with his holds full of spices extracted by force from Muslim and Hindu traders. Columbus sailed to the Indies in three small caravels. Hernán Cortés conquered Mexico in 1518 and 1519 with an expeditionary force of 600 men, seventeen horses, and ten cannon. Later European voyages deployed larger and more heavily armed flotillas, but Columbus's, da Gama's, and Cortés's small forces were sufficient for successful voyages and set the pattern for European mercantilism and colonialism for 300 years.

Portugal and Spain were the early entrants, and later France, the Netherlands, England, and others joined the game. Their colonies, colonial rivalries, and mercantilist activities set the tone for European struggles abroad and at home through the eighteenth century. In 1797 the French colonial historian Moreau de Saint-Méry wrote that the great tall ships of his day carrying African slaves to the colonies and colonial products back to Europe were "the most astonishing machines created by the genius of man." European navies provided the technical means for the West to force itself on the larger world. Experts posit that by 1800 European powers dominated 35 percent of lands, peoples, and resources of the earth.

What role did scientific thought play in this immensely consequential European development? The answer is essentially none. Some of the basic inventions (such as gunpowder and the compass), as we have seen, originated in China, where they were developed independently of any theoretical concerns. In Europe itself, with its established tradition of Aristotle, Euclid, and Ptolemy, none of contemporary natural philosophy was applicable to the development of the new ordnance or any of its ancillary techniques. In retrospect, theoretical ballistics might have been useful, but a science of ballistics had not yet been deduced; it awaited Galileo's law of falling bodies, and even then the applicability of theory to practice in the seventeenth century can be questioned. Metallurgical chemistry could have been useful to foundrymen, but prior to the nineteenth century theory was limited, and alchemy seemingly had nothing to offer. Hydrodynamics, which might have applied to ship design, also lay in the future. The mechanics of materials, which later became a pivotal engineering science, was reconnoitered by Galileo and only applied in the nineteenth century. Scientific cartography probably did play a supporting role in early European overseas expansion, but navigation remained a craft, not a science. The gunners, foundrymen, smiths, shipbuilders, engineers, and navigators all did their work and made their inventions and improvements with the aid of nothing more (and nothing less) than experience, skill, intuition, rules of thumb, and daring.

Indeed, the causal arrow flew from the world of technology to the world of science, as European governments—like their Eastern counterparts—became patrons of science and provided government support for scientific research in the hope of technical and economic benefits. It is no accident, therefore, that the institutionalization and bureaucratization of science by European governments appears first in early modern Portugal and Spain. While now thought of more as a late medieval crusader and less as the great humanist patron of scientific exploration, the Portuguese prince Henry the Navigator (1394–1460) was responsible for the historic series of fifteenth-century Portuguese voyages of exploration along the coasts of West Africa. He did much to promote navigation and to launch Portugal's maritime empire and, driven by the spice trade, Lisbon soon became the world's center of navigational and cartographical expertise. The ruling Portuguese court patronized various royal mathematicians, cosmographers, and professors of mathematics and astronomy, and it created two government offices charged to administer Portuguese trade and to prepare maps. Expert Portuguese cartographers obtained employment internationally.

In Spain, from 1516 to 1598 the Holy Roman emperor Charles V, and his son, Philip II, governed the largest contemporary European empire. Stemming from its colonial rivalry with neighboring Portugal and from technical issues over demarcating the borders of the empire halfway around the world, in the sixteenth century Spain replaced Portugal as the predominant center of scientific navigation and cartography. In Seville the government-sponsored Casa de la Contratación (House of Trade, 1503) kept and continually revised the master-maps of Spanish overseas expansion. The position of Pilot-Major (1508) at the Casa superintended training pilots in navigation; the Cosmographer Royal (1523) was responsible for nautical instruments and charts; and in 1552 Philip II established a royal chair at the Casa in navigation and cosmography. Complementing the Casa de la Contratación, the Council of the Indies, the government body established in 1523 to administer colonial development, came to possess its own corps of royal cosmographers and administrators charged with various scientific and practical duties in the expansion of the Spanish empire. Spanish support for cartography and navigation culminated in the creation of an Academy of Mathematics founded in Madrid by Philip II in 1582 which taught cosmography, navigation, military engineering, and the occult sciences. A sign of the times: the engineering professor of fortifications received twice the salary of the leading university professor of philosophy.

The Spanish government also sponsored systematic and comprehensive geographical and natural history surveys of Spain and the Indies. Officers in the Council of the Indies circulated printed questionnaires and collected useful information on an unprecedented scale, and in the 1570s Philip II sent a historic scientific expedition under Francisco Hernández to the New World to collect geographical, botanical, and medical information. Spain and Portugal were the first European powers to deploy scientific expertise in the service of colonial development. Every succeeding European colonial power—Dutch, French, English, Russian—in turn followed the pattern of state support for science and colonial development established by Spain and Portugal. Thus, by dint of the Military Revolution, the institutionalization of science in early modern Europe began to resemble patterns established in other great civilizations.

CHAPTER 10

Copernicus Incites a Revolution

On his deathbed in 1543 Nicholas Copernicus received the first published copy of his book, *De revolutionibus orbium coelestium (On the Revolutions of the Heavenly Spheres)*. In this seminal work Copernicus proposed a sun-centered or heliocentric cosmology with a moving earth rotating once a day on its own axis and orbiting the sun once a year. In 1543 every culture in the world placed the earth instead at the center of its cosmology. In breaking so radically with geocentrism, received astronomical wisdom, and biblical tradition, Copernicus launched the Scientific Revolution and took the first steps toward the formation of the modern scientific worldview.

The Scientific Revolution represents a turning point in world history. By 1700 European scientists had overthrown the science and worldviews of Aristotle and Ptolemy. Europeans in 1700—and everyone else not long afterwards—lived in a vastly different intellectual world than that experienced by their predecessors in, say, 1500. The role and power of science, as a way of knowing about the world and as an agency with the potential of changing the world, likewise underwent profound restructurings as part of the Scientific Revolution.

The historical concept of *the* Scientific Revolution of the sixteenth and seventeenth centuries emerged only in the twentieth century. The Scientific Revolution was initially thought of as an intellectual transformation of our understanding of nature, a conceptual reordering of the cosmos that entailed, in the felicitous phrase, moving from a closed world to an infinite universe. As scholars have delved deeper into the subject, the unquestioned unity and reality of *the* Scientific Revolution or *a* Scientific Revolution began to break down. The Scientific Revolution as simply an episode in the history of scientific ideas is long a thing of the past. For example, any treatment of the Scientific Revolution must now address not just a triumphant astronomy or mechanics but the "occult" sciences of magic, alchemy, and astrology. Ideological arguments for the social utility of science prove to be a fundamental feature of the Scientific Revolution, and the emergence of new scientific methods—notably experimental science—likewise seems a key property of the "new science" of the sixteenth and seventeenth centuries. Changes in the social and institutional organization of contemporary science are now seen as additional defining elements of the Scientific Revolution. The current interpretative stance rejects any simple notion of the Scientific Revolution as a unitary event with clearly defined chronological or conceptual boundaries. Historians now tend to treat the Scientific Revolution as a useful conceptual tool, setting the episode in a broader historical context as a complex and multifaceted phenomenon to be studied through a variety of approaches.

The New World of the European Renaissance

The social context for science in Europe in the sixteenth and seventeenth centuries had changed in several dramatic ways from the Middle Ages. The Military Revolution, the European voyages of exploration, and the discovery of the New World altered the context in which the Scientific Revolution unfolded. The discovery of the Americas generally undermined the closed Eurocentric cosmos of the later Middle Ages, and the science of geography provided a stimulus of its own to the Scientific Revolution. With an emphasis on observational reports and practical experience, new geographical discoveries challenged received authority; cartography thus provided exemplary new ways of learning about the world in general, ways self-evidently superior to mastering inherited dogma from dusty books. Many of the scientists of the Scientific Revolution seem to have been involved in one fashion or another with geography or cartography.

In the late 1430s Johannes Gutenberg, apparently independently of developments in Asia, invented printing with movable type, and the spread of this powerful new technology after 1450 likewise altered the cultural landscape of early modern Europe. The new medium created a "communications revolution" that increased the amount and accuracy of information available and made scribal copying of books obsolete. Producing some 13,000 works by 1500, printing presses spread rapidly throughout Europe and helped to break down the monopoly of learning in universities and to create a new lay intelligentsia. Indeed, the first print shops became something of intellectual centers themselves with authors, publishers, and workers rubbing shoulders in unprecedented ways in the production of new knowledge. Renaissance humanism, that renowned philosophical and literary movement emphasizing human values and the direct study of classical Greek and Latin texts, is hardly conceivable without the technology of printing that sustained the efforts of learned humanists. Regarding science, the advent of printing and humanist scholarship brought another wave in the recovery of ancient texts. Whereas Europeans first learned of ancient Greek science largely through translations from the Arabic in the twelfth century, in the later fifteenth century scholars brought forth new editions from Greek originals and uncovered influential new sources, notably Archimedes. Similarly, printing disseminated previously recondite handbooks of technical and magical "secrets" that proved influential in the developing Scientific Revolution. And, notably, the technology of printing produced a huge impact on contemporary science without any corresponding input from science on printing technology.

Particularly in Italy, the revival of cultural life and the arts in the fourteenth and fifteenth centuries commonly known as the Renaissance must also be considered as an element of changed conditions of the early modern period. The Italian Renaissance was an urban and comparatively secular phenomenon, aligned with courts and courtly patronage (including patronage by church prelates), but not the university. One associates the great flourish of artistic activity of the Renaissance with such talents as Donatello (1386–1466), Leonardo da Vinci (1452–1519), Raphael (1483–1520), and Michelangelo (1475–1564). In comparison with medieval art, the use of perspective—a projection system that realistically renders the three dimensions of space onto the two dimensions of a canvas—represents a new feature typical of Renaissance painting, and through the work of Leon Battista Alberti (1404–72), Albrecht Dürer (1471–1528), and others, artists learned to practice mathematical rules governing perspective. So noteworthy was this development that historians have been inclined to place Renaissance artists at the vanguard of those uncovering new knowledge about nature in the fifteenth and sixteenth centuries. Whatever one may make of that claim, early modern artists needed accurate knowledge of human muscular anatomy for lifelike renditions, and an explosion of anatomical research in the Renaissance may be attributed to this need in the artistic community.

Indicative of these changing times, the great Renaissance anatomist Andreas Vesalius (1514–64) published his influential anatomical opus, *De humani corporis fabrica (On the Fabric of the Human Body)*, in 1543, the same year that Copernicus published his tome on the heavenly spheres. Vesalius was a military surgeon and his anatomical expertise probably owed as much to the Military Revolution and to the new sorts of wounds inflicted by firearms as it did to any aesthetic requirements of Renaissance art. Other Italian anatomists continued to refine their skills and make anatomical discoveries. Bartolomeo Eustachi (d. 1574) and Gabriel Fallopius (1523–62) gave their names to previously unknown tubes in the body, and in 1559 Realdo Colombo (1520–60) postulated the lesser or pulmonary circulation of the blood from the heart through the lungs. These anatomical developments were capped by the discovery of the circulation of the blood by the English physician William Harvey (1578–1657). He studied in Italy and was elected a fellow of the Royal College of Physicians in London where he lectured on anatomy. By careful observations of the slowly beating hearts of dying animals, along with estimates of the quantity of blood that leaves the heart, Harvey arrived at the conclusion that the arterial and venous blood vessels form a connected circulatory system. The publication of his discovery in 1628 was a revolutionary outcome of the fertile tradition of Renaissance anatomy. Indeed, these revisions of the anatomical doctrines inherited from Galen and Aristotle reflect the comprehensiveness of the Scientific Revolution in the Europe of the sixteenth and seventeenth centuries.

Magic and the occult sciences in the Renaissance formed a defining element of contemporary science and natural philosophy. The occult sciences of the Renaissance included astrology, alchemy, demonology, divination, magic, Neoplatonism, Rosicrucianism (which involved secret societies and occult symbols), and the Cabala (concerning secret mysteries in the Bible). The range of magical activities varied considerably in the early modern period, from proscribed contact with the forces of evil through black magic to "natural" or "mathematical" magic, which had to do with remarkable machines or technical processes (such as burning mirrors or magnets) that produced astounding effects. Despite our prejudices against magic and the occult as irrational delusion and charlatanry, at its highest levels Renaissance magic and associated knowledge systems were serious spiritual and intellectual enterprises that embodied learned understanding of the natural world. The very notion of the occult involved a dual meaning, both as secrets shared among adepts and as secrets hidden in nature.

The occult sciences gained legitimacy and momentum with the recovery and translation of the so-called Hermetic corpus in the mid-fifteenth century. A fundamental principle of the Hermetic philosophy linked the microcosm (or "small world") of the human body with the macrocosm (or "large world") of the universe as a whole through a system of occult (or "hidden") correspondences and relations of "sympathy" and "antipathy." The world, therefore, took on an emblematic quality, replete with hidden meanings, associations, and occult symbolism. In addition to astrological beliefs, Hermeticism affirmed the magical power of an enlightened magician or magus to change the course of nature. (The principle of Renaissance magic that postulated a universe pulsating with "forces" the magus could learn to command flowed into the modern notion, enunciated by Newton, of a universal force of gravity.) Hermeticism thus saw a transcendental, divine order in nature framed by underlying mathematical realities, and held the optimistic vision that humans could both understand nature and, through a technology of magic, operate upon it in their own interests. These characteristics align Renaissance magic with many of the same individuals and historical forces that gave rise to the Scientific Revolution. The anti-Aristotelian and extra-university nature of these movements should also not be overlooked, nor should the opportunities they invited for patron-

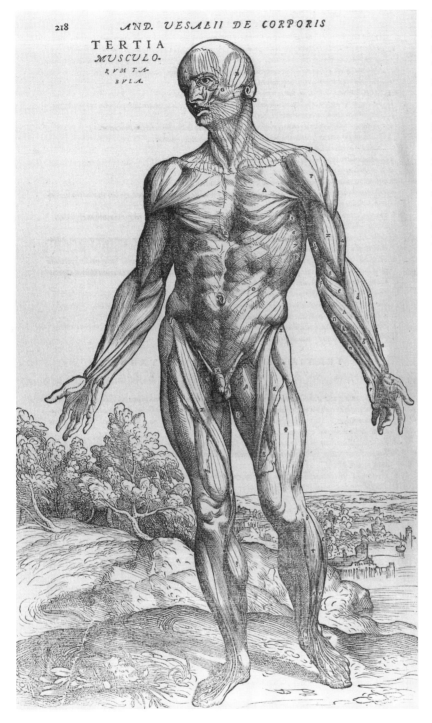

TERTIA
MUSCULO-
RUM TA-
BULA.

Fig. 10.1. The new anatomy. With the creation of gunpowder weapons, physicians and surgeons were confronted with treating more severe wounds and burns. A military surgeon, Andreas Vesalius, produced the first modern reference manual on human anatomy in 1543, the same year that Copernicus published his book on heliocentric astronomy.

age. The relative decline of magic later in the seventeenth century and the transition to more "open" systems of knowledge represent a major transition in the Scientific Revolution, but in the meantime Renaissance magic offered allegedly useful and practical powers.

A monumental historical dislocation, the Protestant Reformation broke the spiritual and political unity of the Catholic Church in the

West in the sixteenth century. The Reformation called into question received religious authority, notably that of the Vatican. In retrospect it represents a major step in the secularization of modern society—that is, the historical shift from ecclesiastical to lay, civil authority governing society. The Reformation began when Martin Luther nailed his Ninety-Five Theses, which were controversial religious propositions, to the door of the church at Wittenberg in 1517, setting off a period of often bloody religious struggle that racked Europe through the Thirty Years' War that ended in 1648. The Scientific Revolution unfolded against the background of the Reformation, and many of its key figures—Johannes Kepler, Galileo Galilei, René Descartes, and Isaac Newton, to name just a few—became deeply affected by religious issues sparked by theological foment.

To this list of changed circumstances facing scientists in the period of the Scientific Revolution, a comparatively minor, yet increasingly irritating problem needs be added: calendar reform. Instituted by Julius Caesar in 45 B.C., the Julian calendar of 365¼ days (with its added full day every fourth February) is longer than the solar year by roughly 10 minutes. By the sixteenth century, the Julian calendar was out of sync with the solar year by some 10 days. Such an inconvenient disjuncture between civil and celestial time exacerbated the already tricky problem of setting the dates for Easter. Pope Sextus IV attempted calendar reform in 1475, but nothing came of it. Pope Leo X raised the issue again in 1512. Consulted on the matter, Nicholas Copernicus expressed his opinion that astronomical theory must be attended to before practical calendar reform was possible.

The Timid Revolutionary

Born in Poland, Copernicus (1473–1543) lived most of his life on the fringes of contemporary scientific civilization, working as a church administrator (canon) in a position gained through family connections. Apparently a timid person, submissive to authority, Copernicus seems an unlikely character to have launched any sort of revolution. In 1491 he matriculated at the University of Cracow, and he spent a decade around 1500 at various universities in Italy, where, in addition to formally studying law and medicine, he developed his interest in astronomy and in general absorbed the cultural ambience of the Italian Renaissance. Indeed, in a typical humanist exercise while still a student, Copernicus translated an otherwise obscure, noncontroversial Greek poet, Theophylactus.

The key to understanding Copernicus and his work comes with the recognition that he was the last of the ancient astronomers, not the first of the moderns. A conservative, he looked backward to ancient Greek astronomy, not forward to some new tradition. He worked as a successor to Ptolemy, not as a precursor of Kepler or Newton. He was at most an ambivalent revolutionary. His object was not to overthrow the old system of Greek astronomy, but rather to restore it to its original purity. In particular, he took seriously the injunction issued nearly 2,000 years earlier to "save the phenomena" and to explain the movements of the heavenly bodies strictly in terms of uniform circular motion. For Copernicus, Ptolemaic astronomy failed to provide a satisfactory account of the stations and retrogradations of the planets; with its elaborate geometrical constructions, it was an astronomical "monster." In particular, he repudiated Ptolemy's equant point—that arbitrary mathematical point in space whence astronomers measured the uniform circular motion of bodies. Uniformity of motion for orbits based on equants was merely a fiction; in fact, as long as astronomers deployed equants, they implied that the planets moved with nonuniform speeds. There had to be a better way, one more consistent with uniform circular motion and ancient tradition.

For Copernicus, that better way turned out to be heliocentrism or placing the sun at (or at least near) the center of the solar system and making the earth a planet. He first proposed heliocentrism in an anonymous manuscript tract, the "Commentariolus," which he circulated among professional astronomers after 1514. But he held off publication of his great work, *De revolutionibus,* possibly because he felt such secrets should not be revealed and certainly for fear, as he put it in his dedication to the pope, of being "hissed off the stage" for such an "absurd" theory. A younger German astronomer and protégé, Rheticus, saw Copernicus's manuscript and published a notice of it, the *Narratio prima* or "First Account," in 1540. With the way apparently cleared, Copernicus approved publication, and his *De revolutionibus orbium coelestium* duly appeared in 1543 just before his death.

Copernicus did not base his astronomy on any new observations. Nor did he *prove* heliocentrism in *De revolutionibus.* Rather, he simply hypothesized heliocentrism and worked out his astronomy from there. In the manner of Euclid's geometry Copernicus posited heliocentrism in a handful of axioms and developed propositions concerning planetary motion under the assumed conditions. He made these bold assumptions for essentially aesthetic and ideological reasons. For Copernicus the heliocentric system possessed greater simplicity and harmony in its proportions; it was intellectually more refined—more "pleasing to the mind"—and economical than what he regarded as the inelegant system of Ptolemy.

The greater simplicity of heliocentrism lay primarily in how it explained the stations and retrogradations of the planets which remained so awkward to explain in geocentric accounts. In the Copernican system such motion is an illusion resulting from the relative motion of the earth and the planet in question against the background of the fixed stars. That is, from a moving earth a moving planet may appear to stop, move backward, and then move forward again, while actually both the observed and the observer circle the sun without any backward motion. With heliocentrism the appearance of the stations and retrogradations of the planets remains, but the problem vanishes: "retrograde" motion automatically follows from the postulate of heliocentrism. The revolutionary character of Copernicus's achievement is nowhere more evident than in the fact that with the adoption of heliocentrism the central theoretical problem in astronomy for two millennia simply disappears.

The Copernican hypothesis was simpler and aesthetically more appealing on additional grounds. It explained why Mercury and Venus never stray farther from the sun than an angular distance of 28° and 48°, respectively. The Ptolemaic system adopted an ad hoc, unsatisfying solution to the problem, while for Copernicus, because the orbits of Mercury and Venus fall within the orbit of the earth, those planets must remain visually in the vicinity of the sun. Similarly the Copernican system dictated a definite order to the planets (Mercury, Venus, Earth, Mars, Jupiter, Saturn), while the matter remained uncertain in Ptolemaic astronomy. Using the Copernican planetary order, observed planetary positions, and simple geometry, astronomers could calculate the relative distance of planets from the sun and the relative size of the solar system.

For Copernicus and like-minded astronomers, the sun occupied a position of paramount importance. In an oft-quoted passage in *De revolutionibus,* one redolent of Neoplatonism if not actual sun-worship, Copernicus wrote:

> In the middle of all sits Sun enthroned. In this most beautiful temple could we place this luminary in any better position from which he can illuminate the whole at once? He is rightly called the Lamp, the Mind, the Ruler of the Universe; Hermes Trismegistus names him the Visible God, Sophocles' Electra calls him the All-seeing. So the Sun sits as upon a royal throne ruling his children the planets which circle round him. . . . Meanwhile the Earth conceives by the Sun, and becomes pregnant with an annual rebirth.

For Copernicus the earth rotates once a day on its axis, thus accounting for the apparent daily motion of everything in the heavens, and the earth revolves around the sun once a year, accounting for the sun's apparent annual motion through the heavens. But Copernicus ascribed not two, but three motions to the earth, and to understand Copernicus's "third motion" reveals the essence of his worldview. In a word, Copernicus held that the planets orbit the sun not in empty or free space but embedded in the crystalline spheres of traditional astronomy. Thus, the spheres in the title of his magnum opus, *On the Revolution of the Heavenly Spheres,* refer not to the spheres of the planets—Earth, Mars, Venus, and so on, but to the crystalline spheres that carry the planets!

That being the case, a serious problem arose for Copernicus, for if

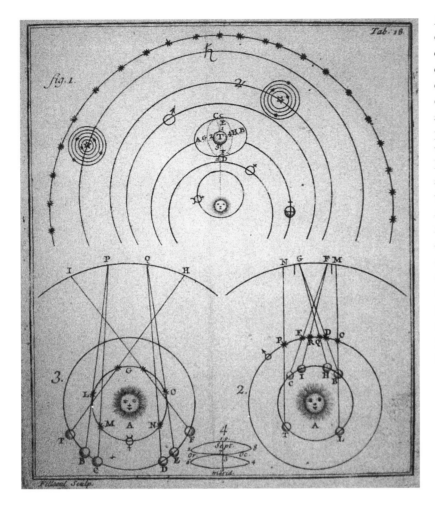

Fig. 10.2. Retrogradation of the planets in the heliocentric system. Copernicus provided a simple explanation for the age-old problem of stations and retrogradations of the planets. In the heliocentric system a planet's looping motion as seen against the background of the fixed stars is only apparent, the result of the relative motion of the earth and the planet in question. In this eighteenth-century engraving, 2 depicts how the superior planets (Mars in this case) seem to retrograde; 3 depicts the inferior plant (Venus in this case).

the earth were carried around the sun in a solid crystalline sphere, the earth's north-south axis would not maintain its constant "tilt" of 23½ degrees toward the pole star (Polaris), and therefore no changes in seasons would occur. Introducing another "conical" motion of the earth's axis, Copernicus kept the earth pointed to the same spot in the heavens and thus accounted for seasonal changes while having the earth carried around by its celestial sphere. In addition, by making this annual third motion of the earth slightly longer than the annual period of the earth's orbit of the sun, Copernicus explained yet another tricky phenomenon, the precession of the equinoxes or the separate motion of the sphere of the fixed stars over a 26,000-year period.

Of course, like Aristarchus before him, Copernicus had to respond to the traditional objections to the idea of a moving earth, and he offered a modified version of standard Aristotelian physics to account for the phenomena. For Copernicus, circular motion is natural to spheres; therefore the earth rotates by its very nature and, like the other planets, is carried around the sun by the inherent natural motion of its crystalline sphere. Material particles naturally aggregate into

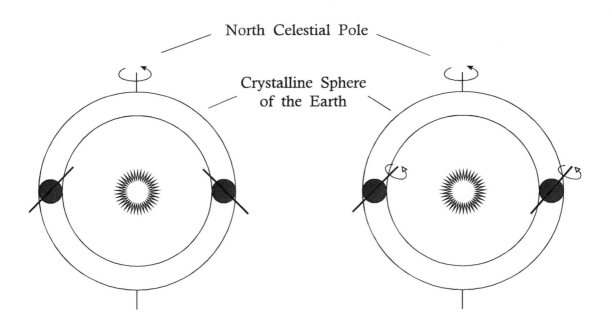

North Celestial Pole

Crystalline Sphere
of the Earth

Without Copernicus'
"Third" Motion

With Copernicus'
"Third" Motion

Fig. 10.3. Copernicus's third ("conical") motion of the earth. To account for the fact that the earth's axis always points in the same direction, Copernicus added a third motion for the earth in addition to its daily and annual movements.

spheres; hence, objects fall downwards on earth, not to the center of the universe, but only to the center of the earth. Bodies do not fly off the earth, given its diurnal and annual motions, because they share in the circular motions of their "mother." Qualitatively, it all works wonderfully well, and the aesthetic superiority shines through in the first twelve folios (twenty-four pages) of the first book of *De revolutionibus,* wherein Copernicus presents the general outline of his system.

The other five books and 195 folios of *De revolutionibus* are quite a different matter. There one finds a highly technical reform of mathematical astronomy, as rigorous and abstruse as Ptolemy's *Almagest.* In fact, superficial comparison can hardly distinguish the works. Copernicus did not intend his work for popular consumption, wanting only to be judged by other professional astronomers. Indeed, he said of his audience that "mathematics is written for mathematicians," and verso to his title page he had printed Plato's motto, "Let no one ignorant of geometry enter here."

Considered as a technical treatise for professional astronomers, therefore, *De revolutionibus* loses much of its aesthetic appeal. As it turns out, the sun stands only near the center of the solar system, not at the center. Copernicus avoided the dreaded equant, to be sure, but, committed to circular motion, he was forced to retain an elaborate apparatus of epicycles and eccentrics in order to explain the remaining irregularities in the apparent speed of the planets as they circled the sun. In the final analysis, as the technical details piled up, Copernicus's astronomy was not any more accurate or more simple than Ptolemy's.

Although he eliminated large epicycles, depending on which circles one counts, Copernicus may have actually employed more epicycles than the corresponding contemporary version of Ptolemy.

Copernican astronomy also faced several nettlesome technical problems that undermined its appeal considerably. The most serious was the problem of stellar parallax, the same problem that subverted Aristarchus and heliocentrism in Greek antiquity. As noted in discussing Aristarchus's proposal, as the earth revolves around the sun, the stars ought to change their apparent relative positions. But astronomers observed no such stellar parallax.

The phenomenon of stellar parallax is in fact a very subtle one, never observable in naked-eye astronomy and not actually demonstrated until 1838. The discovery of stellar aberration by the English Astronomer Royal, James Bradley, in 1729 demonstrated the earth's annual motion but, amazingly, only in 1851 did the physicist J.-B.-L. Foucault definitively prove the daily rotation of the earth by using a giant pendulum. By the eighteenth and nineteenth centuries Ptolemaic astronomy had all but ceased to exist; by that time astronomers universally held to the earth's diurnal motion and to heliocentrism. Can it be that such definitive proofs are not what is needed to persuade converts to a new science?

Be that as it may, the Copernican explanation for the lack of observed stellar parallax resembled Aristarchus's: he assumed the stars were very far away and hence the parallax remains too small to be observed. But this hypothesis produced further problems, notably that the size of the universe ballooned to incredible proportions and the size of the stars (extrapolating from their apparent size) likewise became unbelievably immense. Ptolemaic astronomy had set the sphere of fixed stars at a distance of 20,000 Earth radii. For Copernicus the stars had to lie at least 400,000 Earth radii away, an apparently absurd distance in the context of sixteenth-century astronomy.

The fact that falling bodies do not appear to be left behind as the earth allegedly moves was also a strong impediment to the acceptance of heliocentrism. These and other technical problems meant that Copernican heliocentrism was not immediately hailed as a self-evidently correct or superior astronomical system. But other issues loomed, too, including religious objections that heliocentrism seemingly contradicted passages in the Bible. Copernicus dedicated *De revolutionibus* to Pope Paul III, perhaps to stave off such objections. Pope Clement VII had earlier learned of Copernicus's views in the 1530s and did not object, and Catholic astronomers and churchmen did not take theological exception to the Copernican hypothesis in the second half of the sixteenth century. Some leading Protestants, on the other hand, including Luther and the Danish astronomer Tycho Brahe, did pose such objections, but only in the next century when Galileo fanned the flames of theological controversy did they flare up against Copernicanism.

A spurious prefatory letter attached to *De revolutionibus* explains why Copernicanism did not provoke more strenuous theological reactions. A Lutheran cleric, Andreas Osiander, saw Copernicus's book through the press and on his own authority added an anonymous foreword, "To the Reader Concerning the Hypotheses of this Work." Osiander-cum-Copernicus wrote that heliocentrism need not be true or even probable, merely that it provides a convenient mathematical device that permits astronomers to make more accurate calculations. Copernicus himself held heliocentrism to be a true description of the physical world, but based on Osiander's preface he was taken merely to have created a useful fiction. Paradoxically, Osiander may have helped pave the way for acceptance of Copernicanism by making it superficially palatable for all concerned.

The idea of heliocentrism slowly diffused among astronomers after Copernicus. A new set of astronomical tables—the so-called *Prutenic Tables* calculated on Copernican principles by the astronomer Erasmus Reinhold and published in 1551—represents one practical result forthcoming from Copernicus's work. In 1582, based on these new tables, authorities finally effected calendar reform by instituting the Gregorian calendar in use today. (Named after Pope Gregory XIII, the Gregorian calendar suppresses leap years for centennial years, except those divisible by four.) By the same token, although Copernicus's book was reprinted in 1566 and again in 1617, only a handful of technical astronomers ever read him. A revolution in astronomy is barely discernible even in the second half of the sixteenth century. Not an abrupt transformation of contemporary astronomy or of worldview, the Copernican revolution was, at most, a revolution by degrees.

Tycho's Turn

The great Danish astronomer Tycho Brahe (1546–1601) added to the momentum of the revolution Copernicus had quietly begun. A haughty and arrogant aristocrat, for 20 years from the mid-1570s to the mid-1590s Tycho ruled over the Danish island of Hveen given to him as a fief by the Danish king Frederick II, along with its village, farms, and peasants. There he built and equipped two great astronomical palaces—Uraniborg, the castle of the heavens, and Stjerneborg, the castle of the stars—which together comprised the most magnificent scientific installation of its day. Tycho added his own printing press, a paper mill, a library, and several alchemical laboratories. A practicing alchemist and astrologer, he cast horoscopes for patrons and friends and gave away alchemical medicines. Having lost part of his nose in a duel, Tycho sported a prosthetic replacement, and with his own jester, pets, and coteries of assistants the lord of Uraniborg may seem a virtual self-parody of a Renaissance magus. Tycho had a falling-out with a succeeding Danish king and left Denmark in 1597 to take up a court position as Imperial Mathematician to the Holy Roman Emperor, Rudolph II, in Prague. One report of his death has it that he brought on urinary difficulties because he drank too much at the banquet table and was too polite to get up to relieve himself. He died a few days later.

But Tycho was not merely an eccentric. He was also an adept astronomer who understood the needs of his science. Early in his career he became convinced that the perfection of astronomy depended on accurate and sustained observations of the heavens, and he made it his life's work to undertake those observations. To that end he built large and delicately calibrated naked-eye instruments, such as mural quadrants and armillary spheres—some twenty major instruments at Uraniborg and Stjerneborg. Indicative of the "big science" nature of his enterprise, Tycho received government support totaling some 1 percent of crown revenues, and he boasted that many of his instruments individually cost more than the annual salary of the highest-paid university professors. (Like Copernicus, Tycho's career developed outside the university.) Using these huge and expensive instruments, shielding them from wind stress, minimizing temperature variations, testing and correcting for their intrinsic errors, and adjusting for atmospheric refraction, Tycho produced the most precise naked-eye observations ever, trustworthy down to five or ten seconds of arc in some cases, a minute or two in others, and four minutes of arc in all cases. (A minute of arc is 1/60 of a degree; a second of arc is 1/60 of an arc minute; and, of course, 360° span a circle.) This margin represents an exactitude double that of ancient astronomical observations and one not bested by telescopic observations for yet another century. But the beauty of Tycho's data derived not only from their intrinsic accuracy, but from the systematic character of the observations that Tycho and his assistants methodically compiled night after night over an extended period of years.

Two celestial events further shaped Tycho's astronomy. On the evening of November 11, 1572, as he left his alchemical laboratory, Tycho noticed a "new star" (what we would call a supernova or exploding star) blazing as brightly as Venus in the constellation of Cassiopeia. It shone for three months, and by executing exact parallax observations, Tycho showed that the new star was not located in the earth's atmosphere or in the region below the moon, but stood in the heavens above the sphere of Saturn. In other words, the "new star" really was a new star, even if a temporary one. Tycho thus demonstrated the mutability of the heavens and thereby issued a strong challenge to a central tenet of received dogma in Western cosmology.

Tycho's observations of the comet of 1577 likewise proved unsettling to traditional astronomical theory. Again based on parallax observations, Tycho showed not only that the comet moved in the regions above the moon, but he raised the possibility that it also cut through the crystalline spheres supposedly carrying the planets. In other words, the celestial spheres—those mainstays of Western cosmology and ce-

Fig. 10.4. Tycho Brahe and the mural quadrant. The remarkably accurate set of naked-eye astronomical observations compiled by the sixteenth-century Danish astronomer Tycho Brahe and his assistants depended on large instruments, such as the mural quadrant depicted here. This famous engraving also shows the other components of the research installations erected by Tycho, including an alchemical lab. No contemporary university or university professorship could have paid for Tycho's activities. He depended on major subsidies from the Danish crown.

lestial dynamics from at least the fourth century B.C.—were not real. After Tycho, the only spheres in the heavens were the observed spherical bodies of the sun, the moon, the earth, and the other planets.

Although his work challenged received doctrines, Tycho rejected Copernicus and heliocentrism on strong empirical grounds, especially the lack of observable stellar parallax and because of the consequence that in Tycho's calculations, given the heliocentric system, the fixed stars had to lie an inconceivable 7,850,000 Earth radii distant from the center. The daily or diurnal motion of the earth in the heliocentric system also seemed absurd, and, indicative of a new imagery stemming from the Military Revolution, Tycho introduced a new argument against a spinning earth: a cannon fired toward the west (and a rising horizon) ought to outdistance shots fired toward the east (and a sinking horizon), altogether against experience. Then, too, Tycho, a Protestant, voiced religious objections to the heliocentric system.

In response to deep problems affecting both Ptolemaic and Copernican astronomy Tycho proposed his own system in 1588. In the Tychonic geoheliocentric system the earth remains quiescent at the center of the cosmos, the planets revolve around the sun, and the sun revolves around the earth. This system possessed several advantages: it accounted for stations and retrogradations of the planets without using epicycles, it removed the absurdities of a moving earth, it maintained the traditional scale of the universe, it eliminated the crystalline spheres, and it was mathematically as accurate as its competitors. Holding the earth at rest, the Tychonic system was the equivalent of the Copernican system without the disadvantages of the latter. The Tychonic system represents good, even if conservative, science. But by 1600 with three competing systems and research programs in existence—Ptolemaic, Copernican, and Tychonic—a crisis in astronomy began to mount.

The Music of the Spheres

The case of Johannes Kepler (1571–1630) belies the notion that the internal logic of scientific discovery alone suffices to account for scientific change. Early in his intellectual career Kepler became obsessed with astrology and number mysticism, and more than anything else these obsessions drove his work, shaped his scientific accomplishments, and redirected the course of the Scientific Revolution. Coming from an impoverished and dysfunctional family—his father an errant soldier of fortune, his mother later in life tried as a witch—Kepler attended Lutheran schools and the university at Tübingen as a talented scholarship boy. An unhappy person with bad eyesight and a number of other physical afflictions, Kepler compared himself to a mangy dog. Although he disdained some aspects of astrology, he saw it as an ancient and valid science, and throughout his life he cast horoscopes and wrote up prognostications and calendars (like farmers' almanacs), from which he earned a regular income. On first learning of the Copernican system he became a convert, finding it, like Copernicus did, "pleasing to the mind" and revealing of the workings of the divine in nature.

Kepler did not set out to become an astronomer, but pursued higher studies in theology. However, before granting his degree, authorities at Tübingen nominated Kepler to fill a position as provincial calendar maker and teacher of mathematics in the Protestant high school at Graz

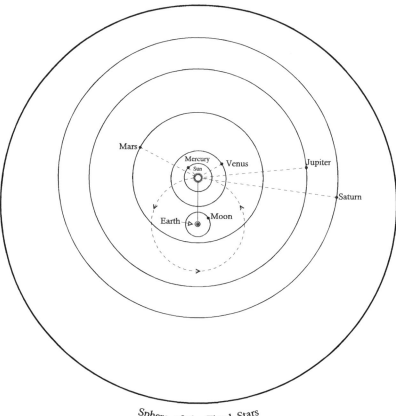

Fig. 10.5. The Tychonic system. In the model of the cosmos developed by Tycho, the earth remains stationary at the center of the universe. The sun orbits the earth, while the other planets revolve around the sun. Tycho's system was good science in that it solved a number of difficult problems in contemporary astronomy, but it did not receive wide acceptance.

Sphere of the Fixed Stars

in Austria, and Kepler accepted. He was a poor teacher, with so few math students that the school assigned him to teach history and ethics. One day—he tells us it was July 19, 1595, in front of his presumably bored geometry class—Kepler had a revelation. He was discussing the cube, the tetrahedron, the octahedron, the icosahedron, and the dodecahedron, which are the five regular solids with identical faces and angles between the faces. (The Greeks had proven that there cannot be more than these five.) Kepler's mystical insight consisted in imagining that these solids might somehow frame the universe, that is, establish the mathematical proportions that spaced the planetary orbits outward from the sun. Thus inspired, Kepler developed his views on the geometrical structure of the universe in a book, *Mysterium cosmographicum (The Mystery of the Universe)*, that appeared in 1596. Kepler's *Mysterium* was the first overtly Copernican work since *De revolutionibus* more than a half a century before, and its origin in pedagogy is one of a handful of exceptions proving the historical rule that nothing of importance for science ever happens in classrooms.

Psychically driven to unpack divine mathematical harmonies of the universe, Kepler was physically driven from Graz in 1600 by the Catholic counter-Reformation and his refusal to convert to Catholicism. He managed to find his way to Prague and a position as Tycho's assistant for the last two years of Tycho's life. The noble, aging Dane assigned the pitiable, younger Kepler data for the planet Mars, in the hope that Kepler could reconcile Tycho's extraordinarily accurate observations with Tychonic theory. The choice of Mars was fortuitous, in that the orbit of Mars is the most eccentric, or noncircular and off-centered, of all the planets. Kepler took up the problem with a vengeance, but only to rescue Mars for Copernican theory and his own intuitions of celestial harmonies. In an intellectual struggle of epic proportions, Kepler worked fervently on the problem for six years, leaving behind some 900 manuscript pages of calculations, testimony to a heroic endeavor and the fact that this exercise in curve fitting proceeded without benefit of mechanical or electronic calculators. In his published account Kepler leads his reader through every tortuous twist and turn. At one point, his circular model for Mars matched the observational data to within 8 minutes of arc, a tremendous achievement, but knowing that Tycho's data were good to 4 minutes, Kepler had to reject his own accomplishment. He made mistakes in his calculations, and then made other mistakes that corrected for them. He got the

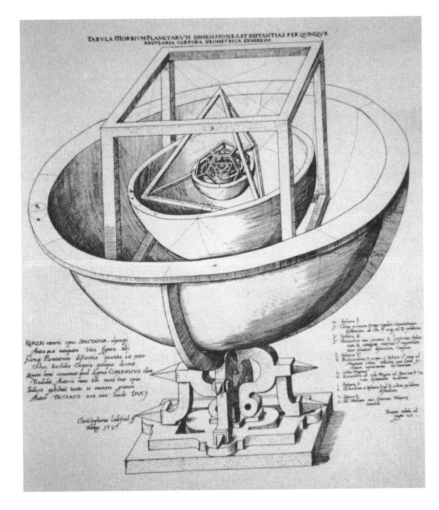

Fig. 10.6. The mystery of the cosmos. In his *Mysterium cosmographicum* (1596) Johannes Kepler conjectured that the spacing of the six known planetary orbits could be explained by nesting them in and around the five regular solids.

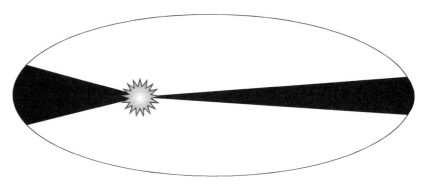

Fig. 10.7. Kepler's elliptical motion of the planets. Based on Tycho Brahe's data, Johannes Kepler broke with the age-old notion that the heavenly bodies move in circles. Kepler reformulated planetary motion in what came to be known as Kepler's three laws: 1) the planets orbit the sun in ellipses with the sun at one focus; 2) planets sweep out equal areas in equal times; and 3) the square of the time of an orbit equals the cube of the mean distance from the sun, or $t^2 \propto r^3$. Kepler reached these conclusions entirely on the basis of astronomical observations and geometrical modeling without providing a convincing physical explanation for why the planets move as they do.

"right" answer and failed to see it. Then, on recognizing an obscure mathematical connection concerning the secant of an angle, Kepler had another flash of insight. "It seemed as if I awoke from sleep and saw a new light break on me," he wrote, and indeed he awoke in a new world.

Kepler concluded that the planets orbit the sun not in circles but in ellipses. This discovery, of course, marks an astonishing turn, in that circles had provided a physics and a metaphysics governing heavenly motion at least since Plato nearly 2,000 years earlier. In his *Astronomia Nova* of 1609 Kepler enunciated the first two of his celebrated three laws of planetary motion: 1) that the planets orbit in ellipses with the sun at one focus, and 2) in what amounts to a planetary speed law, that their radii sweep out equal areas in equal times. Kepler's second law carries the equally disturbing consequence that the planets do not move *uniformly.* As it turns out, Kepler developed his second law before the first, and himself never drew attention to his laws per se. Nevertheless, with the planets moving as Kepler described them and with the sun now indisputably central, Kepler's *Astronomia nova* truly represented a "new astronomy."

Kepler remained in Prague as Imperial Mathematician to Rudolph II until the latter's abdication in 1612. Thereafter he secured a position as provincial mathematician in Linz in Austria, which lasted to 1626, when he moved on to Ulm and Sagan. In the latter period of his life, while the disastrous Thirty Years' War swept over Germany and disrupted his very existence on more than one occasion, Kepler wrote an *Epitome of Copernican Astronomy* (1618–21), which presented his own elliptical vision of the solar system more than the Copernican one, and the *Rudolphine Tables,* a new and highly accurate set of astronomical tables based both on Tycho's data and on Copernican/Keplerian heliocentrism.

In 1619 Kepler issued his *Harmonice mundi,* or *Harmonies of the World.* This work culminated the effort that began with the *Mysterium cosmographicum* and Kepler's meditations and researches on the mathematical order underlying the structure of the cosmos. In the *Harmonice mundi* Kepler calculated astrological relations, correspondences of planets with metals, the music of the spheres—those unheard

tones that he believed were generated by the planets in their motions—
and like connections. Buried in this work lay Kepler's third law, that
the square of the time of a planet's orbit is proportional to the cube of
the mean radius—$t^2 \propto r^3$—at the time an oddly empirical law.

Astronomers and physicists preceding Kepler—even those few who
held to heliocentrism—possessed a traditional dynamics for heavenly
motion: the planets were understood to move uniformly in circles by
their innate natures or carried by crystalline shells. But having dispensed with uniform circular motion, Kepler faced the problem of
providing a dynamics and explaining why the planets move the way
they do through space. Kepler was aware of this obligation and sub-
titled his *Nova Astronomia* of 1609 *The New Astronomy, Based on
Causes, or Celestial Physics*. At an early stage Kepler believed that the
sun possessed an *anima motrix*, a moving soul or spirit akin to the Holy
Ghost, that propelled the planets through their courses. In his more
mature formulations he substituted for this animate spirit a more inanimate moving force or *vis motrix*, a kind of magnetic power. Kepler
derived this latter notion from William Gilbert's influential *De Mag-
nete*, published in 1600, which showed the earth to be a huge magnet.
For Kepler, then, the planets deviate from circular motion as the sun's
and the planets' magnets alternately attract and repel. Kepler's celestial physics provided a plausible explanation for planetary motion, but
not a compelling one, for problems remained. For example, Kepler does
not explain how the force emanating from the sun acts tangentially,
that is, how it "sweeps" the planets along, something like a broom, and
can act at right angles to lines of force emanating from the sun. Also,
paradoxically, he never treats this motive power with the same mathematical rigor and precision he used in determining the orbits of planets. After Kepler, the dynamics of heavenly motion remained an open
question.

Kepler died of a fever in 1630 while traveling to petition for money
owed him. Although he contributed to it mightily, Kepler did not cul-
minate the Scientific Revolution. We extract Kepler's three laws all
too easily from the corpus of his work because we know their historical role and significance for later science, but contemporaries did not
and could not. Few astronomers actually read his works, and by and
large Kepler did not win converts. Indeed, most scientists who became
aware of Kepler's work, notably his great contemporary, Galileo,
rejected his views. As an eccentric mystic, Kepler enjoyed the reputation, rather, of a great astronomer gone slightly mad.

CHAPTER 11

The Crime and Punishment
of Galileo Galilei

Galileo Galilei (1564–1642) looms as a pivotal figure in the Scientific Revolution and in the history of modern science. His great renown and importance for this story derive from several sources and accomplishments. His improvement of the telescope, his astronomical discoveries, and his research on motion and falling bodies brought him international fame and an enduring place in the annals of science. No less striking is Galileo's career as a Renaissance scientist, reflecting as it does deep changes in the social character of science in the sixteenth and seventeenth centuries. And his infamous trial and recantation of Copernicanism at the hands of the Catholic Inquisition is a notorious chapter in relations between faith and reason, which contributed to the slowly emerging recognition of the value of intellectual freedom.

Galileo, Court, and Telescope

Galileo's life and career unfolded in clearly demarcated stages. Born in Pisa and raised in Florence, throughout his life Galileo maintained his identity as a Tuscan. His father served the Medici court as a professional musician. Galileo attended the university at Pisa as a medical student, but he secretly studied mathematics, and eventually the father consented to his son's pursuit of the more socially vague career of mathematician. After a brief apprenticeship, and through patronage connections, at age 25 Galileo secured a term appointment at the University of Pisa in 1589.

At this stage Galileo followed the medieval model of a career in mathematics and natural philosophy institutionalized in universities. He toiled diligently as a university professor of mathematics for three years at Pisa and then, again through patronage connections, from 1592 at the University of Padua in the republic of Venice. As professor of mathematics at Padua, Galileo endured a lowly status in the university, earning only an eighth as much as the professor of theology. He lectured daily and disconsolately during university terms on a variety of subjects—astronomy, mathematics, fortifications, surveying—and came to feel that teaching interfered with his ambitions to conduct research: "It is a hindrance not a help to my work." To make ends meet, Galileo boarded foreign students and took in others for private tutoring. He employed an artisan to manufacture a "geometric and military compass"—the proportional dividers that he had invented and sold to engineers and architects. In Padua he took on a long-term mistress, Marina Gamba, with whom he had three children. Galileo was a devoted father. He had red hair, a short temper, a mastery of language, a gift for mockery in debate, and a liking for wine. All in all, until he stumbled onto the telescope—or better, until the telescope stumbled onto him—Galileo was a hardworking, low-paid, disgruntled, relatively undistinguished professor at a second-rate university. He was already 45 years old in 1609 when he suddenly achieved fame and immortality.

A Dutchman named Hans Lipperhey invented the telescope in Holland in 1608. In Padua Galileo heard a report of the "toy" and that was evidently enough for him to understand the concept of the telescope and to craft one. Galileo ground his own lenses of Venetian glass. His first attempt resulted in an eight-power telescope, which he soon bettered with models of 20 and 30 magnifications. Galileo's renown stems from his boldly turning his improved spyglass to the heavens and discovering a fabulous new celestial world. He rushed into print in 1610 with the forty-page pamphlet *Sidereus nuncius (Starry Messenger)* and, with a job-seeker's instincts, he dedicated it to Cosimo II de' Medici, the grand duke of Tuscany. In the *Starry Messenger* Galileo announced the existence of myriads of new stars, never before seen by anyone, making up the Milky Way. He showed that the moon, far from being a perfect sphere, was deformed by huge mountains, craters and valleys, and that it may well have an atmosphere. Most spectacularly, Galileo revealed that four moons circled Jupiter. The previously unknown moons of Jupiter indicated that other centers besides the earth or the sun existed around which bodies could orbit. Recognizing their importance and their potential for his career ambitions, Galileo unashamedly named these four moons the Medicean Stars.

The first telescopic discoveries did not involve simply pointing the telescope to the heavens and instantaneously seeing what Galileo reported. We should not underestimate the difficulties or the early disputes that arose over interpreting the visual images presented by the telescope, conceptualizing new astronomical entities, and the acceptance of the telescope as a legitimate tool in astronomy. Galileo could "see" the mountains on the moon only by interpreting the changing shadows they cast over a period of weeks; and he could "see" the moons of Jupiter only by similarly observing their changing positions by careful and protracted observations. As a result, Galileo's mar-

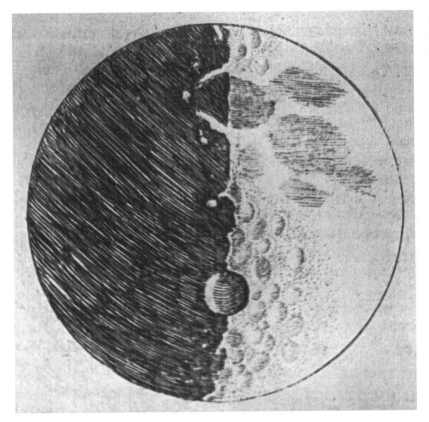

Fig. 11.1. Galileo's *Starry Messenger*. In his famous pamphlet published in 1610, Galileo depicted the results of the observations he made with the telescope. He showed, contrary to accepted dogma, that the moon was not perfectly spherical and had mountains.

velous discoveries soon became incontrovertible, and they brought to the fore the question of the true system of the world.

Galileo craftily parleyed his new fame as an international celebrity into a career move from the University of Padua to the position of Chief Mathematician and Philosopher (with a handsome salary) at the Medici court in Florence back in his native Tuscany. He had long wooed the Medicis, tutoring the crown prince (turned Grand Duke Cosimo II) for a number of summers. The appointment culminated a ritualized negotiation between client and patron. Having presented his telescope to the Venetian Senate, receiving a salary increase and life-tenure at the university in Padua in turn, Galileo burnt his bridges to Venice by accepting the court post in Florence. He wanted the more prestigious position and the time to pursue his own work, free from the burden of teaching undergraduates. For its part the Medici court added another star to the heavens of its reputation, and someone with nominal engineering expertise who might do useful work. Galileo's title of philosopher elevated his status, putting him on par with his professorial adversaries, and it raised the status of his mathematical natural philosophy. With his move from Padua to the court at Florence, Galileo fashioned himself as a scientific courtier.

Galileo's career displays a new pattern for the organization and pursuit of science at the turn of the seventeenth century. He was a Renaissance scientist, albeit a late one, the equivalent in science of Michelangelo in art, and like his Renaissance compeers, Galileo Galilei is known universally by his first name. That Galileo left traditional university life underscores how much this new "Renaissance" mode differed from medieval university-based precedents. His Aristotelian opponents remained intellectually and institutionally rooted in universities, while his own science found its arena among the public and in the courts of the great.

Universities were not active seats of change in the Scientific Revolution, and Renaissance courts and court life provided the key new setting for science. One can usefully distinguish between the court itself, where the ruler and his retainers did business, and full-fledged states that later developed more elaborate bureaucratic apparatus, of which the court formed only a part. An entire patronage system arose, especially in the courts of Renaissance Italy, which provided a new, historically significant means of social support for science. Medici court patronage shaped Galileo's career and his science, while patronage forthcoming from the Holy Roman Emperor, Rudolph II, had supported Tycho and then Kepler at Prague as Imperial Mathematicians. European courts employed varieties of experts: artists, physicians, surgeons, alchemists, astronomers, astrologers, mathematicians, engineers, architects, projectors, surveyors, and cartographers. In offering court-based patronage and support, patrons certainly sought useful services and were motivated mainly by the hope of gaining practical results. But Renaissance patronage represents a social and cultural system that involved much more than patrons "purchasing" useful services. Such exchange constituted a small part of a patronage system that flourished in aristocratic cultures and that entailed hierarchies of "clients" and patrons. For example, patrons accrued glory and enhanced their reputations by supporting glorious clients. Patrons also stirred up controversies, and Galileo became embroiled in several that began as arguments at court. As a social institution, the court patronage system legitimated and helped define the social role of science and scientists in the seventeenth century. The pattern of Renaissance court patronage and court-science seen in Galileo's career did not die out in the seventeenth century, but continued into the eighteenth. Even Isaac Newton became drawn into a controversy (over historical chronology) through his role as a courtier attending the Hanoverian Princess Caroline.

Renaissance academies complemented courts as new institutions for science. Appearing in the fifteenth century as outgrowths of the humanist movement and the spread of printing, hundreds, indeed thousands, of literary and fine arts societies sprang up wherever the educated gathered across Europe over the next three centuries. Private salons and informal associations of amateurs proliferated not only in opposition to university Aristotelianism, but also because of a limited number of university positions. Renaissance-type academies characteristically operated with formal constitutions, but they usually lacked official charters certified by the state. The patron often played the essential role, and many Renaissance academies proved unable to survive without active promotion by the patron.

Renaissance science academies represent a late manifestation of the humanist academy movement. Two early anti-Copernican academies, dating to 1550 and 1560 and both called Accademia degli Immobili (the Academy of the Unmoved), may have been the first to concern themselves directly with science or natural philosophy. Naples was a center of esoteric knowledge, and the Neapolitan magus Giambattista Della Porta (1535–1615) organized an early experimental society, the Academy of the Secrets of Nature (Academia Secretorum Naturae or Accademia dei Secreti) in Naples in the 1560s. A volume of curiosities and wonders, *Magia naturalis (Natural Magic)*, published by Della Porta in 1558 and again in 1589, probably reflects the interests and activities of Della Porta's academy. (More than fifty editions and translations of this influential work appeared in the sixteenth and seventeenth centuries.) Although Della Porta's dabbling in magic brought the scrutiny of the Inquisition, which forced him to disband his academy, he enjoyed a sizable reputation for his learning, and he received patronage offers from numerous courts.

The next Renaissance academy to concern itself with science was the Accademia dei Lincei (1603–30). The Accademia dei Lincei (or Academy of the Lynx-Eyed) appeared in Rome in 1603, patronized by Roman aristocrat Federico Cesi. Della Porta was an early member, but after 1610 the Lincei, led by Cesi, shifted to Galileo's more open approach and program for science. Galileo became a member of the Accademia as part of a triumphal trip to Rome in 1611, and afterwards he valued his title of Lincean academician and used it proudly in print. In turn, Cesi and the academy published several of Galileo's works, including his *Letters on Sunspots* (1613) and the *Assayer* (1623). The Accademia provided a key extra-university institutional prop upon which Galileo erected his career as Renaissance courtier and scientist, and Cesi's untimely death in 1630 and the collapse of the Accademia dei Lincei left Galileo without important support when he came to trial in 1633. In the long run the Renaissance-style social pattern exemplified by Galileo's career gave way to one centered on the nation-state and on national scientific academies. In the meantime, particularly in Italy, the Renaissance court provided a notable home for scientists.

Galileo, Copernicus, and the Church

Galileo's was always a contentious personality, and as soon as he moved from university to court he became embroiled in disputes with adversaries in Florence. He brought with him controversies surrounding the telescope and what it revealed, and he promptly quarreled with Aristotelian opponents over the physics of floating bodies. In these controversies university academics were Galileo's main opponents, and at issue were points of science and Aristotelian natural philosophy. The telescopic discoveries soon put the Copernican question at center stage, and all too quickly theological objections emerged, bringing a whole new class of enemies for Galileo—the theologians. Already in 1611 his name had come up in Inquisition proceedings. Some Dominicans publicly preached against him in 1614, and in 1615 zealous souls actively denounced Galileo to the Inquisition. Although nothing came of these first accusations, the bureaucracy of the Inquisition opened a file with his name on it. Well before his trial and conviction in 1633, Galileo in some sense had already become a prisoner of the Inquisition.

As his star began to shine more brightly at the Medicean court, Galileo preached Copernicanism evermore strongly. The *Starry Messenger* promised that a later book on the system of the world would "prove the Earth to be a wandering body." His *Letters on Sunspots* followed in 1613 with further telescopic novelties concerning spots at or near the surface of the sun, the changing shape of Venus as it makes its orbit, and curious news about the planet Saturn. The discovery of sunspots in particular challenged the purported incorruptibility of the sun, and Galileo would later use the phases of Venus as definitive evidence against the Ptolemaic system. In the *Letter on Sunspots* Galileo asserted that his observations "verified" Copernicus's *De revolutionibus*. A dispute arose at the Medici dinner table in late 1613 over the religious implications of Copernicanism and apparent conflicts with literal interpretations of the Bible. Out of this dispute came Galileo's courtly but provocative "Letter to the Grand Duchess Christina"—the mother of his patron—"Concerning the Use of Biblical Quotations in Matters of Science" (1615). In this work Galileo took the position that faith and reason cannot be in contradiction since the Bible is the word of God and nature is the work of God. However, in instances where there *appears* to be a contradiction science supersedes theology in questions concerning nature, for, as he put it, the Bible was written to be understood by the common people and can readily be reinterpreted, but nature possesses a reality that cannot be altered. For Galileo, if scientists demonstrate some truth of nature that seems to contradict statements found in the Bible, theologians must then articulate reinterpretations of the literal sense of Holy Writ. (This is essentially the position of the Catholic Church today.) Galileo's postulate that science and the human study of nature should take priority over traditional theology represents a radical step, much removed from the medieval role of science as handmaiden to theology, and one almost calculated to provoke the animosity of theologians. Especially offensive was his insufferable arrogance in counseling theologians in the conduct of their business.

Galileo actively defended Copernicanism and led a vigorous campaign to persuade church authorities to accept Copernican heliocentrism over the Aristotelian-Ptolemaic worldview to which religious and scientific thought had long been wedded. In 1616 Galileo's position lost out when the Inquisition ruled Copernicus's opinion erroneous and formally heretical, and the Congregation of the Index placed Copernicus's *De revolutionibus* on the list of books banned by the church. Robert Bellarmine (1542–1621), ranking cardinal of the Inquisition and elderly defender of the faith, wrote that he had understood (from Osiander's preface) that Copernicus posited heliocentrism merely as a mathematical conceit to facilitate astronomical calculations. Since God in his omnipotence could make the heavens go around in any of a thousand ways He pleased, it would be dangerous, said Bellarmine, to set human reason above divine potency and clearly stated biblical language unless clear proof existed that the Bible was mistaken.

On the surface Galileo managed to keep his name out of the proceedings that condemned Copernicanism in 1616, and indeed Bellarmine and the Inquisition afforded Galileo the honor of prior notification of the decision. Different versions have been proposed for what actually transpired at Galileo's meeting with Bellarmine on February 26, 1616. The Inquisition secretly specified steps to be taken (up to and including imprisonment) if Galileo refused to accept the verdict of church authorities as communicated to him not to hold or defend Copernican views. It seems evident that Galileo must have immediately acquiesced, but nevertheless—either in 1616 or at some later time— an irregular, possibly forged, notarial document, one that would come back to haunt Galileo, found its way into the Inquisition's files indicating that a specific personal injunction had been delivered to him "not to hold, teach, or defend [Copernicanism] in any way whatsoever, either orally or in writing." For his part, Galileo obtained a written certificate from Bellarmine later in 1616 confirming that he, Galileo, merely could not "hold or defend" Copernicanism, leaving the option in Galileo's mind to "teach" the disputed doctrine, even while not exactly holding or defending it.

Galileo lost in 1616, but life went on. He was over 50, but still famous and still a Medici luminary. Copernicanism remained off-limits, but other scientific subjects engaged his interest and other quarrels developed. In 1618 a controversy erupted over three comets observed that year, with Galileo drawn into a bitter intellectual brawl with a powerful Jesuit, Orazio Grassi. Their pamphlet war culminated in Galileo's *Assayer* of 1623, sometimes labeled Galileo's manifesto for the new science. The timing for the publication of the *Assayer* proved propitious, for in 1623 the old pope, Gregory XV, died and a new pope was elected, Urban VIII—Maffeo Barberini, a Florentine himself and a longtime friend of Galileo. Prospects looked bright, and the *Assayer* seemed to be the perfect vehicle for Galileo to establish patronage ties at the highest level in Rome. The Accademia dei Lincei published the *Assayer* with an effusive dedication to Urban, who had the book read to him at mealtimes. The new pope was delighted. He showered presents on Galileo, who was sojourning in Rome for six weeks in 1624, and invited him for promenades and discussions in the Vatican gardens. Apparently in that context Galileo asked for and received permission to revisit Copernicanism. Regarding the projected work, Urban insisted, however, that Galileo provide an impartial treatment of the Ptolemaic and Copernican systems and that he stress the point that God could move the heavenly bodies in numberless ways regardless of appearances, and, hence, humans cannot detect the true causes of observed events. Urban's involvement in shaping Galileo's opus extended even to its title. Galileo had wanted to call it *On the Tides,* to highlight his theory that the earth's motion causes the tides and that, hence, the tides confirm that motion. Urban immortalized the work with the title, *Dialogue on the Two Chief World Systems,* a title meant to imply an impartial review of the two planetary theories.

Eight years followed before the *Two Chief World Systems* appeared in print in 1632. By then in his 60s, Galileo was ill during much of the time it took to write the work. Delays ensued in securing the required approvals and licenses, as anxious censors and officials in Rome and Florence gingerly processed the manuscript. When it finally issued from the press in Florence in late February of 1632 Galileo's book was a bombshell for more reasons than one.

First and foremost, the *Dialogue on the Two Chief World Systems* made the clearest, fullest, and most persuasive presentation yet of arguments in favor of Copernicanism and against traditional Aristotelian/Ptolemaic astronomy and natural philosophy. Galileo wrote the work in Italian for the largest popular audience he could reach, and he cast it as a literary dialogue among three interlocutors, Salviati (who in essence spoke for Galileo), Segredo (who represented the interested, intelligent amateur), and Simplicio (the "simpleton" who staunchly voiced Aristotelian views). A spirited and accessible work of literature, the "action" unfolds over four "days." In day 1, using evidence concerning the moon and other new telescopic discoveries, Galileo developed a devastating critique of traditional Aristotelian notions of place, motion, up and down, and the venerable distinction between the celestial heavens and the earth. In day 2 he treats the earth's daily rotation on its axis and deals with apparent conundrums, such as why objects do not go flying off a spinning earth, why we do not experience winds constantly out of the east as the earth spins, why birds or butterflies have no more difficulty flying west than east, why a dropped ball falls at the base of a tower on a moving earth, and why a cannonball flies the same distance east and west. His explanations hinged on the idea that earthbound bodies share a common motion and seem to move only relatively to one another. In day 3 Galileo moves on to consider

DIALOGO
di
GALILEO GALILEI LINCEO
AL SER.ᵐᵒ FERD. II. GRAN. DVCA DI
TOSCANA

Fig. 11.2. The crime of Galileo. In his *Two Chief World Systems* (1632) Galileo presented strong arguments in favor of Copernicanism. Written in Italian as a set of informal conversations, the work was and is accessible to readers not trained in astronomy. The book led to Galileo's arrest by the Roman Inquisition and his subsequent imprisonment.

the heliocentric system and the annual motion of the earth around the sun. Among an array of arguments in favor of Copernicanism and heliocentrism, Galileo introduces his "smoking gun" against Ptolemaic astronomy, the phases of Venus. Seen through a telescope, Venus changes its shape like the earth's moon: a new Venus, a quarter Venus, "horned" Venus. The upshot of Galileo's technical point is that the observed phases of Venus are incompatible with the geocentric Ptolemaic system. The phases did not prove Copernicanism, however, for the observations are consistent with the Tychonic system, too; but Galileo brushed Tycho aside. Last, in day 4 of the *Dialogue,* Galileo offers what in his own mind represents positive proof of the Copernican system, his idiosyncratic account of the tides. His explanation: a spinning, revolving earth induces sloshing motions in the seas and oceans and thereby causes the tides. He gives an elegant mathematical explanation for seasonal variations in tides, and he introduces William Gilbert's work concerning the earth as a giant magnet.

In assessing the merits of the astronomical systems of Ptolemy and Copernicus, Galileo adopted, at least superficially, the impartial posture urged upon him by Urban VIII, pretending to treat indeterminately "now the reasons for one side, and now for the other." He sprinkled his text with the appropriate nonpartisan caveats, saying here and there that he had "not decided" and that he merely wore Copernicus's "mask." But the sharply partisan character of the work could not be denied. Not only was the superiority of Copernicus argued at every turn, Galileo repeatedly refuted Aristotle and made Simplicio seem ignorant and foolish throughout. To make matters worse, Galileo inserted the language dictated to him by the pope concerning God's divine omnipotence and the limits of human reason, but only at the very end of the work and, most provocatively, in Simplicio's voice as "heard from a most eminent and learned person." If Urban had winked at Galileo back in 1624 when they discussed an impartial treatment of Copernicanism and the Ptolemaic system of the world, then Galileo had duly maintained the appearances. If Urban was, or later became, serious about a balanced treatment, then Galileo mocked him and was in deep trouble.

Once the *Dialogue* began to circulate, the reaction was immediate and harsh. In the summer of 1632, on the pope's orders, sales stopped, copies retrieved, and materials confiscated from the printer. In an unusual move designed either to shield Galileo or simply to frame his downfall, Urban convened a high-ranking special committee to evaluate the situation. The matter then passed formally to the Inquisition, which in the fall of 1632 called Galileo to Rome to answer charges. At first Galileo, now 68, resisted the summons, at one point pathetically going so far as to send a "doctor's note" testifying to his inability to travel. But the Inquisition remained adamant; Galileo would be brought in chains if necessary. But it was not; Galileo made the trip to Rome on a litter.

The raw facts concerning Galileo's trial have been well known for more than a century, but radically differing interpretations and explanations of the trial continue to be offered down to the present. An early view—now dismissed—envisioned the episode as a great battle in the supposed war between science and religion, with Galileo as hero-scientist beaten down by theological obscurantism for having discovered scientific truth. Another interpretation, reflecting an awareness of twentieth-century totalitarian regimes, uncovered the essence of Galileo's trial in the bureaucratic state apparatus of the Inquisition. From another perspective, Galileo was simply insubordinate and questioned biblical authority at a time when the church was locked in a deadly embrace with the Reformation. A further documented but more conspiratorial version sees the formal charges stemming from Galileo's Copernicanism as a ruse by church authorities to hide a bitter dispute with the Jesuits as well as other more serious theological accusations having to do with Galileo's atomism and with difficulties reconciling atomism with the miracle of transubstantiation in the Catholic mass. Others argue that Urban's papacy had lost much of its luster and was in political trouble by 1632, and recent accounts evoke a patronage crisis for Galileo and his fall as a courtier. Galileo's trial is a tablet on which historians write and rewrite improved versions, similar to the way legal scholars review and argue old court cases.

Galileo's trial did not proceed in a wholly straightforward manner. Galileo first stayed at the Medicean embassy in Rome, and although treated preferentially, he then had to enter the Inquisition prison like all persons prior to questioning. He appeared before the Inquisition for the first time on April 12, 1633, but he was not told of any charges against him. He rather cockily defended his *Dialogue* by claiming that, in fact, he had not sided with Copernicus against Ptolemy, but had instead shown Copernicus's reasoning to be "invalid and inconclusive." The Inquisitors possessed the tainted notarial minute of 1616, so they thought they had caught Galileo violating a personal injunction not to deal with Copernicanism in any way. When confronted with this document, Galileo produced the certificate he had from the late Cardinal Bellarmine that only prohibited him from "holding or defending" Copernicanism, not from teaching it or treating it altogether.

The Inquisition procured expert opinions that in fact Galileo did defend and hold Copernicanism in his book, but Galileo's genuine certificate from Bellarmine remained an obstacle to resolving the neat case the Inquisition initially envisioned. To avoid a potentially embarrassing outcome, an Inquisition official visited Galileo extrajudicially in his cell to discuss a compromise: Galileo would be led to see and admit the error of his ways, for which some slap on the wrist would be imposed. Appearing next before the Inquisition, Galileo duly confessed to inadvertence and "vainglorious ambition." Shamefully, after he was dismissed he returned to the Inquisition's chambers to volunteer to write yet another "day" for his *Dialogue* to truly set the matter right.

Alas for Galileo, the compromise fell through, and having obtained his confession, Urban rejected the Inquisition's solution and insisted that formal charges of heresy be pressed against Galileo. Hauled before the Inquisition once again, Galileo had nothing to say. Threatened with torture, he said only that he was not a Copernican and had abandoned the opinion in 1616. "For the rest, here I am in your hands; do as you please."

The Inquisition found Galileo guilty of "vehement suspicion of heresy," just a notch below conviction for heresy itself, for which punishment was immediate burning at the stake. Galileo's *Two Chief World Systems* was put on the Index of prohibited books, and on June 22,

1633, this once-proud Michelangelo of Italian science, 69 years old, kneeling in a public ceremony dressed in the white gown of the penitent and with a candle in hand, was forced to "abjure, curse, and detest" the Copernican heresy and to promise to denounce all such heretics. Galileo remained a formal prisoner of the Inquisition under house arrest for life. Arriving for confinement at Siena in July of 1633, myth has it that, stepping out of his coach, Galileo touched his finger to the earth and uttered the words, "Eppur si muove"—"And yet it moves." The History of Science Museum in Florence displays the bones of Galileo's middle finger, a scientific relic gesturing defiantly at us today.

The trial and punishment of Galileo are sometimes invoked to support the claim that science functions best in a democracy. This claim is demonstrably false, as some of the least democratic societies have been and continue to be successful in the pursuit of science and technological advance. The issue, rather, is the independence of scientific communities, regardless of the political context. Where it has occurred, the intervention of political authorities—whether the Catholic Church or the Communist Party—has inhibited scientific development. Fortunately, political authorities rarely have any interest in the abstractions of theoretical science. In the Christian tradition only the movement of the earth and the origin of species pitted natural philosophy against biblical sovereignty. Whether science takes place in democratic or antidemocratic societies has little to do with its development.

Galileo, Falling Bodies, and Experiment

Transferred to his home outside of Florence in December of 1633 and attended by his daughter Virginia, Galileo turned 70, a prisoner of the Inquisition, humiliated by his recantation, and already half blind. Remarkably, he did not simply give up the ghost. Instead, Galileo turned to producing what many judge to be his scientific masterpiece, *Discourses on Two New Sciences* (1638). In his *Two New Sciences* or the *Discorsi,* as it is known, Galileo published two remarkable discoveries—his mathematical analysis of a loaded beam or cantilever and his law of falling bodies. The work represents Galileo's greatest positive contribution to physical science and to the ongoing course of the Scientific Revolution of the sixteenth and seventeenth centuries. The *Two New Sciences* also shows other dimensions of Galileo's genius as an expert mathematician and experimenter.

In writing the *Two New Sciences* Galileo did not suddenly begin a new research program. As might be expected, he went back to his old notes and to scientific work he had done before 1610, before his first telescope, before international scientific fame, and before the astronomy, the polemics, and the condemnation. The technical and esoteric topics Galileo deals with in the *Two New Sciences* are politically safe and theologically noncontroversial subjects involving how beams break and how balls roll down inclined planes.

The *Two New Sciences* was published somewhat covertly in Protestant Holland by the Elsevier press in 1638. Like its companion masterpiece, *Dialogue on the Two Chief World Systems,* the *Two New Sciences* is set in dialogue form and divided into four "days." The same three interlocutors appear as in the *Dialogue:* Salviati, Segredo, and Simplicio, although this time they act less antagonistically and play somewhat different roles. Whereas previously Salviati clearly represented Galileo, Simplicio Aristotle, and Segredo the interested amateur, the three characters in the *Two New Sciences* more likely represent the chronological stages Galileo himself went through in arriving at his mature views concerning mechanics. Simplicio represents his initial, Aristotelian phase, Segredo an Archimedean middle period, and Salviati his late views. In the *Two New Sciences* the Aristotelian Simplicio, in particular, seems much more flexible, even admitting, "If I were to begin my studies over again, I should try to follow the advice of Plato and commence from mathematics."

The *Two New Sciences* was far more of a mathematical treatise than the previous *Two Chief World Systems.* At one point Salviati reads from a Latin mathematical text written by "our Author," Galileo himself. The book opens with a discussion among Salviati and his two friends at the Arsenal of Venice, that famous center of technology, the largest and most advanced industrial enterprise in Europe, where craftsmen and artisans built ships, cast cannon, twisted rope, poured tar, melted glass, and worked at a hundred other technological and industrial activities in the service of the Venetian Republic. Galileo set the scene at the Arsenal of Venice in order to juxtapose, rhetorically and self-consciously, the enterprises we designate as science and technology. While the Arsenal was clearly an important device for Galileo as an extra-university setting for his new sciences, experts disagree over whether Galileo came there to teach or to learn.

Historians of ideas have tended to skip over the first two "days" of the *Two New Sciences* where Galileo treats the mundane field of strength of materials. They prefer to concentrate on days 3 and 4, where Galileo presents his original findings on the more abstract study of motion, of such high importance for Newton and the Scientific Revolution. But the strength of materials is directly relevant to engineering and to the connection of science and technology.

In days 1 and 2 of the *Two New Sciences* Galileo explores the general topics of cohesion of bodies and the breaking strength of materials. In day 1 he considers a potpourri of technical and theoretical problems, some old and some new. He wonders, for example, about size effect (the theory of scaling) and why one cannot build a wooden boat weighing a million tons. He asks what makes a marble column hold together. He presents an extraordinary matter theory (involving infinities of infinitesimals), and he ingeniously tackles the coherence of bodies, surface tension, the nature of fluids, condensation and rarefaction, gilding, the explosion of gunpowder, the weight of air, the propagation of light, geometrical propositions about cylinders, mathematical paradoxes concerning infinity, and discoveries about the constancy of the swing of the pendulum. (Galileo supposedly discovered the isochronism of the pendulum back in the 1580s while at Pisa.) The discussion is always brilliant and entertaining.

In the second "day" of his *Discourse* Galileo extends the range of ancient mechanics in considering the loaded beam. He mathematically determines the effects of the internal stresses induced by external loads and by the intrinsic heaviness of the beam itself. This problem had previously received little theoretical attention and might be of interest to architects, stonemasons, carpenters, shipwrights, millwrights, and engineers. Galileo conducted no experiments. Instead, he applied theoretical statics to the problem and, despite a misguided assumption about the distribution of internal stresses in the beam, he arrived at the fundamentally correct conclusion that the strength of the beam (its flexural strength) is proportional to the *square* of the cross-sectional depth. (*AB* in the figure.) In the end, however, artisans and craftsmen greeted Galileo's results with indifference. Contemporary engineers were fully capable of solving their problems using traditional and hard-won rules of thumb without adopting the meager theoretical principles that contemporary science could offer.

In the more closely studied days 3 and 4 Galileo unveiled the second of his *Two New Sciences,* the study of local motion, that is, motion in the neighborhood of the earth. In a word, Galileo overthrew the traditional Aristotelian conception held by almost all contemporary scientists that the rate at which a body falls is proportional to its *weight*. The medium through which bodies fall, rather than playing an essential resisting role per Aristotelian interpretations, became merely an accidental "impediment" to ideal fall that would occur in a vacuum. To so reconceptualize motion and fall was to strike at the core of Aristotle's physics, and Galileo's work in these areas proved central to breaking down the Aristotelian worldview.

In examining the historical development of Galileo's thought, one must recognize a "process of discovery" as Galileo fought his way through the maze of factors involved in motion and fall. How to conceptualize the factors, much less how to relate them, was not clear to Galileo at the outset. As a student he adhered to Aristotle's views. Early on he became convinced that a rigid Aristotelian interpretation of fall was false, and his skepticism may have led to a demonstration concerning falling bodies at the Leaning Tower of Pisa. At one point in the long and complex process of working out his ideas, Galileo thought that the density of the medium through which bodies fell (such

Fig. 11.3. The strength of materials. While under house arrest in 1638 Galileo, at the age of 74, published the *Two New Sciences* in which he formulated the law of falling bodies and deduced that the strength of a loaded beam is proportional to the square of the depth of its cross-section.

as air) was a key factor in determining how fast they fall. By 1604 he had come to believe that all bodies would fall at the same speed in a vacuum and that the distance a body covers in free fall is measured by the square of the time. But in 1604 he envisioned the right law for the wrong reason, thinking (mistakenly) that the velocity of fall is proportional to the distance covered, rather than time passed (as he finally concluded). Only after returning to his old work in 1633 did Galileo arrive at his mature view that velocity is proportional to time elapsed, not distance covered; the distance a body falls in free fall remains proportional to the *square* of the time of fall (distance $s \propto t^2$). This is Galileo's law of falling bodies. For Galileo, all bodies (independent of their weight) fall at the same accelerated rate in a vacuum.

Our knowledge of the "correct" answers may obscure Galileo's intellectual achievement. Heavy objects do seem to fall faster than light ones, as Aristotelian theory would predict—a heavy book reaches the ground before a light sheet of paper, for example. Many factors are involved in fall: the weight of a body, or, as we would say, its "mass" and "momentum," the latter measured in several ways; the medium a body moves in, the density or specific gravity of an object, the buoyancy of the medium, the shape of the falling body, the resistance it might offer (and different measurements of same), the distance covered, time elapsed, initial velocity (or speed), average velocity, terminal velocity, and accelerations of various types. Which factors are essential? Galileo faced formidable conceptual problems in coming to terms with free fall.

Two further points need to be made concerning Galileo's fundamental contribution to theoretical mechanics. First, his law is a *kinematical* law, that is, it renders a *description* of motion, not an account of the causes of motion. Galileo's law *describes how* bodies fall; it does not *explain why* they fall. In this way Galileo self-consciously avoided any discussion of *cause*. As a methodological maneuver, the beauty and power of the move derive from what can be gleaned from kinematics alone without inquiring into causes. In effect, Galileo is saying, look what we can accomplish by concentrating on a mathematical description of phenomena without confusing matters by arguing over causes of phenomena.

The second point concerns the fact that all of the kinematical rules Galileo announces in the *Two New Sciences,* including the germ of his law of falling bodies, were, as previously noted in chapter 9, discovered and enunciated three centuries earlier by Nicole Oresme and a group of late medieval scholastics at Oxford University known as the Mertonians or the Calculators. There were differences, however; the most important was, as Galileo himself was quick to point out, that while the Mertonians speculated about abstract possibilities for motion, Galileo believed what he had discovered applied to the real world and to the way bodies actually fall here on earth.

Days 3 and 4 of the *Two New Sciences* also speak to the question of the role of experiment in Galileo's science and about how he satisfied himself that his mathematical formulations concerning motion apply in nature while the Mertonian-like ideas remained merely speculative. Galileo is often referred to as the "father of experimental science," and indeed, doesn't the popular stereotype portray Galileo "experimenting" by dropping balls from the Leaning Tower of Pisa? Alas, it has proven all too easy to (mis)cast Galileo as the father of the "experimental method" or the "Scientific Method." The advent of experimental science marked an important development in science in the seventeenth century to which Galileo contributed mightily, but simplistic and uncritical readings of Galileo's work misplace the role of experiment in his science and they reinforce myths about how science works. Galileo decidedly did not operate according to some cliché of the "Scientific Method" that, stereotypically, has scientists formulating hypotheses, testing them through experiment, and deciding their truth or falsity on the basis, simply, of experimental results. The reality of Galilean experiment is more interesting, more complex, and historically more important.

With regard to the Leaning Tower experiment, as a junior professor at the University of Pisa between 1589 and 1592, Galileo supposedly dropped balls from the Tower before an audience of students and professors in order to demonstrate the falsity of Aristotle's theories about heavy bodies falling faster than light bodies. One can question whether such an experiment actually took place because the first written record that Galileo performed any demonstration at Pisa dates only from 1657, 15 years after Galileo's death. The popular image is surely wrong that Galileo was trying to prove his law of falling bodies experimentally. We know that Galileo did not arrive at a formulation of his law until 1604, so he could not have been "proving" it in an experiment a decade before in Pisa. But he may well have performed a demonstration to illustrate problems with the Aristotelian analysis of fall.

The precise role of experiment in Galileo's science and the reality of *how* Galileo came to reject Aristotle are much more complex and nuanced than the cliché of Galileo dropping balls from the Leaning Tower of Pisa would suggest. The way Galileo arrived at his law of falling bodies and associated kinematics did indeed involve considerable experimentation in the sense that he repeatedly made all manner of tests and trials in coming to grips with the phenomena. He left voluminous manuscript records of such experiments. Galileo was a deft experimenter, manually dexterous, and expert with equipment, as his facility for building telescopes also demonstrates. Experiment in this sense figures prominently in Galileo's approaches to research. But formally, Galileo reserved experiment not to test his propositions, as we might think retrospectively, but to confirm and illustrate his principles. In a word, Galileo's experiments do not confirm hypotheses, they demonstrate conclusions previously arrived at through analytical reasoning.

In a key passage in the *Two New Sciences,* after laying out his rules concerning motion, Galileo has Simplicio inquire, "But I am still doubtful whether this is the acceleration employed by nature in the motion of her falling heavy bodies [and I ask you to present] some experiment . . . that agree[s] in various cases with the demonstrated conclusions." To which Salviati replies, "Like a true scientist, you make a very reasonable demand, for this is usual and necessary in those sciences which apply mathematical demonstrations to physical conclusions, as may be seen among writers on optics, astronomers, mechanics, musicians, and others who *confirm their principles with sensory experiences.*"

With that Galileo turns to his celebrated inclined plane experiment. He first describes his experimental equipment: a wooden beam 24 feet long, three inches thick, with a channel chiseled in one edge, smoothed, and lined with parchment. One end of the beam is raised two to four feet, and a rounded bronze ball allowed to roll down the channel. Two paragraphs later he describes his timing method: he collected water running from a container and measured its weight to determine a time interval. It hardly needs to be pointed out that a not-perfectly-spherical or uniform bronze ball rolling and rattling down and over a vellum-lined channel no matter how smooth could not exactly produce the predicted results. Too many "impediments" were at play in the experiment. Under the control of the human eye and hand, Galileo's ingenious water device doubtless slipped a drop or two here and there, and inaccuracies stemming from measurement even in a delicate balance had to throw off the results still further. Yet Galileo goes on unashamedly to claim that "by experiments repeated a full hundred times, the spaces were always found to be to one another as the squares of the times . . . these operations repeated time and again never differed by any notable amount." But he presents no data, and he does not tell us what "any notable amount" means. The community of scientists in France later had cause to doubt the validity of Galileo's claimed results when they tried to replicate his experiment based on his text. Galileo, however, took his own report of the experiment as sufficient evidence to confirm what mathematical analysis previously demonstrated as true. Beyond indicating the distinctive role of experiment in Galileo's science, the inclined plane experiment illustrates the complexities of how experiments work in practice rather than according to some abstract theory of "scientific method."

Finally, in day 4 of the *Two New Sciences,* Galileo extends his considerations of motion to cover projectiles and projectile motion. The figure shows the conceptual model Galileo used to analyze projectile motion. For Galileo the motion of a thrown or shot object is compounded of two other motions. On the one hand, the projected body falls downward according to the law of falling bodies presented in day 3. On the other hand, it moves *inertially* along a horizontal line, meaning that it moves of its own accord without any separate mover acting on it. Galileo's concept of inertia, first presented in his *Sunspot Letters* in 1613, requires further clarification, but its revolutionary implications should already be clear. As we recall, for "violent" motion Aristotle required a mover. How to provide such a mover in the case of a projectile after it separates from its launcher had been a nagging problem for Aristotelian mechanics for 2,000 years. Galileo offered a revolutionary reformulation that eliminated the problem altogether. For Galileo, and later for Descartes and Newton, no mover is required, for there is nothing to explain in accounting for the natural inertial motion of bodies. Such is the stuff of scientific revolutions.

A crucial and revealing difference separates Galileo's view of inertia from the view later adopted by Descartes and Newton. Whereas the latter (and modern science generally) adopts *rectilinear* or straight-line inertia, Galileo held to horizontal or so-called *circular* inertia. He believed that bodies moving inertially would travel, not in straight lines, but in curves following the horizon, in fact, in circles around the earth. For Galileo the horizontal line is not straight, but a segment of a circle around the earth's center, that is, the "horizon." Galileo's revolution-

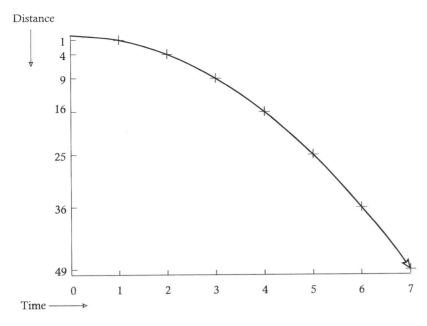

Distance

1	
4	
9	
16	
25	
36	
49	

0   1   2   3   4   5   6   7

Time ⟶

Fig. 11.4. Parabolic motion of projectiles. By separating projectile motion into two components, Galileo could show that projected bodies follow parabolic paths. Along the vertical axis a body falls with uniformly accelerated motion; it moves inertially with a uniform, constant speed on the horizontal axis; when these two motions are combined, the body traces the curve of a parabola. Galileo thus reconceptualized the centuries-old problem of projectile motion, recasting it in something like its modern form.

ary "discovery" (or "invention," if you prefer) of inertia removed a major objection to Copernicanism, for if objects move inertially they will not appear to be left behind as the earth moves, and the concept contributed forcefully to the overthrow of Aristotle and the Aristotelian worldview. That Galileo held to circular inertia would be an oddity of history, except that it provides yet another instance of the continuing power of circles on the imagination of scientists well into the seventeenth century.

Galileo drew the corollary from his analysis of the compound motion of projectiles that the resulting curve is a parabola—at least in theory. (Ironically, it would be a parabola only if the earth were flat.) Galileo's discovery of the parabolic motion of projectiles represents another notable achievement, and this one offered obvious practical possibilities for artillery and ballistics. He recognized such applied uses, and in day 4 he published detailed mathematical tables of elevation and range for gunners derived wholly from theory. The case would seem a golden instance of theoretical science being turned to an applied end but, alas, Galileo's theoretical understanding had no impact on gunnery in practice. By the time Galileo published his tables, cannons and artillery had reshaped the face of Europe for over 300 years. Skilled gunners and military engineers had long since worked out artillery "rules," tables, and procedures for hitting their targets. The technology of cannonry may have been more influential on Galileo's science than the other way around.

Galileo knew that problems remained in the realm of mechanics. He understood very well, for example, that a ball of cotton behaves differently from a ball of lead, and he groped his way toward notions of what we would call "force" and "momentum." But Galileo never got beyond initial speculations about these matters, and at one point in day 4 of the *Two New Sciences* he says, almost nostalgically, "I should still like to find a way of measuring this force of impact." That work—the measurement of force—would come after Galileo, with Isaac Newton. Galileo went completely blind, and doctors withheld his beloved wine. He died in 1642, the very year Newton was born. His erstwhile friend, Pope Urban VIII, prohibited any monument to his memory.

Post-Galileo

The trial and punishment of Galileo did not end scientific activity in Italy in the second half of the seventeenth century, but the case severely dampened the level and quality of contemporary Italian science. The atmosphere in Italy remained repressive, and church authorities were vigilant. Copernicanism and grand cosmological theorizing remained off-limits, and Italian scientists avoided them in favor of safer endeavors like strictly observational astronomy. A tolerance toward Galileo emerged only a hundred years after his death with an Italian edition of his works sanctioned by the liberal Pope Benedict XIV. The Catholic Church authorized the teaching of Copernicus only in 1822, and Copernicus was finally taken off the Index in 1835. Galileo himself was not fully rehabilitated until the 1990s.

Galileo and the patronage system are partly to blame for the failure of a Galilean school to take hold in Italy. Especially in the initial controversies in the 1610s, Galileo had followers and, as something of a patron in his own right, he succeeded in placing a number of them, including Benedetto Castelli (1578–1643) as professor of mathematics at the University of Pisa. Because he was a courtier, however, he did not train students. Vincenzio Viviani (1622–1703) and Evangelista Torricelli (1608–47) joined the master as copyists and assistants only in the last few years of his life. The younger mathematician Francesco Bonaventura Cavalieri (1598–1647) was a true pupil, and Galileo's son, Vincenzio Galilei (1606–49), also carried on his father's work, especially in the development of the pendulum clock. But Galileo's few direct scientific descendants, excepting Viviani, had passed from the scene by 1650. The patronage system robbed him of offspring.

With the waning of Italian science after 1633, a characteristic feature of the period of the Scientific Revolution is the geographical movement of scientific activity northward out of Italy and into the Atlantic states—France, Holland, and England. An active scientific community of independent amateurs arose in France and included such luminaries as Pierre Gassendi (1592–1655), Pierre Fermat (1601–65), Blaise Pascal (1623–62), and René Descartes (1596–1650). Although they were insulated from the direct power of the church in Rome, Galileo's trial by the Inquisition produced a chilling effect on French scientific intellectuals. Descartes, for example, suspended publication of his Copernican treatise, *Le Monde*, in 1633.

René Descartes inherited the mantel of intellectual leadership for the new science. Trained by the Jesuits, Descartes was a polymath genius and soldier of fortune who retired at the age of 32 to take up a contemplative life devoted to philosophy and science. Part of Descartes's fame stems from his achievements in algebra and analytical geometry, including the introduction of "Cartesian" coordinates. He also produced original work in optics and meteorology, and he demonstrated his concern for how scientific knowledge is produced in his historic *Discourse on Method* (1627). Descartes likewise wrote on theology and metaphysics and is often hailed as the father of modern philosophy. For our purposes Descartes's importance derives from the fact that he developed a complete cosmology and world system to replace Aristotle's and competing alternatives that were at play in the early decades of the seventeenth century.

In coming to grips with the state of science and philosophy in his day, Descartes proposed a completely mechanical view of the world, and his mechanization of the universe represented a radical departure. For Descartes, the world and everything in it functions as a great machine linked and governed by the laws of mechanics and of impact. On a cosmological scale he pictured moons carried around planets and planets carried around the sun in great whirlpools of aetherial matter; his *Principles of Philosophy* (1644) elaborated this heliocentric vortex theory. In the realm of physiology and medicine Descartes provided a rational, mechanical alternative to traditional Aristotelian-Galenic accounts. Although his system was mathematically vague and open to critical challenges, Descartes may be fairly said to have capped the Scientific Revolution, in that Cartesian natural philosophy subsumed all the controversies raised over the century since Copernicus and encompassed all the discoveries of the new science. Even more, Descartes provided a comprehensive explanatory alternative to Aristotle and all other competing systems. Whether Descartes was right was the only issue in science after his death in 1650.

Descartes lived and worked for two decades in the Netherlands, that Protestant republic famous for its social and intellectual tolerance. The Dutch republic produced its own outstanding contributors as part of the northward movement of the Scientific Revolution, including the mathematician and engineer Simon Stevin (1548–1620), the atomist Isaac Beeckman (1588–1637), and most notably Christiaan Huygens (1629–95), perhaps the foremost Cartesian and spokesman for the new mechanical science in the second half of the seventeenth century.

The Low Countries likewise became the locus of pioneering work using the microscope. The drapier-turned-scientist Anton van Leeuwenhoek (1632–1723) achieved international renown for uncovering

Fig. 11.5. Descartes's system of the world. The constitution of the universe was an open question. The great French philosopher and mathematician René Descartes responded by envisioning a universe filled with an aetherial fluid sweeping the planets and other heavenly bodies around in vortices. The figure depicts a comet crossing the vortex of our solar system.

a heretofore hidden and unknown "world of the very small." The novelties he discovered included observing blood corpuscles, spermatozoa, and other minute "animalcules." His countryman Jan Swammerdam (1637–80) likewise pushed the microscope to new limits, particularly in his delicate dissections and preparations of plants and insects. These trailblazing Dutch microscopists were joined by the Italian Marcello Malpighi (1628–94) and the Englishman Robert Hooke (1635–1703), whose *Micrographia* appeared in London in 1665. All of these early investigators used single-lens beads for their microscopes, and technique proved crucial for success. But unlike its cousin instrument, the telescope, which became universally accepted and an essential tool in astronomy, the microscope raised more questions than it answered for seventeenth-century observers and theorists. What one "sees" through the microscope represents a complex interplay of ideas and images, and agreement over what was seen and what the images said about insect anatomy, capillary circulation, or embryology, for example, was not forthcoming. The diverging fates of the microscope and the telescope in the seventeenth century suggest that shared intellectual frameworks are required for establishing research traditions, not the instruments themselves. Only in the nineteenth century, and under a different set of conditions, did the compound microscope become a standard piece of laboratory equipment.

England was another Protestant maritime state that fostered a community of men who pursued science after Galileo. We have already mentioned the court physician William Gilbert (1544–1603) and his influential work on the magnet. We saw, too, that the English physician, William Harvey (1578–1657) made the revolutionary discovery of the circulation of the blood in 1618. One can likewise point to Francis Bacon (1561–1626), the lord chancellor of England who proved such an effective spokesman for the new science; the aristocrat and experimental chemist Robert Boyle (1627–91); and, of course, Isaac Newton (1642–1727), just to name a few of the luminaries of seventeenth-century English science. Institutions facilitated the growth of science in contemporary England, notably, the Royal College of Physicians (1518), Gresham College (a new institute with salaried professors founded in 1598), and, later in the seventeenth century, the Royal Society of London (1662) and the Royal Observatory at Greenwich (1675). Royal funding for new scientific chairs at Oxford (geometry-astronomy in 1619 and natural philosophy in 1621) and later at Cambridge (1663) likewise helps explain the flourishing of English science in the later seventeenth century.

Ideology and Utility

Although not absolutely new, claims for the social utility of science began to be widely asserted in the seventeenth century, the conviction that science and scientific activities can promote human welfare and should therefore be encouraged. The ideology was activist and contrasted with the Hellenic view of the practical irrelevance of natural philosophy and the medieval view of science as the subservient handmaiden to theology.

The ideology for the social utility of science sprang from more than one historical source. Renaissance magic and Hermeticism, with their belief in the possibility of controlling forces that permeate the universe, represent one route from which emerged the doctrine that knowledge can and should be applied. Alchemy, in both its medicinal and metallurgical forms, exemplifies another. The Neoplatonist and humanist Pico della Mirandola (1463–94), for example, saw magic as simply the practical part of natural science. Although he walked a tightrope in these regards, Giambattista Della Porta also favored the idea that natural magic held powers useful to princes and governments. Astrology and the occult formed a notable part of the reward systems of patrons: Tycho turned out astrological forecasts, and Kepler made a career as a court astrologer. Philip II, the level-headed ruler of Spain and the Spanish Empire from 1556 to 1598, became deeply involved in the occult. He patronized numerous alchemists, and he built a huge alchemical laboratory capable of manufacturing alchemical medicines in volume. Charles II of England possessed his own alchemical laboratory. Reliable reports circulated of the alchemical multiplication of gold, and the utility of occult studies remained widely accepted through the 1600s.

The perception of science as useful knowledge found its foremost ideologue in Francis Bacon. Bacon pointed to gunpowder, the compass, silk, and the printing press as examples of the kind of worthwhile inventions potentially forthcoming from systematic investigation and discovery. (Bacon neglected to say that these technologies arose independently of natural philosophy but, no matter, for future scientific research promised similarly useful devices and techniques.) Among the castes of laborers Bacon envisioned for a scientific utopia, he set aside one group, the "Dowry men," especially to search for practical benefits. In categorizing different types of experiments, Bacon likewise specified that "experiments of fruit" must be combined with "experiments of light" to produce practical outcomes. His influence in the world of science was largely posthumous, but it proved no less powerful for that.

Descartes, too, was influential in advocating what he called a "practical philosophy" and the idea that knowledge should be applied "for the general good of all men." Descartes considered medicine a principal arena where useful advances and practical applications of theory might be found. Later in the seventeenth century Robert Boyle likewise enunciated the goal of new medical therapies derived from experimental philosophy. Notwithstanding the fact that ties between contemporary scientific theory and effective medical technique were tenuous at best and largely remained so until the twentieth century, the seventeenth-century ideologues of the new science were quick to hitch their wagons to the draught horse of medical practice.

Deep in Book II of the *Principia* even Isaac Newton made an argument for utility. After completing a complex demonstration concerning hydrodynamics and the shape of bodies producing the least resistance while moving in fluids, he commented dryly that "this proposition I conceive may be of use in the building of ships." Newtonian theory—the result of pure science—stood far removed from economic reality and any practical application, but the case brings home the distinction between what the new ideology claimed and what it could deliver.

As part of their ideology, seventeenth-century thinkers likewise expressed new attitudes about nature and the exploitation of nature. Bacon and Descartes separately voiced the view that humans should be the master and possessor of nature, that nature and the world's natural resources should be vigorously exploited for the benefit of humankind—that is, those who own or control knowledge. The notion that nature was subject to human dominion possessed biblical authority and was already operative in the Middle Ages. But a distinctive imagery of the violent rape and torture of nature as an aspect of scientific practice came to the fore in seventeenth-century thought on these matters. Bacon, for example, asserted bluntly that "Nature must be taken by the forelock."

The notions that science is useful, that science is a public good, and that knowledge is power have ruled as cultural leitmotifs in the West since the seventeenth century and everywhere since the nineteenth. The further implication was twofold: science and scientists deserved support, and the power they brought should be used for the commonweal. The status of the old ideas that natural philosophy was for natural philosophers or subservient to theology diminished. Evidently the new notions were more consistent with the interests of the new centralized states and the development of merchant capitalism in Europe.

CHAPTER 12

"God said, 'Let Newton be!'"

The Scientific Revolution was a complex social and intellectual affair that was much more than a collection of individual scientific biographies. Nevertheless, Isaac Newton (1642–1727) so dominates the intellectual landscape of the latter seventeenth and early eighteenth centuries that his life and works compel close scrutiny. Newton's career speaks volumes about the end of one era and the beginning of another, and in following the trajectory of Newton's biography, we can simultaneously survey the thematics of the Scientific Revolution raised in previous chapters.

Newton's *Principia Mathematica Philosophia Naturalis* or *Mathematical Principles of Natural Philosophy* (1687) effectively concluded a line of theoretical inquiry into cosmology and the underlying physics of the world that stretched back through Descartes, Galileo, Kepler, and Copernicus and ultimately back to Aristotle. With universal gravitation and his laws of motion, Newton's physics finally unified the celestial and terrestrial realms separated since Aristotle. Newton buried not only the moribund Aristotelian world, but also the newer mechanical alternative worked out by Descartes. He is also acclaimed as a mathematician and the co-inventor (along with Leibniz) of the calculus; and he did fundamental work in optics as well. The renown of Newton's genius stems from the many domains of his accomplishment as well as from the profundity of his contributions to each.

But Newton's overriding importance in the history of modern science derives not simply from what he did for contemporary science, but also from his abiding role in molding scientific traditions that came afterward. Newton culminated the Scientific Revolution, and at the same time set the agenda for scientific research in astronomy, mechanics, optics, and a range of other sciences. In so doing he profoundly, indelibly shaped the history of science over the two centuries following his death.

Newton's professional career likewise is of general import and repays examination. That he metamorphosed from the reclusive Lucasian Professor of Mathematics at Cambridge University into a bureaucratic civil servant at the Royal Mint, president of the Royal Society of London, and finally *Sir* Isaac is revealing of the social history of science and reflects larger patterns of social and institutional changes in European science in the seventeenth century.

Newton also comes to us as a popular stereotype. Who has not absorbed the cliché of Newton discovering gravity by being hit on the head by a falling apple? In his own day and variously afterward Newton and Newtonian science have been the objects of mythological development and political exploitation. One especially influential image crafted in the eighteenth century held Newton to be the epitome of the rational scientist. Recent work—the product of a substantial scholarly industry—now portrays the great man not only as a brilliant natural scientist but also as a magus, deeply involved in alchemy, religious fanaticism, and investigating occult forces at play in nature and in history. The different Newtons revealed in the historical record not only raise the perennial question about historical truth, but the case highlights no less pointed issues concerning the cultural meaning of Newtonian science and the social uses to which it was put in his own day and afterwards.

From Lincolnshire to Cambridge

Isaac Newton was born on the family farm in Woolsthorpe, Lincolnshire, England, on Christmas day, 1642. He was a posthumous child, his father having died earlier that fall. (The absence of the father and the coincidence of his birthday with Jesus' may have confirmed for Newton his sense of having a special relationship with God.) On his father's side the Newtons were a modestly rising yeoman family of English farmers and herders; he was better connected on his mother's side through a wealthier and more educated family of rural gentility and clergy. Underweight and premature, the newborn was not expected to survive. He reportedly fit in a quart jar and was too weak to lift his head to the breast.

Newton experienced an awkward and unpleasant childhood. His mother remarried when he was only 3 and left him at Woolsthorpe with his grandparents. All readings of Newton now acknowledge that he was an alienated and tortured neurotic, emotionally disfigured early in life. As a widow with other children, Newton's mother returned to her son when he was 10, and at 12 Newton began grammar school in nearby Grantham. Leaving the Grantham school at 17, he appeared headed for an unhappy life as a rural farmer and property manager, but the obviously clever lad so failed at farming and so rebelled against his fate that the only alternative was to ship him off to university. And so, through a family tie and after a bit of brushing up back at Grantham, Newton headed up to Cambridge in 1661.

England had entered a new period of its history just the previous year when it restored the monarchy under Charles II after two decades of civil and religious strife. Although not directly affected by the turmoil, Newton's Lincolnshire youth had coincided with civil war in England (1642–49) and the subsequent republican period of the English Commonwealth (1649–60) which had overthrown the monarchy and the official Anglican church. The peaceful installation of Charles II as Anglican king of England in 1660 reinstituted both monarchy and the state church, but on more constitutional terms and with a measure of religious and political toleration. Further changes lay ahead, and tensions ran high in Restoration England as Newton proceeded to Cambridge.

Newton matriculated at Trinity College as a subsizar or scholarship boy, and for his first years he suffered the indignity of serving the senior students. (Newton resented—but always retained—his somewhat peasant roots.) Cambridge University was a sleepy intellectual backwater that continued to embrace Aristotle even in the 1660s. Fortunately for Newton, its relaxed rules and lack of oversight left him free to study on his own. He soon made his way to the cutting edge of contemporary mathematics, mechanics, and natural philosophy. Symptomatic of the changing state of science in the later seventeenth century, Newton mastered Descartes's work before he studied Euclid. After he took his B.A. degree at the university in 1665, he stayed on at Trinity and was soon elected a permanent fellow of the college. In 1669, on the retirement of Isaac Barrow (his own professor), Isaac Newton, M.A.—26 years old—became the second Lucasian Professor of Mathematics, the new chair for science established in 1663 to invigorate the university. In the space of eight years, but as part of a medieval career pattern that had gone on for four centuries, Newton slipped into the role of university don.

In 1665, while Newton was still a student, the plague struck at Cambridge. The university closed for almost two years, and Newton returned to Woolsthorpe. Earlier biographical accounts made much of the so-called Miracle Year that followed in 1666, wherein Newton supposedly discovered gravity, invented the calculus, and developed his theory of light and colors. Historians today take a more nuanced view of Newton's development at this stage, setting 1666 in the context of the preceding period of intense study of science and mathematics and the leisure at Woolsthorpe of thinking through problems on his own. Unknown to the rest of the world, in 1666 Newton was in fact the world's leading mathematician and was as knowledgeable as anyone about science or natural philosophy (new or old). He thought about gravity and calculated in a rough way the effects of gravity extending to the moon. Using prisms, he investigated light and colors, discovered new phenomena, and toyed with new explanations. He also gained his fundamental insight into the calculus by seeing a relation between tangents to curves and areas under curves (what we call the derivative and the integral). What emerged in 1666 was not fully formulated and completed work, as legend has it. Instead, Newton's initial insights into mechanics, optics, and mathematics developed further after he returned to Cambridge in 1667, and indeed these subjects occupied much of his life's work in science.

Newton published first on optics. In a paper appearing in 1672, he offered his views that light is composed of rays, that different rays are refracted to different degrees through lenses or prisms, that each ray corresponds to a different color, and that white light represents an amalgamation of all rays and colors. Newton adduced these views using carefully controlled experiments. In his famous "crucial experiment," for example, he passed a beam of light through a prism to create a spectrum. Refracting portions of that spectrum through a second prism failed to produce another spectrum or other changes, thus demonstrating to his satisfaction that colors are properties of light and not produced by refraction. The conclusions Newton drew embodied radical new concepts of the nature of light and colors, contrary to both Aristotelian and Cartesian theory, but he naively believed he was simply making public straightforward facts about nature. Newton noted a practical, technological spinoff of his discoveries: to avoid chromatic aberration from light refracting through lenses he designed a *reflecting* telescope that used a mirror to focus light. Newton presented his reflecting telescope to the Royal Society of London, which in turn elected him a Fellow in 1672.

Newton's 1672 paper on light proved a tour de force, but disputes soon arose. Aristotelian and Cartesian opponents attacked Newton's findings, and he became entangled in arguments that continued doggedly for decades over precise details of experimental procedure and the interpretation of results. These events led Newton to shun a public life in science. Some of his mathematical work circulated privately in manuscript, but after his initial public notice Newton retreated as much as possible into his own world at Cambridge. There, spurred by the requirement of his professorship to become a priest in the Anglican Church, he took up serious study of theology and biblical prophecy in the 1670s and early 1680s. Religion fervently occupied Newton all his life, but he rejected Christian orthodoxy, seeing, for example, the Christian trinity as a hoax foisted on the early church. He developed other heretical theological views (known as Arianism, something like a fiery unitarianism) which put him dangerously out of touch with English society around him. Newton's fanaticism extended to the belief that there existed bodies of secret, pristine, and arcane knowledge concerning matters religious and scientific that God had initially conveyed to Noah, which then passed to Moses and Pythagoras, and which had come down to Newton's day in an esoteric oral tradition of prophets and the elect few like Newton himself, or so he believed, who could read hidden codes in nature and the Bible. He seems to have taken seriously the poetic line later written about him by Edmond Halley, "Nearer the gods no mortal may approach." But Newton kept these dissident views very private, and in 1675 he was spared, one might say miraculously, having to resign the Lucasian professorship when the requirement for taking religious orders was dropped.

In the quest after secret knowledge, alchemy occupied the major portion of Newton's time and attention from the mid-1670s through the mid-1680s. His alchemical investigations represent a continuation and extension of his natural philosophical researches into mechanics, optics, and mathematics. Newton was a serious, practicing alchemist—not some sort of protochemist. He kept his alchemical furnaces burning for weeks at a time, and he mastered the difficult occult literature. He did not try to transmute lead into gold; instead, using alchemical science, he pried as hard as he could into forces and powers at work in nature. He stayed in touch with an alchemical underground, and he exchanged alchemical secrets with Robert Boyle and John Locke. The largest part of Newton's manuscripts and papers concern alchemy, and the influence of alchemy reverberates throughout Newton's published opus. This was not the Enlightenment's Newton.

Science Reorganized

Newton's 1672 paper on light was published in the *Philosophical Transactions* of the Royal Society as a letter to its secretary, Henry Oldenburg. The sponsoring organization, the Royal Society of London for Improving Natural Knowledge, was a new scientific institution present on the English scene. Founded in 1660 and granted a royal charter by Charles II in 1662, the Royal Society was a state scientific society that maintained itself through the dues of its members.

The Royal Society (1662) and the Paris Academy of Sciences (1666) were the flagships of an organizational revolution of the seventeenth century. They created a new institutional base for science and scientists, and they ushered in a new age of academies characteristic of organized science in the following century. Major national academies of science subsequently arose in Prussia, Russia, and Sweden, and the model of a state academy or society of science spread throughout Europe and to its colonies around the world. Scientific academies and societies coordinated a variety of scientific activities on several levels: they offered paid positions, sponsored prizes and expeditions, maintained a publishing program, superintended expeditions and surveys, and rendered a diversity of special functions in the service of the state and society. These institutions, incorporating a broad array of scientific interests, dominated organized science until the coming of specialized scientific societies and a renewed scientific vitality in universities in the nineteenth century.

Growing out of Renaissance and courtly precedents, these new learned societies of the seventeenth and eighteenth centuries were creations of nation-states and ruling governments. State-supported scientific societies possessed a more permanent character than their Renaissance cousins, in that they received official charters from government powers incorporating them as legal institutions and permanent corporations. Given the increasing separation of government operations from royal households, official state scientific societies became detached from court activity and integrated into government bureaucracies. Society members tended to act less as scientific courtiers and more as expert functionaries in the service of the state. The state academies and societies were also institutions specifically concerned with the natural sciences; they were not subservient to other missions, they largely governed themselves, and, unlike universities, they did no teaching. The growth and maturation of state academies and societies of science in the eighteenth century provide impressive evidence of the greater social assimilation of science after the Scientific Revolution.

Whereas upper-class women reigned in the literary and intellectual *salons* of the seventeenth and eighteenth centuries, the world of science and the learned society remained largely a man's world. The French marquise Madame de Châtelet (1706–49) is a notable exception, being a member of several academies, a contributor of original science, and Newton's underrated French translator. Several Italian towns with universities and academies produced a string of female scientific luminaries—the experimental Newtonian Laura Bassi (d. 1778), for example. Bassi and her fellow female scientists brought credit on themselves and their gender, on their towns, and on the contemporary world of learning.

Other extra-university patterns and alternatives developed for organizing and communicating science in the seventeenth century. Up until that point personal travel, the private letter, and the printed book represented the mainstays of scientific communications. As the Scientific Revolution unfolded, informal circles of correspondence began to effect new lines of communication among people interested in the new science. Then, in the second half of the seventeenth century, coincident with the new state-supported learned societies, the periodical journal, the main form for the publication of scientific research ever since, made its appearance. The *Philosophical Transactions* of the Royal Society of London and the *Journal des Sçavans* in France both appeared in 1666 and other influential early scientific journals followed. Journals provided a new mode of communicating and disseminating scientific knowledge and research. They allowed for comparatively speedy publication,

Fig. 12.1. Louis XIV visits the Paris Academy. This fanciful seventeenth-century engraving, showing King Louis XIV of France visiting the Paris Academy of Sciences, illustrates the growing role of state support for the sciences in seventeenth-century Europe.

and they created the scientific paper, which became the unit of production in the scientific world.

To better administer commerce at home and abroad, European states, following the precedent established in Islam, also founded royal and national astronomical observatories in France (1667), England (1675), Prussia (1700), Russia (1724), and Sweden (1747). Similarly, states began to sponsor and maintain national botanical gardens: the Jardin du Roi in Paris dates from 1635 and the Royal Gardens at Kew from 1753. These state gardens and hundreds like them arose either by fiat or by central governments coopting older university pharmacy gardens. The state botanical gardens of Europe became centers for scientific study, even as mercantilistic policies spread networks of Dutch, English, and French botanical gardens worldwide.

From the sixteenth century onward, European courts and governments institutionalized science, and, in turn, scientific specialists assisted governments in botanical gardens, in observatories, in the scientific societies, in specialized professorships, and in various nooks and crannies of government bureaucracies all across Europe. The River Commission of Medici Florence, for example, possessed a technical staff of fifteen, and the court called on Galileo for his opinion in engineering matters. Newton later served the English crown at the Mint. Like their predecessors in hydraulic civilizations, European scientists increasingly became state functionaries. Scientific experts and the institutions they represented performed useful services for the state. The Paris Academy of Sciences, for example, administered patent applications as an arm of the royal government of France, and, like the Astronomical Bureau in China, it controlled publication of official astronomical tables.

The ideology of utility cemented new relations between science and government in early modern Europe. For the first time European governments struck a firm deal with science and natural philosophy, a reciprocal exchange of useful services and powers in return for recognition, support, and self-government. Science actively sold itself to government, and, to a degree at least, the state—first the courts of absolute princes, then the bureaucracies of nation states—began to buy into science. The historical significance of the new contract established between science and the state in early modern Europe is that European governments began to imitate ancient hydraulic civilizations in the use of scientific experts.

One should not exaggerate the extent of government support for science in Europe which developed by the end of the Scientific Revolution. Mathematicians, scientists, and technical experts did not generally enjoy a high social status. Charles II famously ridiculed his own Royal Society for the useless "weighing of air" with its air-pump experiments. Often, too, receiving payment for technical services or promised stipends proved difficult for scientific and technical personnel in court or state employ. Even the great Paris Academy of Sciences received meager funds in its early decades and at a level far lower than its sister fine arts academies. Tellingly, Louis XIV visited his science academy only once and then reluctantly. European governments insisted that science be cheap as well as useful. Until the twentieth century they saw to it that the first condition obtained. As for the second, they were only partially successful.

Reframing the Universe

In August of 1684 Edmond Halley traveled to Cambridge to ask Isaac Newton a question. Earlier that year at the Royal Society in London, Halley, Robert Hooke, and Christopher Wren toyed with connections between Kepler's law of elliptical planetary motion and a force of attraction emanating from the sun—clearly the idea of such a force was in the air. Newton and Hooke, curator of experiments at the Royal Society, had earlier corresponded over the idea in 1679 and 1680. Wren had even proposed a prize for a mathematical demonstration. Halley visited the reclusive Newton and asked him about the motion of a planet orbiting the sun under a $1/r^2$ attractive force. Harking back to 1666, Newton immediately replied that the shape of the orbit would be an ellipse and that he had calculated it. After fumbling through his papers, Newton promised the awestruck Halley that he would send him the calculation. Three months later Halley received a nine-page manuscript, "On the Motion of Orbiting Bodies," that outlined the basic principles of celestial mechanics. Everyone who saw it immediately recognized the significance of Newton's little tract.

The *Principia* remained to be written, however. When Newton returned to his initial calculations of 1666, he discovered errors and conceptual ambiguities. His initial response to Halley in 1684 merely sketched the new physics, and an intense two-year period followed as he struggled with the conceptual, mathematical, and authorial obstacles in expanding his insight into a rigorous, complete treatment. Halley shepherded the great work through the press, nominally under the aegis of the Royal Society. The *Principia* appeared in 1687 with the imprimatur of the Royal Society.

The *Principia* is a highly mathematical or, better, geometrical text, and Newton begins it with definitions and axioms. He defines his terms (e.g., mass, force) and states his historic three laws of motion: 1) his inertial law that bodies in motion remain at rest or in straight-line motion unless acted upon by an outside force; 2) that force is measured by change in motion (although he never wrote $F = ma$); and 3) that for every action there is an equal and opposite reaction. In his frontmatter he introduces his ideas about absolute space and time; and in a scholium—a mere footnote—he shows that Galileo's hard-won law of falling bodies ($s \propto t^2$) follows as a consequence of his, Newton's, laws. At the outset it is evident that Newtonian dynamics, in explaining motion with forces, subsumes Galilean descriptive kinematics.

The body of the *Principia* consists of three books. The first provides an abstract treatment of the motion of bodies in free space. Section 1 actually continues the preceding prefatory remarks in presenting the analytical techniques of the calculus (integration and differentiation) used in the rest of the *Principia*. Newton states these techniques in the language of geometry because virtually he alone knew the calculus while virtually all his potential readers understood only geometry.

Newton gets down to cases in section 2 of Book I, "The Determination of Centripetal Forces." There he proves that a body orbiting an attracting central force obeys Kepler's second law and sweeps out equal areas in equal times. That is, if a body at point A orbits S and is drawn toward S by a centripetal or gravitating force of some variety, then the line AS will sweep out equal areas in equal times. Newton also shows the converse: if an orbiting body obeys Kepler's second law, then an attracting or gravitational force of some sort may be seen to be operating.

Newton then turns to a deceptive Proposition IV, Theorem 4. In this proposition he considers the following simplified situation. Acted on by a centripetal (or attracting) force, a body sweeps out an arc. The illustration shows a body (b) moving with a velocity of (v) from A to A' along arc (a) in a certain time (t) and at a radius (r) from the center of attracting force at F. Newton is interested in the abstract mathematical relations between and among these parameters: (a), (r), (F), (t), and (v). In the proposition and then in a staccato of corollaries, he reels off those relations. In the sixth corollary Newton slips in: if $t^2 \propto r^3$, then $F \propto 1/r^2$. This seemingly meek proposition masks a deep insight into nature, for in it Newton is saying that if Kepler's *third* law holds true, then bodies are maintained in their orbits by a gravitational force that varies inversely as the square of the distance, which is Newton's law of gravity. And conversely, Newton shows that bodies orbiting under a $1/r^2$ law of gravity must obey Kepler's third law. In other words, Kepler's third law demonstrates Newtonian gravitation and vice versa. In a footnote Newton owns up to the explosive significance of what he has done, saying, "The case of the sixth Corollary obtains in the celestial bodies."

The remainder of Book I develops the full mechanics implied in these initial propositions. Newton extends the analysis to all the conic sections; he proves that the attractive force of an extended solid body (like the earth) can be mathematically reduced to attraction from its center; he shows off his virtuosity in discussions of pendulums; he explores the abstract mathematics of universal gravitation; and he provides the mathematical tools for determining orbits from observations and vice versa.

Rather than pressing the astronomical implications of Book I, like the second act of a three-act play or the slow second movement of a concerto, the second book of the *Principia* pauses to consider the motion of bodies, not in free space, but in resisting media. In essence, in Book II Newton provides a mathematical treatise on hydrostatics and hydrodynamics. At first glance this diversion away from the main theme of gravitational celestial mechanics seems peculiar, until one recalls that in the Cartesian system a thick aether fills the cosmos and that, for Descartes, planets are carried around in vortices or whirlpools which are, in effect, hydrodynamical systems. In exploring the physics of these systems, Newton seeks to smash the Cartesian system, and Book II concludes with the devastating remark, "Hence it is manifest that the planets are not carried round in corporeal vortices. . . . Let philosophers then see how [Kepler's third law] can be accounted for by vortices."

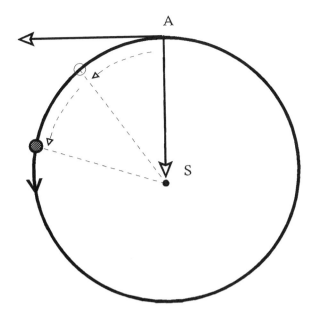

Fig. 12.2. Newton's *Principia*. In Book I of his *Mathematical Principles of Natural Philosophy* (1687) Isaac Newton linked gravity to Kepler's laws of planetary motion. In the proposition shown, Newton demonstrates that a body moving inertially and affected by a gravitational force will obey Kepler's Second Law and sweep out equal areas in equal times. That is, given Kepler's Second Law, orbiting bodies can be understood to be moving under the influence of some kind of gravitational force.

The *Principia*'s denouement occurs in Book III, "The System of the World." Newton first presents the "phenomena" of a heliocentric solar system where orbiting bodies obey Kepler's three laws. In particular, he provides reliable observational data coupling Kepler's third law with the motions of the moon around the earth, the planets around the sun, and satellites around Jupiter and Saturn, respectively. Geocentrism is shown to be absurd and inconsistent with the known facts. Using Kepler's third law and Book I of the *Principia*, Newton then proposes that the forces holding the world's planets and moons in their orbits are $1/r^2$ attracting forces and, in particular, "that the Moon gravitates towards the Earth."

Newton's celestial mechanics hinges on the case of the earth's moon. This case and the case of the great comet of 1680 were the only ones that Newton used to back up his celestial mechanics, for they were the only instances where he had adequate data. With regard to the moon, Newton knew the rough distance between it and the earth (60 Earth radii). He knew the time of its orbit (one month). From that he could calculate the force holding the moon in orbit. In an elegant bit of calculation, using Galileo's law of falling bodies, Newton demonstrated conclusively that the force responsible for the fall of bodies at the surface of the earth—the earth's gravity—is the very same force holding the moon in its orbit and that gravity varies inversely as the square of the distance from the center of the earth. In proving this one exquisite case Newton united the heavens and the earth and closed the door on now-stale cosmological debates going back to Copernicus and Aristotle. In proving this and the comet case, Newton simultaneously opened the door on a whole new world of problems to solve.

The remainder of Book III spelled out areas for research in the new

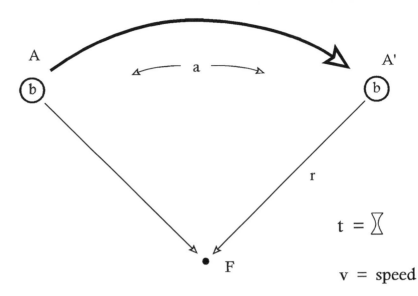

Fig. 12.3. Proposition linking gravitation and Kepler's Third Law. In this proposition from Book I of the *Principia* Newton analyzed various factors affecting the motion of a body, *b*, while it moves along the arc (*a*) from *A* to *A'* while under the influence of an attracting gravitational force *F*. He was able to show that if the force of gravity decreases as the square of the distance, then $t^2 \propto r^3$, or Kepler's Third Law applies, and vice versa. In other words, given Kepler's Third Law, then Newton's $1/r^2$ law of gravity necessarily follows, and given Newton's law of gravity, then Kepler's Third Law must also apply. Newton thus provided a new physics to explain the motion of the heavenly bodies.

era to follow the *Principia:* the moon's precise orbit, astronomical perturbations, the shape of the earth, gravity and the tides, and especially comets. With regard to comets, Newton introduced his other smoking gun: observations of the comet of 1680 and calculations proving its Keplerian solar orbit under the influence of the sun's $1/r^2$ gravity. Newton's would be the model solution for every comet to follow.

The first edition of the *Principia* in 1687 tailed off in an alchemical allusion to cometary vapors transmuting into terrestrial substances through a "slow heat." Twenty-five years later in 1713—a very different period in Newton's life—the author added a General Scholium to the second edition of the *Principia.* Commensurate with the grandeur of the subject, in the General Scholium Newton offered a disquisition on God, the "intelligent and powerful Being" evident in "this most beautiful system of the Sun, planets, and comets." Newton's natural philosophy led him to natural theology and the conclusion that God can be known through His contrivances in nature. Newton's God becomes the great Clockmaker who watches his great machine tick away according to His established laws of nature. In the General Scholium Newton also explains his approach to treating gravity mathematically without inquiring into its cause; about the cause of gravity, Newton noted, in his famous phrase, that he "feigns no hypotheses" *(hypotheses non fingo).* The Scholium itself ends with an incongruous last paragraph about a "subtle spirit" correlated with light, heat, cohesion, electricity, physiology, and sensation. These matters, indeed, as Newton wrote, were "not to be explained in a few words."

Virtually overnight the *Principia* made Newton, at 44, a famous man. But in the short run, it made little difference in his reclusive life at Cambridge. After the *Principia* Newton returned to his private meditations and alchemical fires. In 1693 he suffered a serious mental breakdown and remained incommunicado for four months. He went without sleep for days on end; he wrote paranoid letters to acquaintances, such as John Locke whom he accused of "endeavoring to embroil me with women." Newton himself later spoke of the "distemper" that seized him at that time. He remained what we would call clinically depressed for at least 18 months. The year 1693 also marks the end of the period of creative scientific work in Newton's life.

Several factors combined to precipitate Newton's breakdown. He had exhausted himself in his alchemical program and possibly became disillusioned with it. Perhaps he had poisoned himself with heavy metals (like mercury vapor) and for a while became "mad as a hatter." A priority dispute with the German philosopher G. W. Leibniz over the invention of the calculus was beginning to heat up, and his early efforts to secure a post-*Principia* appointment in London had come to naught. Still and all, historians are loath to say directly that another reason behind Newton's disintegration in 1693 stems from the breakup of his relationship with the 25-year-old Swiss mathematician Nicolas Fatio de Duillier. The climax of their presumably chaste affair in June of 1693 proved the proximate cause of Newton's psychic collapse.

Newton's mental disorders of 1693 underscore an oft-noted affinity between creativity and madness. His salvation from the abyss came from an unexpected quarter: politics. The Glorious Revolution in England in 1688 deposed the Catholic King James II and installed the Dutch Protestants William and Mary of Orange as joint king and queen. The period prior to this revolution was tense and affected the university at Cambridge where James tried to impose Catholic Fellows contrary to university rights and privileges. Somewhat out of character, Newton stood firm publicly on this issue, and the university elected him one of its representatives to the king. His election can be explained in part by Newton's eccentricity, which served as a front for those dons who had more to lose, and doubtless also by Newton's abiding hatred of Catholicism as the institution of the Antichrist. Newton went on to represent the university at the Convention Parliament that greeted William and Mary. He had shown himself courageous politically and had sided with the winning party. He stood to be rewarded.

After some disappointment and delay, Newton succeeded in obtaining an appointment. He left Cambridge in 1696 and moved permanently to London to become Warden of the English Mint, and in 1699 he succeeded to the top post of Master of the Mint. Newton applied himself assiduously at the Mint in recoining efforts and in the grisly task of prosecuting counterfeiters. Most of his income derived from his position at the Mint, which he held until his death in 1727.

Newton had spent 30 years sequestered at Cambridge, and his passage from the university to the government, from don to functionary, deserves to be highlighted. Indicative of the changing social circumstances of science in the seventeenth century, Newton's employment as a government civil servant—and notably one with technical expertise—illustrates again that contemporary universities were not active scientific centers and that new careers in science were pursued elsewhere. As the century went on individual princely patronage waned, and the new social setting for science in Newton's day saw scientists more and more institutionalized in the bureaucracies of central governments and in national scientific societies. Christopher Wren, for example, was president of the Royal Society before Newton, became the king's chief architect and, after the fire of London, rebuilt St. Paul's Cathedral.

That Newton became president of the Royal Society of London comes as no surprise. Newton's association with the Royal Society was longstanding, but he stood for president only in 1703 after the death of his longtime enemy, Hooke. Although not without opposition, Newton was reelected annually, and he remained sitting president until his death. With one foot at the Mint and the other at the Royal Society, Newton thus stood well planted at the center of the small but growing world of official English science. (That world also included the Royal Observatory at Greenwich.) In his new role Newton played the autocrat of English science. At the Royal Society he packed the Fellowship and stacked the governing Council. He maneuvered to control the observatory at Greenwich, and his voice was decisive in scientific appointments of protégés throughout Great Britain. Not for nothing, he became Sir Isaac, knighted by Queen Anne in 1705.

Newton shamefully abused his authority and position at the Royal Society in the notorious priority dispute with Leibniz over the invention of the calculus. Both prodigies seem to have independently invented the fundamental theorem of the calculus, Newton in 1665–66 and Leibniz around 1676, but they clashed mightily over who should get the credit. Leibniz published first (in 1684) and was recognized first, but in claiming sole credit he was not candid about what he learned of Newton's mathematics on a visit to London in 1676. Newton kept his mathematical accomplishments more or less private, but a paper trail existed, and as word leaked out and as material began to be published about Newton's priority in the discovery of the calculus, Newton's minions leveled veiled charges of thievery, fraud, and plagiarism against Leibniz. In 1711 Leibniz made the mistake of appealing to the Royal Society for justice in the matter, thereby falling into Newton's hands. Newton created a kangaroo court within the Society, and he himself furiously penned the inquisitorial report, the *Commercium epistolicum,* that "reckoned Mr. Newton the first Inventor." Newton further pursued Leibniz in 1714, writing and anonymously publishing a long "Account of the *Commercium epistolicum*" in the *Philosophical Transactions* of the Royal Society. Even after Leibniz's death in 1716 Newton actively pushed his claims for sole credit. The unfortunate episode sparked some worthwhile philosophical discussion about the worldviews of Newton and Leibniz, but otherwise it proved a disaster. The priority dispute over the calculus says much about Newton's character, about simultaneous discovery in science, and about the still-nascent state of modern social norms of science, such as establishing priority and sharing credit for discoveries. Ironically, Newton's version of the calculus became a dead end, while Leibniz's notation and concepts of *d* (for derivative) and $\int$ (for integral) prevailed.

Coincident with his ascent to the presidency of the Royal Society, Newton published the first edition of his *Opticks* in 1704. The *Opticks* ranks alongside the *Principia* as the other major work through which Newton and Newtonian science became widely known. The *Opticks* did not represent any new scientific work. Rather, the book set out completed work stretching back to 1666. The *Opticks* was more readable and accessible than the *Principia*, being experimental in approach and without complicated mathematics. A notable aspect of Newton's experimentalism in the *Opticks*, likewise differentiating it from the more deductive *Principia*, was Newton's offering "proof by experiments," by which he presented whole collections of experiments to argue about the phenomena they revealed.

Newton wrote the *Opticks* in three books. He began by summarizing the body of optical knowledge developed from antiquity to his own day. He then recapitulated in nearly 200 pages his short paper of 1672, presenting experiments concerning the refraction of light, the heterogeneity of white light, the nature of colors, colored bodies, and the rainbow, and ways to improve the telescope. In Book I and elsewhere in the *Opticks* Newton revealed a Pythagorean and Neoplatonic side, by linking the seven colors of the rainbow with the seven notes in the Western musical scale and with underlying mathematical ratios. In Book II Newton moved on to consider thin-film phenomena such as the colored patterns that arise on soap bubbles or when a lens and a plate of glass are pushed together, phenomena now known as Newton's rings. Newton had difficulty reconciling the phenomena of the rings with his particle theory of light, and he had to introduce awkward explanations to get his rays to behave as required. Also in Book II, Newton speculated on matter theory and the atomic constitution of matter, expressing the hope that microscopes may be improved to where we could see atoms but, using the language of alchemy, he feared "the more secret and noble Works of Nature within corpuscles" must remain hidden. In Book III Newton took up the newly discovered phenomenon of diffraction, or the bending of light around edges and the appearance of fringes in shadows, but he abruptly broke off discussion, saying that since he first performed the experiments he had become "interrupted and cannot now think of taking these things into farther consideration."

Rather than a grand theoretical conclusion analogous to the General Scholium of the *Principia*, Newton concluded the *Opticks* by proposing a series of general scientific research questions, his famous Queries. Newton's queries were rhetorical, in that even if they ended with a question mark they expressed his views on a variety of scientific subjects. He explicitly intended that they spark "a farther search by others," and they had that effect. In three different editions of the *Opticks* Newton posed a total of thirty-one queries. In the 1704 edition he offered sixteen queries concerning light, heat, vision, and sound. He added seven more (later numbered 25–31) for the Latin edition of 1706 dealing with his particle theory of light, the polarization of light, and an attack on the wave theory of Huygens (and Descartes). In the last two queries of this set Newton ventured a frank discussion of (al)chemical and biological change ("very conformable to the course of Nature, which seems delighted with Transmutations"), and he set forth a lengthy disquisition on chemical phenomena. In these Queries his model for interpreting phenomena consisted of "massy, hard, impenetrable" atoms endowed with attractive and repulsive powers operating over a distance, atoms and powers which he called upon to explain gravity, inertia, electricity, magnetism, capillary action, cohesion, vegetable fermentation, combustion, and many other phenomena. Over a decade later, in the second English edition of 1717, Newton inserted eight more queries between the first and second sets. In these he retreated from the idea of atoms invested with active principles, and instead deployed various superfine, self-repulsive ethers to account for optical phenomena, gravity, electricity, magnetism, heat, and physiology. The collection of queries as a whole may have lacked intellectual coherence, but it set up research traditions and provided something of a general conceptual umbrella, particularly for the less well-developed sciences mentioned.

As in the *Principia,* Newton concluded the *Opticks* on a note of natural theology, by finding God in nature and claiming that it appears "from phenomena that there is a Being incorporeal, living, intelligent [and] omnipresent." Studying nature can reveal God's providence or design in nature and "our duty towards him." For Newton, natural philosophy again leads to natural theology.

Natural theology in Newtonian science resonated with the surrounding culture in ways that need to be emphasized. In particular, the content of his natural theology fit the larger interests and designs of English religious and political moderates (Latitudinarians) at and after 1688. Newton's Latitudinarian followers extracted from his science theological and sociopolitical points concerning God's existence, his providence, the sacredness of property, and the legitimacy of social hierarchy, duty, and enlightened self-interest. Newtonian cosmology and natural philosophy, in other words, supported, indeed stood at the center of, the reigning social and political ideology in England.

Nowhere is this coincidence of science and ideology more plain than in the noted series of Boyle lectures given from 1692 "for proving the

Christian Religion against Infidels." Newton's followers controlled the administration of the lectures, and the chosen lecturers were all Newtonians. Newton himself had a hand in the appointment of the first lecturer, Richard Bentley, who pontificated on *The Folly and Unreasonableness of Atheism,* and Newton actively advised Bentley over points of science and natural theology. A not-untypical offering was that by William Derham, Boyle lecturer in 1711–12: *Physico-Theology: or, A Demonstration of the Being and Attributes of God from the World of Creation.* The Boyle lectures provided an influential means of spreading the word not only about God but about the Newtonian world picture, too. The case of the Boyle lectures epitomizes the social use of scientific knowledge, but it does not exhaust Newton's great influence in extrascientific, social realms.

Newton grew old and fat. He superintended the Mint and ruled at the Royal Society. With the help of younger disciples he attended to new editions of his *Opticks* in 1717 and 1721 and the *Principia* in 1713 and 1726. In his latter years Newton turned again to theology and biblical prophecy. His heretical views had not changed in the decades since Cambridge, but he continued to hide them. (A sanitized version of his observations on prophecy saw a posthumous publication.) Late in life and through his connections to the royal English court, Newton became drawn into a controversy over systems of historical chronology, which brought the aging giant into conflict with the learned scholars at the Académie des Inscriptions et Belles-Lettres in Paris. Circulated in manuscript and published posthumously in 1728, Newton's *Chronology of Ancient Kingdoms Amended* proposed an astronomically based dating system, but the French academicians bested Newton in erudition and showed up the shortcomings of his chronology. In the last years of his life Newton became incontinent and stuck more and more to his rooms, which he had decorated in crimson. Finally true to his religious beliefs, he refused the rites of the Anglican Church and died on March 20, 1727, at the age of 85.

The Newton who died refusing the comfort of the church was not the same Newton England buried in Westminster Abbey—the heart and soul of English Anglican rule. Newton, the closet Arian, the last of the magicians, the emotionally warped and scarred human being in life, underwent a resurrection of sorts as a national hero and the model of the rational scientist. Alexander Pope immortalized this glory to England in his apotheosis:

> Nature, and Nature's Laws lay hid in Night.
> God said, *Let Newton be!* and All was *Light.*

Newton's effect on English natural theology has been mentioned. The social influence of Newtonian science on the French Enlightenment of the eighteenth century was of a similar character but of even greater historical impact. Ironically, with his mystical speculations largely hidden until the twentieth century, Newton may be fairly said to be a founding father of the Enlightenment, that campaign of reason against superstition and irrationality that arose in France and then spread across eighteenth-century Europe and America. Inspired by an idealized Newton, Newton's clockwork God, and the success of the Scientific Revolution generally, influential Enlightenment *philosophes* such as Voltaire, Montesquieu, Condillac, Diderot, La Mettrie, and others sought to extend to the social and human sciences the progress exemplified in the natural sciences in the previous century. Indeed, the first Enlightenment document may well be the ode penned by Edmond Halley to preface the first edition of the *Principia,* wherein Halley wrote of understanding the world: "In reason's light, the clouds of ignorance/ Dispelled at last by science." Voltaire attended Newton's funeral and brought back the famous anecdote of having left France and a universe filled with the Cartesian ether to arrive in England and a universe of empty Newtonian space. By the mid-eighteenth century, aided by Voltaire and Madame du Châtelet, Newton's science conquered France and won out over Descartes among French intellectuals and scientists.

The forces associated with Newtonian science and the Newtonian Enlightenment were liberal, progressive, reformist, and even revolutionary, and they played major roles in the prehistory and history of the American Revolution of 1776 and the French Revolution of 1789. Indeed, as evidenced in the Declaration of Independence, with its proposition that "all men are created equal," the political realm can be represented as a Newtonian system of politically equal citizen-atoms moving in law-like patterns under the influence of a universal political gravity and a democratic impulse toward civic association.

Theory and Practice

The new ideology of the seventeenth century strengthened the claim that science should be useful and applied. What, then, were the real connections between science and technology in the period of the Scientific Revolution? By and large no technological or industrial revolution paralleled the Scientific Revolution in sixteenth- and seventeenth-century Europe and, although inventions like the printing press, cannons, and the gunned ship had epoch-making impact, their development proceeded without the applications of science or natural philosophy. With the possible exception of cartography, no technological application or spinoff from science produced a significant economic, medical, or military impact in the early modern period. Overall, European science and technology remained the largely separate enterprises, intellectually and sociologically, they had been since antiquity.

Cannonry and ballistics again prove revealing. As already noted, cannon technology became perfected independently of any science or theory. Governments created artillery schools, but instruction was almost entirely practical, and all commentators agree that the science of the day was irrelevant to gunnery. Experience alone counted. The same might be said about fortification and building techniques in the seventeenth century; theory mattered little. Indeed, as we saw, Galileo's ballistic theory and published tables of range *followed* the maturation of early artillery techniques. In this case, once again reversing the usual logic of applied science, technology and engineering posed problems for scientific research and so shaped the history of science.

Cartography—the art and applied science of making maps—may well have been the first modern scientific technology. Spurred by the voyages of discovery, the printing press, and the humanist recovery of Ptolemy's *Geographia,* sixteenth-century European cartographers quickly surpassed Ptolemy and all ancient and medieval predecessors. Cartography and mathematical geography were certainly scientific in that practitioners had to know trigonometry, spherical geometry, the gnomon, cosmography, and the practical mathematics associated with geodesy, surveying, and drafting. Gerardus Mercator (1512–94) was a Flemish professor of cosmography and court cosmographer, and notably his Mercator projection can only be constructed mathematically. Early Portuguese and Spanish support for cartography has been mentioned. In France an ambitious project to map the kingdom scientifically began in 1669. Conducted by the Cassini family from the Paris Observatory and the Paris Academy, the project lapsed from time to time for want of adequate funding, but for over a century the French cartographical establishment—also in coordination with the French navy—turned out whole series of highly accurate maps of France, Europe, and the overseas colonies, many unsurpassed until the twentieth century. Accurate maps formed a nontrivial adjunct to the pursuit of trade and the government of nations and empires. Maps and mapping projects not only concerned navigation and surveying, but they also connected with resource identification and economic development. In these ways cartography as an applied science made a difference in the development and expansion of early modern Europe and indicated what might be forthcoming from state support of science.

Other efforts to exploit or somehow apply science were less fruitful than the case of cartography. The practical problem of finding one's location at sea—the famous problem of longitude—illustrates the more characteristic distance between the promise of early modern science and its practical payoff in applied technologies. For nearly 300 years European maritime activity proceeded in the face of this unsolved navigational problem. In the Northern hemisphere the determination of *latitude* is relatively simple: it requires only the measurement of the angle to the North Star, a procedure that presents no great problem even on a moving ship. But the determination of *longitude* was an intractable problem for ships' captains, trading houses, and Atlantic maritime nations that had to rely on fallible experience and navigators' reports for east-west voyages. Already in 1598 Philip III of Spain offered a monetary prize for a practical solution to the problem of longitude; the Dutch republic put up 25,000 florins in 1626; the Royal Observatory at Greenwich in England was created with the specific charge of "perfecting navigation"; in 1714—propelled by merchant interests—the specially created British Board of Longitude offered the extraordinary sum of £20,000 for the solution of the longitude problem; and in 1716 the French government responded with a substantial prize of its own.

In principle, contemporary astronomy offered several means of solving the problem. The key to the solution lay in determining the differences in time between a known locale such as Greenwich or Paris and the unknown locale. In 1612 Galileo himself proposed that his newly discovered satellites of Jupiter could function as a celestial clock against which the required times could be measured, and he worked (unsuccessfully) to develop practical procedures for the Spanish government. In 1668 the French-Italian astronomer J. D. Cassini published tables of Jupiter's satellites for longitude observation purposes, and other such tables followed in the eighteenth century. Astronomers also tried simultaneous lunar observations. Successful for determining land-based longitude, lunar observations were impractical for observations at sea.

The ultimate solution to the problem of longitude finally came not from the world of science but from the world of the crafts—the improved technologies of clockmaking. In the early 1760s the English clockmaker John Harrison perfected a chronometer (a time-measuring instrument) which employed counteracting balances to compensate for the pitching and rolling of a ship and thermocouples to compensate for temperature variations. With Harrison's chronometer sailors could carry Greenwich time on their voyages, and by comparing it with easily ascertained local time they could at last determine their longitude at sea. After considerable wrangling, the clockmaker finally received the prize from the Board of Longitude.

Practical efforts undertaken in the early Royal Society likewise typify the general failure of applied science in the seventeenth century. The early Royal Society expressed its Baconian commitments through committees it established to pursue useful inquiries into navigation, shipbuilding, reforestation, and the history of trades, but very little came from this extended collective effort. In response to a request by the Royal Navy, for example, the Royal Society conducted experiments on the strength of beams. Fellows tested specimens of various types of wood and different cross-sectional shapes and arrived at a conclusion at odds with Galileo's theoretical determination. A decade later one of them, William Petty, recognized their error and described how Galileo's result could be used by builders to size beams, but here again practicing engineers, in possession of reliable rules of thumb, had no need of Petty's reassurance.

The case of scientific instruments in general and the telescope in particular provides a minor but revealing example of more complex interactions possible between science and technology which began to emerge in the seventeenth century. Scientific instruments, especially the telescope and the microscope, became more common in seventeenth-century research, and they demonstrate once again the historical impact of technology on contemporary science. The development of the telescope, however, shows a further pattern of sequential and reciprocal interactions over time, back and forth, between scientific theory and technological practice. The first telescopes were crafted entirely without the benefit of any theory, Galileo's claims to the contrary notwithstanding. Once the telescope existed, new discoveries were made in optics, notably chromatic and spherical aberration, distortions introduced in the passage of light through lenses. These discoveries in turn posed practical issues for improving telescopes and theoretical ones for the science of optics in accounting for the phenomena. On the practical front scientists and lens grinders responded by trying to grind nonspherical lenses. Newton in his 1672 paper on light provided a theoretical explanation for chromatic aberration and, based on this discovery, he went on to propose a remedy with his reflecting telescope. Ultimately, the solution to the problem of chromatic aberration—the creation of compound lenses with compensating indices of refraction—emerged only after the 1730s and came from the world of technology and the glassmaking craft. All the while, of course, European astronomers made groundbreaking astronomical discoveries with technologically improved telescopes, a reaffirmation of the principle that science often follows, rather than leads, technology.

By the same token, as notable as the telescope case was—and it certainly was for optics and astronomy—in general science and technology did not interact strongly in the era of the Scientific Revolution. Where scientific insights had (at least potentially) practical import, natural philosophers and theoreticians were ignored in favor of engineers, builders, architects, craftsmen, and others with practical empirical experience. Indeed, contemporary technology seems to have had a greater effect on science than the other way around, and one needs to be wary of concluding that the modern alliance of science and technology appeared with the Scientific Revolution. Not until the nineteenth century did ideological promise bear substantial fruit.

Matter and Methods

Competing visions of science and nature jostled among different groups of European intellectuals in the sixteenth and seventeenth centuries: old-fashioned Aristotelianism, various occult traditions, and the "new science," also known as the mechanical or experimental philosophy. The new science reintroduced doctrines, dormant since antiquity, of atoms and corpuscles, and it offered mechanical explanations for nature's actions in terms of matter in motion. The new natural philosophy tended to sustain scientific claims through experiments and, unlike occult learning, it espoused the open communication of knowledge. By 1700, the new science—especially in its new institutional settings—had won out over alternatives whose validity was unquestioned a century earlier. Such "open" systems of knowledge proved of broader appeal to princes and states, and that fact may help account for the relative decline of "closed," magical, or secret doctrines which could possibly be politically dangerous.

Proponents of the new science defended atomism and the mechanical philosophy aggressively against charges of atheism and determinism. Descartes handled these issues with his well-known dualism between matter and spirit that preserved all the formalisms of official religion. We have seen how natural theology formed so important a part of Newtonian science. As a standard operating procedure, the new academies and societies refused all discussion of politics and religion. Scientific research was directed at nature, not society or divinity.

The seventeenth century witnessed the rise and spread of experimental science. Gilbert played with magnets; Galileo rolled balls down inclined planes; Torricelli toyed with tubes of mercury to arrive at the principle of air pressure; Pascal sent a barometer to the top of a mountain to confirm the conjecture that the atmosphere forms a great sea of air; William Harvey dissected and vivisected countless animals in his quest to understand the heart; Newton ran beams of light through prisms and lenses. Although seventeenth-century scientists experimented, they used a variety of approaches and they did not agree on an "experimental method" or the precise role of experiment in generating new knowledge. In particular, what is usually thought of as the hypothetico-deductive method was not historically prominent in seventeenth-century science. That is, one finds little of the procedure later adduced by philosophers of science of formally posing a hypothesis, testing it through experiment, and accepting or rejecting the initial hypothesis based on the results of experiment. Galileo, as we saw, utilized experiments to confirm points already established; Newton used experiment one way in his "crucial experiment" of the 1672 paper and in another way in his "proof by experiment" in his *Opticks*. Harvey undertook systematic experimentation in his dissections, but only as an old-style Aristotelian concerned with "animality" and not to demonstrate the circulation of the blood. "Thought experiments"—experiments impossible to carry out in practice—likewise formed part of seventeenth-century science; consider, for example, Galileo's proposal to view the earth from the moon. Given these different approaches to experiment, including the hypothetico-deductive—experiment has to be seen not as a fixed entity or technique, but as historically contingent:

various approaches to experiment and experimental science emerged, and from heterogeneous roots.

Alchemy doubtless formed one route to modern experimental science. So, too, did handbooks of "secrets" that mushroomed along with printing. These sixteenth- and seventeenth-century "how-to" texts contained practical empirical formulas and recipes for making inks, paints, dyes, alloys, wine, perfumes, soaps, and the like. In their Academy of Secrets Della Porta and friends did not undertake experiments to test scientific hypotheses, but to verify the effectiveness of the practical "secrets" of which they had heard. The Renaissance model of experimentation has been likened to a noble "hunt" for secrets of nature with successful hunters sometimes receiving big rewards from patrons.

As a component of his inductive program, Francis Bacon argued for a multifaceted role for experiments. In the first instance he intended that they produce new phenomena artificially, to "twist the lion's tail," as the expression has it. Such experiments were not to test anything, but to add facts and examples for later inductive analysis. A classic example of a Baconian experiment is Boyle's sustained observations of a rotting piece of meat that glowed. So, too, is the story of Bacon himself stuffing a chicken with snow to observe the outcome. (He died from the resulting chill he suffered.) But Bacon also envisioned a role, something like theory testing, for experiments conducted at a higher level in the inductive process, "experiments of light," experiments designed to question nature more specifically and in the manner of an experimental test as we commonly envision it. Then, Bacon also combined "experiments of light" with practical applied research in his "experiments of fruit."

Descartes, on the other hand, rejected Baconian induction and Bacon's approach to experiment. Descartes's deductive method of philosophizing from first, mechanical principles led him to downplay the role of experiment altogether. Instead he wanted to deduce effects from causes and not to derive causes experimentally through effects. The masses of experiments performed by others confused rather than clarified matters for Descartes, although he did allow a restricted role for experiments at an advanced stage of deduction as a means of testing not *a* single theory but between and among plausible theoretical alternatives.

As experimental science matured and developed, experiment came to be used as a refined tool to test theories or hypotheses and advance scientific discourse. Newton's "crucial experiment" proving light to be composed of rays of differing degrees of refraction had this character. Robert Hooke arrived at the generalization that stress and strain are proportional—Hooke's Law—by tests conducted with springs. Robert Boyle is also widely credited for promoting experimental science as a powerful new technique. With increasing, if diverse, attention given to

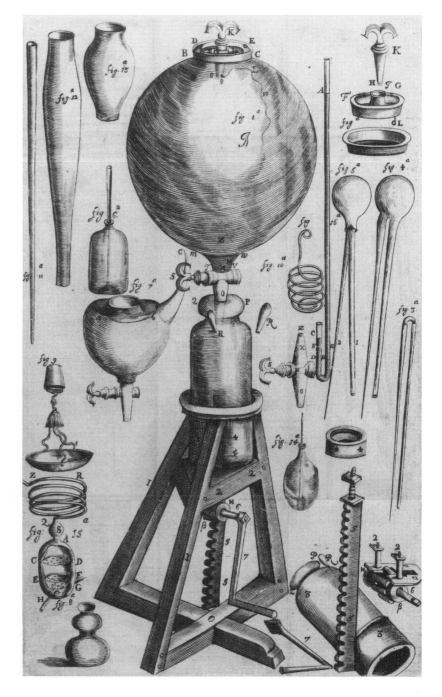

Fig. 12.4. The air pump. This invention led to many experiments in which objects were observed under conditions of a near vacuum. The realities of such experiments were more complex and difficult than apparent in the reported results.

experiment in the seventeenth century, the scientific enterprise naturally became more instrument based and technology dependent. Generations of amateur and professional inquirers into nature applied themselves with telescopes, microscopes, thermometers, beakers, scales, barometers, clocks, inclined planes, prisms, lenses, mirrors, and, later, static electricity machines. (The increasing demand for instruments also provided employment for a growing number of expert instrument makers.) Some of these technological implements extended the range of the senses, and instruments became essential tools for the production of knowledge in a number of arenas of science.

To conclude with one prominent example, Boyle, assisted by Hooke, invented the air pump or vacuum pump in 1658–59. With an evacuated receiver providing a new "experimental space," Boyle's pneumatic pump added a spectacular new instrument to the armamentarium of seventeenth-century experimental science. Boyle used the apparatus to investigate the elasticity or "spring of the air," and he came to discover the law correlating pressure and volume of confined gases. Textbook accounts of this discovery make it seem misleadingly simple and the straightforward product of the experimental apparatus. Newer research has revealed the social and instrumental complexities of producing new facts about nature through the use of instruments and experiments. In the decade after Boyle's invention, for example, a mere handful of costly pumps existed in just a few locales. They all leaked and were difficult to maintain in working order. Replicating experiments and effects proved elusive. Only a practiced operator could demonstrate claims, and claims were regularly contested. A demonstration setting like that of the Royal Society became a public yet restricted forum where effects were tried and where social standing and prestige mattered a great deal. Word of experiments spread through written reports of trustworthy observers allowing others to "witness" the experiment at the removes of time and space. The intricate ways in which these factors played themselves out in the discovery of Boyle's Law embodied subtle social conventions for establishing matters of fact, and the case underscores the fundamentally social nature of the process that is the production of new knowledge.

PART IV

A Brave New World

In the eighteenth century a movement unfolded in England that transformed human existence: industrialization. In essence, the historical process of industrialization—or the Industrial Revolution as it is commonly known—entailed a shift from agriculture as the primary human activity to the mechanization of the production of goods in factories. The consequences of that shift proved profound, at least the equivalent of the earlier Neolithic and Urban Revolutions of the prehistoric and early historic eras.

The sciences provided an example of what reason and experiment might accomplish, but the technology of the Industrial Revolution remained in classical independence of the world of science; only during the nineteenth and twentieth centuries did thinkers and toolmakers finally forge a common culture. Then, new sciences with clearly practical potential made their appearance—electricity, thermodynamics, kinematics, industrial chemistry, molecular biology, and aerodynamics. While science was thus offering a handshake to technology, engineers founded the first professional engineering societies and entered the university to be trained in the new sciences. This nineteenth-century merger of the theoretical and craft traditions has produced the scientific-technological culture in which, for better or worse, we are immersed today.

With regard to the sciences themselves, the elaboration of Darwin's theory of evolution has had profound effects on the modern worldview. Darwin's insights about the diversity of life shifted scientific thought from the work of a divine Designer to a blind process of trial and error and natural selection. Subsequent discoveries in the life sciences continue to produce evermore nuanced views of the nature and history of life on Earth, including ourselves. In the physical sciences, too, the ground has shifted, with both theoretical and practical consequences. The strictly causal Newtonian world enclosed in a fixed universe has given way to relativity, indeterminacy, and speculations about multiple universes. And theoretical research, favored by contributions from technology, has in turn spun off useful applications in nuclear energy, medicine, pharmacology, biochemistry, agriculture, computers, and artificial intelligence, to name just a few fields.

Occasionally technological change still happens on its own terms independently of science, and many scientists today likewise continue to pursue abstract and useless inquiries into the arcana of nature just as their natural philosophical predecessors did 2,500 years ago in ancient Greece. But the merger that has been forged between technology and science has created the unprecedented world we live in today. Our attitudes toward that world are not unmixed: we glorify the conveniences that the modern scientifico-technological revolution has brought us, yet we are fearful of consequences like nuclear or biological warfare or ecological catastrophe. From their separate origins and historically occasional contact, thinking and toolmaking—science and technology—have indeed combined to give us a brave but insecure new world.

CHAPTER 13

The Industrial Revolution

Many factors have been at play in the making of the modern world over the last two or three hundred years, but changes in technology stand at the center of all accounts, notably the Industrial Revolution, that epoch-making technological transformation that took off in the eighteenth century and that gave birth to a whole new mode for human existence, Industrial Civilization.

At the beginning of the eighteenth century, as the Scientific Revolution receded into history, Europe remained the scene of agrarian societies. The bulk of the population, more than 90 percent, lived in rural settings and engaged directly in agricultural activities. Of the urban dwellers few worked as factory labor. Manufactured goods were for the most part the products of either cottage industries in farming communities or of skilled craftsmen. The physical resources that characterized those traditional societies were wood, wind, and water. Then a radical transformation began to sweep first England and, during the next century, Europe and North America.

The Industrial Revolution saw a demographic shift away from traditional agriculture and trade to the mechanization of production, the elaboration of the factory system, and the development of global market systems to support industrial production. Iron, coal, and steam became the emblematic resources.

The changes wrought by the Industrial Revolution are of a magnitude not seen since the Neolithic Revolution 12,000 years ago when humans first turned away from foraging to food-producing, or the great Urban Revolution that occurred with the rise of cities and fully civilized life in the pristine societies at the dawn of written history 5,000 years ago. Largely as the result of the Industrial Revolution, the technical, economic, political, and social bases of life have become transformed virtually everywhere in the last 200 years. The point applies not only to strictly industrial societies, but also to traditional agrarian societies, remaining groups of pastoral nomads, and surviving hunter-gatherers—all humanity has been effected by the coming of industrial civilization.

Industrialization unleashed processes of fundamental social change as well as technological innovation and economic growth. People migrated from the countryside to cities, expanding urban populations of low-paid factory workers; factory labor increased and class conflict intensified; new coercive institutions such as public schools and well-regulated prisons came into being as agents of social control; the family ceased to be a center of production, and a new division of labor took hold—typically men secured employment in factories while women were mainly restricted to domestic duties. A further demographic upheaval accompanied industrialization as mass migration saw millions of Europeans head westward across the Atlantic and eastward into the expanding Russian empire. The process surged and became a global tidal wave that continues to transform every corner of the world, often with unsettling results.

Ecological Stimulus, Technological Response

The history of industry in Europe since the rise of European civilization in the tenth century is the history of long-term economic development and technological innovation against a background of environmental constraints and pressures. Indeed, the industrialization of Europe generally marched in step with a growth of population that produced a constant threat of scarcity. From a low of 2 million in the middle of the fifteenth century, after a hundred years of repeated ravages of the Black Death, the population of England and Wales rose to about 5.5 million by the end of the seventeenth century and, with increasing rapidity, to 9 million by the end of the eighteenth. (This increase resulted from a lowering of mortality rates, probably through improved hygiene, and from changing agricultural practices.) As the population quintupled during a span of 350 years, pressure on resources increased and in some cases became severe.

Perhaps the most serious shortage that developed was in land itself. In England, in many ways a typical agrarian society, land was put to many uses—as cropland, as pasture for cattle, horses, and sheep, as forest for timber, and, increasingly, for expanding towns and cities as the burgeoning population sought nonagricultural means of subsistence in urban centers. During the sixteenth and early seventeenth centuries English towns, trade, and industry had been growing, but by the mid-seventeenth century growth began to be checked as critical bottlenecks formed. Because economic life is a process of interlocked activities, a shortage or restriction in one area can disrupt the entire system. During the eighteenth century several of these constraints were successively broken by technological innovations, and the British economy began to grow again at an unprecedented rapid pace. Iron-making, the spinning and weaving industries, mining, and transportation were all improved by the application of new technologies. Many of the innovations were radical and ingenious departures from traditional methods. The main effect of these technological novelties, however, was to expand the economy in the face of a population that was outpacing limited resources and inadequate methods. Behind the process of industrialization a growing population pressed against its economic and ecological limits.

Indicative of these pressures and constraints, the emergence of a set of new farming techniques known as the Norfolk system provided the necessary agricultural surplus to support the coming of industrialization to England. The new practices replaced the medieval three-field system with a new four-field system of crop rotation; the added production of turnips and clover permitted the over-wintering of greater numbers of cattle, with consequent and consequential rises in meat production. The Norfolk system succeeded in part by enclosing public land (the Common) and making it subject to cultivation and private ownership. The enclosure movement increased agricultural productivity, but it also rendered landless sizable numbers of marginal villagers and farmers who were then "freed" for an industrial labor pool.

The English "timber famine" may also serve as an instructive case study of ecological and economic tensions that propelled change in eighteenth-century England. The British Isles were never heavily endowed with forest, and with the advent of Neolithic farming thousands of years earlier timber reserves became further depleted by the expansion of cropland and pasturage. In the early modern era military and naval requirements along with the beginnings of industrial intensification placed increasing strains on a dwindling supply of timber. Shipbuilding, for example, a major industry in a maritime nation, consumed vast quantities of timber. By the beginning of the eighteenth century construction of a large man-of-war devoured 4,000 trees. And just prior to the American War of Independence one-third of the British merchant marine had to be built in the American colonies where timber remained plentiful. The smelting of iron ore, another major industry, depleted whole forests, with each furnace annually consuming the equivalent of four square kilometers of woodlands. And, like the smelting and refining of iron, the making of bread, beer, and glass likewise depended on wood as fuel in the form of charcoal (charred wood). In none of these production processes could coal be substituted since, using contemporary techniques, the fuel or its fumes came into direct contact with the product, which would then be ruined by coal's impurities, notably sulfur. Also, in the heating and lighting of buildings wood was preferred as fuel since coal fires produced noxious fumes. As the scarcity of timber spread, its price inevitably rose. From 1500 to 1700 while general prices rose fivefold in England the price of firewood rose tenfold. As a result of this energy crisis, by the beginning of the eighteenth century

British iron production actually declined, owing to a shortage of fuel.

The increasing scarcity of wood created a bottleneck in several industries. Under these conditions the incentive to conserve timber did not derive primarily from a desire to increase efficiency. Rather, it was a response to a threatening decline in the standard of living. It sprang from a problem whose roots lay in a wasting ecological imbalance, intensified by population growth and the conversion of woodland to other uses. The upshot of the timber famine and the outcome of the Industrial Revolution generally were neither foreordained nor foreseen. The fundamental historical processes that unfolded in eighteenth-century England resulted from an unpredictable interaction of various industries and technologies, including the most important industry, the making of iron.

While iron-making consumed extravagant quantities of timber, it was an industry that seemed to lend itself to the substitution of plentiful coal for scarce wood. During the seventeenth century many attempts were made to smelt iron ore using coal as fuel, but all of them proved unsuccessful. In processes like cooking, for example, where the product could be separated from the fuel by the use of pots the substitution of coal for wood caused no problems. But in the traditional method of smelting iron the fuel and the iron ore had to be physically mixed in order to bring the ore to the required temperature. In the eleventh century Chinese ironmasters had already developed methods of smelting that employed coal instead of wood as fuel, but in Europe a comparable development did not take place until the eighteenth century. In 1709 Abraham Darby, a Quaker ironmaster, succeeded in using coke (charred coal) instead of charcoal in the blast furnace, although not until mid-century did the new process come into common use.

Darby arrived at his discovery strictly through tinkering with contemporary methods. Neither scientific theory nor organized or institutionalized science played any role in the process. Applicable theoretical principles of metallurgy had not come into being, and even "carbon" and "oxygen" were entities yet to be defined. As a typical artisan-engineer, Darby left no record of his experiments or, rather, tinkering, and we can only guess at how he might have achieved his success. As both the size of the blast furnace traditionally used for smelting iron ore and the strength of the blast slowly and incrementally increased, higher temperatures may have proved capable of burning off the impurities in the coal which had ruined the iron in earlier attempts.

In 1784 English inventor Henry Cort developed the "puddling" process for converting pig (or cast) iron to wrought iron using coal, a technique that involved stirring the melt and that increased productivity. These changes rendered English iron production geographically and materially independent of the forest. With the lag in iron production thus relieved, the world entered a new Iron Age. In the course of the eighteenth century British iron production multiplied more than ten-fold from a low point of fewer than 25,000 tons per year. And from 1788 to the middle of the nineteenth century, with railroad construction booming production increased another fortyfold.

Another key industry of the Industrial Revolution, coal mining, displayed a similar pattern of development. It too had been growing in step with population growth, and it too encountered a production bottleneck. As superficial deposits became depleted, mine shafts were sunk deeper and therefore filled with groundwater at a more rapid rate. Traditional methods of removing water from the mines employed pumps of various designs driven by animal power. By the end of the seventeenth century it became clear that a more effective source of power was required to drive the pumps. "Fire engines" soon resulted—devices that would employ fire in one way or another to raise water. In 1712 an obscure English ironmonger, Thomas Newcomen, invented the first practical steam engine.

The steam engine was a technological innovation that changed the course of industrial development. The origin of the steam engine lies in independent craft traditions. Through intuition, tinkering, and a stroke of luck, Newcomen and his plumber assistant, John Cawley (or Calley), hit on the method of condensing steam in a cylinder and creating a partial vacuum so that atmospheric pressure would drive the piston. Even if the idea of atmospheric pressure as a potential motor was, as it were, "in the air," the actual design of the steam engine owed nothing to science. The complicated valve mechanisms, the technique of injecting cold water into the cylinder to condense the steam, and the mechanical linkages to the pump evolved through trial and error. The alternate heating and cooling of the large cylinder made the engine highly inefficient and a profligate consumer of coal, but Newcomen engines proved sufficiently economical and were widely adopted because they operated primarily at coal mines where coal was cheap. But the heavy consumption of coal remained a shortcoming of the design, especially when employed in other applications, and it provoked attempts to increase the efficiency of the engine.

Around the mid-eighteenth century two English craftsmen, John Smeaton and James Watt, using entirely different approaches, improved the Newcomen engine. Their work also remained essentially within the craft tradition without applying scientific abstractions. Smeaton, who later became president of the Society of Civil Engineers ("the Smeatonians"), employed strictly empirical methods and systematically tested model steam engines, varying the dimensions of the parts without modifying the basic design; he thereby achieved a doubling of efficiency of the Newcomen engine. Watt introduced a fundamental novelty that resulted in a radical improvement in efficiency. In a flash of insight during a Sunday stroll in 1765 he arrived at the idea of condensing the steam in a separate vessel kept cold outside of the cylinder, thereby

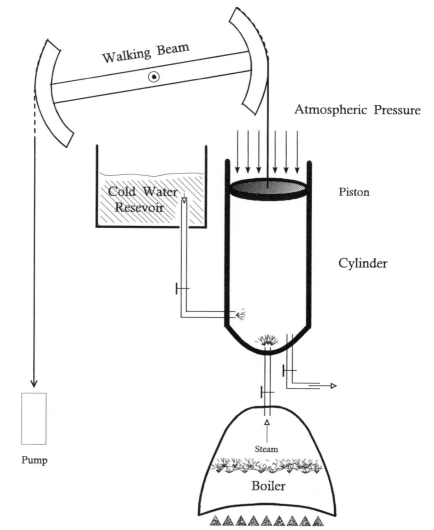

Fig. 13.1. Steam power. As mine shafts went deeper it became increasingly difficult to remove ground water. The problem was initially solved by Thomas Newcomen's invention of the atmospheric steam engine in 1712. Because it alternatively heated and chilled the cylinder Newcomen's engine was inherently inefficient and was only cost-effective when deployed near coal mines where fuel was inexpensive.

leaving the cylinder hot throughout the cycle. By eliminating the alternate heating and cooling of the cylinder, the saving in coal was considerable. Constructed and sold in a famous partnership with the Birmingham manufacturer Matthew Boulton, Watt engines were widely adopted and were soon used in industries other than coal mining. The success of the Watt engine stemmed in part from the commercial strategy of leasing the machines and charging only for a percentage of the savings in the cost of coal over traditional Newcomen engines. Freed from a dependence on cheap coal at the coal mouth, the Watt engine could be set up and used to drive mills virtually anywhere, thereby promoting the expansion of manufacture in urban centers. Five hundred steam engines puffed away in Britain in 1800, and the number rose rapidly after that.

Since the early steam engine relied on atmospheric pressure it was necessarily a large machine (sometimes called a "stationary" engine). By the end of the eighteenth century stationary engines were used to drive machinery in factories, and since boats were sufficiently large to hold them, early steamboats were also powered by atmospheric engines. But the railroad locomotive had to await the invention of a compact, high-pressure design, developed in 1800 by another Englishman, Richard Trevithick. By employing pressures much higher than atmospheric, Trevithick's engine was much smaller and could thus be placed on a carriage of reasonable size. At first Trevithick intended it as a replacement for the atmospheric engines used primarily in mines and mills, but he found that mine operators and manufacturers were unwilling to scrap their atmospheric engines for the still-questionable benefits of the new design, especially since the high-pressure engine was reputed to be unsafe (a reputation publicized by Watt who attempted to fend off the competition of the new design). Trevithick then took his invention to Peru where he hoped that at the reduced atmospheric pressures of high-altitude mining his engine would have a competitive advantage. When that venture failed he returned to London and did what inventors have often done before and since—he turned his invention into a novelty by running a locomotive on a circular track as an amusement for which he charged admission.

But the railroad could not be denied. England was becoming the bustling workshop of the world, and the need to transport bulky goods was rapidly increasing. Despite contemporary efforts to build improved "turnpikes," animal-drawn wagons on the primitive road system proved inadequate, especially for moving large quantities of coal on inland routes. At first, the solution seemed to be river and canal transportation, and in 1757 and 1764 the first two canals were constructed to link coal fields with Manchester via the River Mersey. Canal mileage and the associated number of locks and crossings increased dramatically thereafter. But Trevithick's high-pressure steam engine altered the economics of transportation by making the railroad possible. In 1814 British engineer George Stephenson unveiled his first steam locomotive. The railroads initially hauled coal for short runs from the mines, but the railroad age truly dawned with the first public line which opened between Liverpool and Manchester in 1830. Indeed, a railroad mania ensued, as the world's land areas began to be encrusted with a dense network of iron rails. In Britain, the railroad-building boom peaked in the 1840s; in 1847 alone nearly 6,500 miles of railroad track were under construction. As railroad transportation developed, it interacted with the iron industry in a pattern of mutual stimulation: the rapid growth of railways was made possible by the availability of cheap iron, and, in turn, rail transportation facilitated—indeed demanded—the further growth of iron production.

The multistage production of textiles represents a case where the interdependence of subsidiary technologies stimulated rapid growth. In a pattern of challenge and response, alternating technical innovations in spinning and weaving machinery propelled development. In 1733

Fig. 13.2. Watt's steam engine. In 1765 James Watt hit upon a way to greatly improve the efficiency of steam engines: condense the steam in a condenser separated from the main cylinder. That way the cylinder could remain hot through the entire cycle, thereby increasing efficiency and lowering operating costs. As important as Watt's technical innovation was, the success of his steam engine depended as much on the manufacturing partnership he established with the early industrialist Matthew Boulton and the marketing strategies they devised.

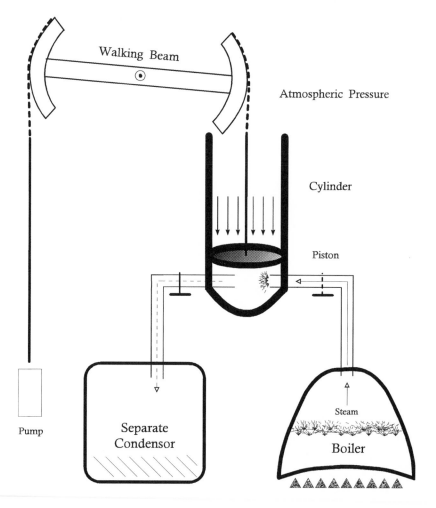

John Kay, initially a clockmaker, invented the "flying shuttle," which improved weaving but thereby created a lag in the spinning of thread. A combination of technical developments elaborated by a series of artisans and engineers in the 1760s and 1770s then mechanized spinning; these improvements left weaving as the constriction point in the production process, an imbalance intensified by the invention of cylinder carding in 1775 which made spinning even more efficient. After 1785, with the mechanically powered loom, to which the steam engine would ultimately be applied as an independent power source, weaving became mechanized. Not unexpectedly, therefore, between 1764 and 1812 worker productivity in the cotton industry increased by a factor of 200. In 1813 2,400 power looms operated; by 1833 the number had skyrocketed to 100,000. (The power loom replaced not only the hand loom but also hand-loom weavers.) The mechanization and industrialization of textile production marked the arrival of industrial civilization in England.

The process that began in Britain in the 1780s—what economic historians have called the "takeoff into sustained growth"—can be attributed to the mutually reinforcing effects of major industries as they developed separately and jointly. Thus, as iron began to be smelted with coal, it stimulated the growth of the coal industry, which led to the creation of steam engines to clear the mines, while the need to transport large quantities of coal led to the railroad, which then led back to an enormous increase of iron production in an upward-spiraling, symbiotic process. The result was a nation, and eventually a world, transformed. A population of rural farmers became a population of urban factory workers. Locomotives and iron tracks replaced horses and dirt roads. Iron and steel increasingly replaced wood and stone as structural materials, and the steamship replaced the sailing ship. As with the Neolithic and Urban Revolutions previously, once these fundamental processes of change were under way, there was no going back to earlier modes of social or economic life.

Industrial Civilization

Not every aspect of emerging industrial civilization can be treated here, but four features require attention: new energy sources fueling the Industrial Revolution, the new organization of labor in the factory system of production, new means of financing industrial development, and ideological changes that accompanied industrialization.

Premodern societies depended overwhelmingly on human and animal muscle power and to some extent on wind and water power. For fuels they used renewable resources, notably wood. In turning to the steam engine and nonrenewable fossil fuels like coal and, later, oil, the advent of industrial civilization brought about significant changes in available energy resources and patterns of energy consumption. The production of coal and then oil rose exponentially from the eighteenth century, to the point where per capita energy consumption in industrialized societies today is 5 to 10 times greater than in traditional preindustrial cultures. With its sharply increased consumption of energy, industrialization produced not merely a reordering of traditional society but, rather, a new type of society in which industrial production represents the major economic activity.

Factories were not unknown before the eighteenth century, but the predominant type of manufacturing remained the domestic or cottage system of production, which was household-based or sited in artisanal craft shops. The new factory system that arose with the Industrial Revolution came to involve centralized and standardized production using machines, wage labor, and an organization of the production process that involved rigid hierarchies of supervisors governing workers. A series of power-driven textile mills created by Richard Arkwright in the 1770s and 1780s, employing hundreds of workers, represents the coming into being of the first factories in the modern sense. The so-called American system of manufacturing with interchangeable parts—developed in Britain, but widely applied in the United States—represents a key innovation that later emerged in the mid-nineteenth century. The formal assembly line, perfected by Henry Ford in the automobile industry in the second decade of the twentieth century, culminated the evolution of the modern factory.

The factory produced radical social transformations. An industrial, urban-based labor force formed a new working class that rivaled the traditional rural peasantry, while a money economy replaced traditional exchanges of goods and services. For workers, the factory imposed an unprecedented alienation of work from home and family life. Bosses were a novelty, and the clock, the medieval device to tell time, became the industrial master that governed time and the workplace. Especially in its early phases in England, the factory system entailed a severe exploitation of labor. Two-thirds of Arkwright's 1,150 workers in 1789, for example, were children. The Parliamentary Act of 1799 made it illegal to unionize and prescribed a three-month prison sentence for anyone seeking to ameliorate working conditions by means of organized action. The Bill of 1825 recognized worker "combinations" but severely restricted union activity; comparable restrictions on business organizations and price-fixing were not enforced. The doleful effects on labor sound as one of the traditional themes in the study of the Industrial Revolution. Quite apart from the political and moral issues, the exploitation of labor formed a systemic component of industrialization. During the second quarter of the nineteenth century, for example, as cotton production quadrupled and profits doubled, wages remained almost unchanged.

Like labor, capital and new means of financing were equally essential for industrialization. The history of European capitalism extends back to the late Middle Ages, and the development of the new industrial capitalism emerged out of the background of merchant capitalism and the success of overseas trade in commodity products through the eighteenth century. Profits from British colonial trade in sugar and slaves in large measure provided the accumulated capital needed to fund industrial development. Although authorities created the Bank of England in 1694, the official state bank did little to promote industry. Rather, hundreds of private banks arose in the English Midlands to handle capital requirements of nascent industries. Interest rates fell steadily through the seventeenth and eighteenth centuries, reaching a low of 3 percent in 1757 and remaining low thereafter, despite some ups and downs (as during the American war). A low interest rate made available large amounts of cheap money, without which the capital requirements of early factories could not have been met. Growing out of prior groups of commodity brokers and insurance agents, the London Stock Exchange opened in 1773, and in 1803 it offered its first list of traded stocks.

The ideological effects of industrialization proved no less powerful.

Mercantilism, or the idea of a state-controlled economy restricting free trade in the interests of bolstering exports and accumulating gold and silver for the state's coffers, stood as the reigning economic theory and the basis of the economic policies of contemporary European governments. Not coincidentally, new ideas about open markets and free enterprise or "laissez-faire" capitalism emerged with the coming of the Industrial Revolution. Adam Smith's pathbreaking work, *The Wealth of Nations,* appearing in 1776, signaled this new ideology of the marketplace. As the Industrial Revolution gained momentum, however, and its unanticipated labor strife and social costs emerged more clearly, other voices began to speak in opposition to free-market capitalism. In particular, the work of Karl Marx (1818–83), *Das Kapital* (three volumes from 1867), provided an analytical critique of the new economic relations. Marx underscored the inevitable exploitation of labor for profit by the owners of factories and the means of production; class warfare between workers and owners, according to Marx, would result in a transformation of society. In this way Marx provided the ideological underpinnings of socialism and communism.

The later eighteenth and nineteenth centuries also witnessed the flowering of the Romantic movement. In poetry, literature, music, and other of the fine arts artists turned away from classical styles of the preceding decades and toward themes associated with the simplicity of nature, the family, and matters of the human heart. This blossoming of romanticism needs to be seen first and foremost as a reactionary response to the ravages of industrialization.

The processes of industrialization continued to gain momentum in England in the nineteenth century. Worker productivity doubled from 1830 to 1850. Iron production jumped from 700,000 tons in 1830 to 4 million tons in 1860. Coal production soared from 24 million tons in 1830 to 110 million tons in 1870. In 1850 for the first time the urban population of England topped 50 percent. The first "world's fair"—the Great International Exhibition—opened in London in 1851. The machines on display within the magnificent iron and glass "Crystal Palace" vividly exemplified the power of industrialization and the new technologies then transforming the world. Britain, at least, was a very different country than it had been a century earlier.

Science and the Industrial Revolution

All of the technical innovations that formed the basis of the Industrial Revolution of the eighteenth and the first half of the nineteenth centuries were made by men who can best be described as craftsmen, artisans, or engineers. Few of them were university educated, and all of them achieved their results without the benefit of scientific theory. Nonetheless, given the technical nature of the inventions, a persistent legend arose that the originators must have been counseled by the great

Fig. 13.3. The Industrial Age. The Crystal Palace, erected in 1851 for an international exposition in London, was a splendid cast-iron and glass structure that heralded the new industrial era.

figures of the Scientific Revolution. During the eighteenth century John Robison, a professor at the University of Edinburgh, publicized the myths that Newcomen had been instructed by Robert Hooke, one of the leading savants of seventeenth-century English science, and that Watt had applied Joseph Black's theory of latent heat in arriving at the separate condenser. These claims have been discredited by historical research. The French physicist Sadi Carnot, for example, produced the first scientific analysis of the operation of the steam engine in his work *Reflections on the Motive Power of Fire* published in 1824, long after steam engines had become commonplace. And Watt's cleverly designed parallel motion could not even begin to be studied scientifically until kinematic synthesis developed the appropriate analytical techniques in the last quarter of the nineteenth century, partly in an attempt, in fact, to analyze parallel motion. These examples rank among the many instances where, contrary to the claims that eighteenth-century engineers benefited from scientific theory, technical developments provoked the interests of scientists and led to theoretical advances. In this context it is worth noting that industrialization spread to South Asia and the Far East long before the Western scientific tradition took hold in those regions.

The myth that the theoretical innovations of the Scientific Revolution account for the technical inventions of the Industrial Revolution finds reinforcement in the common belief, which has been challenged repeatedly in these pages, that technology is inherently applied science, a belief only partially true even today when research and development are indeed often conducted in close contact. In the eighteenth and early nineteenth centuries it was almost never true. This is not to say that science played no social or ideological role in promoting industrialization. On the contrary, as the Industrial Revolution unfolded in England, science permeated the social and cultural fabric of European civilization. Multitudes of learned societies and academies dotted the European map, where scientists and literate engineers occasionally rubbed shoulders. Public lectures alerted large, nonspecialist audiences to the triumphs of scientific discovery and the analytical potency of experiment and the scientific method. Natural theology, the doctrine that the study of nature is an act of piety, strengthened the concordance between science and religion and reinforced the notion of the useful exploitation of nature. Science raised the status of the reasoned life and was honored as a cultural and intellectual enterprise. The rational sciences offered a new outlook and worldview. In this sense, scientific culture was important and perhaps essential to the Industrial Revolution. But the scientific enterprise itself continued to be shaped in a Hellenic mold, largely divorced from the practical affairs of society, except where patronized by the state; technologists and engineers proceeded without tapping bodies of scientific knowledge.

Although technology developed along traditional lines without the benefits of scientific theory, several eminent craftsmen made social contact with the world of science in eighteenth-century Europe. In England, engineers James Watt and John Smeaton and the potter Josiah Wedgwood became members of the Royal Society and contributed to its *Philosophical Transactions*. But, in fact, their publications had little or nothing to do with their contributions to industry. Watt published letters and articles on the composition of water and on the "medicinal use of factitious airs." Both were stated in terms of phlogiston chemistry, and neither had any bearing on his steam engineering. Wedgwood became keenly interested in chemistry, performed chemical experiments, discovered that clay shrinks when heated, and in 1782 invented the pyrometer based on that discovery. He also corresponded with eminent chemists, including Joseph Priestley and Antoine Lavoisier. But his novel and successful ceramic ware, which goes under the name of Wedgwood, came *before* his entrance into the world of chemistry. His father and brother were potters; Wedgwood was unschooled, and he came to pottery through a craft apprenticeship in his brother's factory. His career as a potter led him to an interest in chemistry. And Smeaton, who coined the term *civil engineer* to distinguish civilian consultants from the military engineers graduating from the newly founded Royal Military Academy at Woolwich, distinguished himself as a builder of large-scale public works. He won the Copley Medal of the Royal Society for a paper he published in the *Philosophical Transactions*. The paper, a report on his empirical demonstration that overshot waterwheels are more efficient than undershot, contained a keen insight, and he applied it in his own hydraulic projects where he consistently preferred the overshot wheel. (Other engineers continued to use undershot wheels because they were less costly to build.) But Smeaton's scientific preference regarding waterwheels had an insignificant effect, if any, on the industrialization of Britain.

In 1742, the English government, recognizing the need for formally trained artillery (and, soon afterwards, engineering) officers, established the military academy at Woolwich. There, cadets were schooled in the "fluxions" (Newton's form of the calculus) and the elements of statics, among other subjects. However, the scant knowledge of the graduates combined with their lack of craft experience disqualified them as working engineers, and during the eighteenth century only civilian engineers, unschooled in any science, contributed to the program of industrialization. The period of the Industrial Revolution did indeed see a revision in the traditional affiliation between technology and the crafts in favor of new links between technology and science: the perceived rational methods of experimental science began to be applied in industry, and some of the leading engineers drew closer socially to the world of science. But the gulf between practical applications and theoretical research remained to be bridged.

An episode in Watt's career reflects the dimensions of that gulf. During the 1780s he performed experiments on the commercial application of the chlorine process for bleaching textiles, a process discovered by the French chemist C. L. Berthollet. For Berthollet the research was purely scientific, and he published his results without regard for commercial possibilities or financial gain. But Watt's father-in-law, James MacGregor, was in the bleaching business, and Watt hoped that the three of them—Watt, MacGregor, and Berthollet—could, by keeping their improvements secret, acquire a patent and reap substantial profits. When Watt complained to Berthollet that by "making his discoveries . . . publick" he undermined the possibility, Berthollet replied, "If one loves science one has little need of wealth." Berthollet stood almost alone among scientists of the eighteenth century called upon to defend the ethos of pure science. For the others the occasion to apply theoretical research to practical problems almost never arose.

Almost never, but not quite. A curious instance occurred at the turn of the nineteenth century when English scientists were called upon to apply their knowledge to industry. The resulting failure points conclusively to the features of applied science that were still lacking. The Port of London determined that another bridge across the Thames was required to serve the growing metropolis. Proposals were solicited, and among them Thomas Telford (who later became the first president of the Institution of Civil Engineers) submitted a spectacular design in the form of a single cast-iron arch with a span of 600 feet. Since cast-iron bridges were still a novelty, with only three or four having been built,

no established tradition or rules of thumb existed to guide the design. Nor was there any applied science that could be brought to bear on the project, or any professorships of what would nowadays be called engineering science at universities. The parliamentary committee considering plans for improvement of the port recognized the difficulty and proposed to consult "the Persons most eminent in Great Britain for their Theoretic as well as Practical Knowledge of such Subjects."

Since such knowledge, theoretical and practical, was not combined in any individuals in 1800, Parliament created two committees, one of mathematicians and natural scientists and the other of practicing builders. Each was asked to respond to a questionnaire about Telford's design in the hope that applied science would emerge from the combined answers. The results illustrate the futility of the proceedings, based on the misconception that applied science could be derived from the combined expertise of mathematicians who possessed little knowledge of building and builders who were largely ignorant of mathematics and theoretical mechanics. Among the answers provided by the "practitioners" a few sensible suggestions emerged, but engineers were still defeated by their lack of any theory of structures (which had not yet been formulated) and by the complexity of the design, which would present a formidable theoretical problem even today. But the inability to bring contemporary scientific knowledge to bear on the solution of practical problems was especially evident in the replies of the "theoreticians," who included the Astronomer Royal and the Professor of

Fig. 13.4. Thomas Telford's iron bridge across the Menai Straits (completed 1824). In the eighteenth century, iron was adapted to structural engineering. The cast-iron arch bridge was followed by wrought-iron suspension bridges and wrought-iron tubular bridges. Organizational and managerial skills proved no less important for the success of bridge building than the technical mastery of iron as a construction material.

Geometry at Oxford. The Astronomer Royal's views on mechanical engineering were ridiculed not long afterwards as "a sentence from the lofty tribunal of refined science, which the simplest workman must feel to be erroneous." The astronomer's incompetent testimony was enriched only by his knowledge of heavenly phenomena: he suggested that "the Bridge be painted White, as it will thereby be least affected by the Rays of the Sun," and that "it be secured against Lightening." The contribution of the Savilian Professor of Geometry was equally silly: he calculated the length of the bridge to ten-millionths of an inch and its weight to thousandths of an ounce.

To their credit, some of the theoreticians on the committee recognized that science was not yet prepared to minister to technology. Isaac Milner, the Lucasian Professor of Mathematics at Cambridge (Newton's professorship), observed that theory would be useless in such applications until it was combined with practical knowledge. The theoretician, he observed, "may . . . appear learned, by producing long and intricate Calculations, founded upon imaginary Hypotheses, and both the Symbols and the Numbers may be all perfectly right to the smallest Fraction, and *the Bridge be still unsafe*." And John Playfair, professor of mathematics at Edinburgh, ended his report by noting that theoretical mechanics "aided by all the Resources of the higher Geometry, [has] not gone farther than to determine the Equilibrium of a Set of smooth Wedges." At the beginning of the nineteenth century the design of a structure as complex as Telford's fixed arch was still to be left to the intuition and experience of craftsmen—"Men bred in the School of daily Practice and Experience." It would be another half century before John Rankine's *Manual of Applied Mechanics* and similar works would show the way toward the engineering sciences.

Whatever cultural influence the Scientific Revolution exerted on the Industrial Revolution, it did not extend to the application of scientific theory to technical invention. Although the governments of Europe were rational in their Baconian hope that science would assist society, their interests were more narrowly focused on governance, and the technical dimension of the Industrial Revolution was left to be crafted by the ingenuity of unschooled artisans working without benefit of theoretical knowledge. That knowledge was not yet compacted into textbooks. The universities had no programs or even courses in the engineering sciences. There were no professional engineering societies. The constants and tables that would convert abstract mathematical principles into engineering formulas had not yet been determined and compiled. And no research laboratories had yet come into being. Those developments and applied science awaited a later day.

CHAPTER 14

The Road to Modern Science: Pure and Applied

The Scientific Revolution that culminated in the work and worldview of Isaac Newton did not automatically create modern science. Indeed, we today stand several social and intellectual removes from Newton, and to understand the maturation of science since his day, its trajectory, particularly in the physical sciences, needs to be sketched. This chapter surveys the period to the end of the nineteenth century; later chapters bring the story down to the present. At the same time, coincident with the spread of industrialization, science and industry forged crucial new ties in the nineteenth century that established the pattern for applied science today. At long last, the application of scientific knowledge is clearly seen in the world of the toolmaker.

The Legacies of Bacon and Newton

The exponential growth of knowledge since the seventeenth century presents a difficulty in thinking about the history of science after the Scientific Revolution. The scale of the scientific enterprise and its output have increased by orders of magnitude since then, making it hard to review in any detail the sciences of the past two centuries. One way around these difficulties involves modeling the phenomena, that is, to reduce a complex reality to a more simple model by identifying essential elements, linking them conceptually, and seeing how they interact.

It is thus possible to envision two distinct scientific traditions developing in the aftermath of the Scientific Revolution. One, often labeled the "Classical sciences," includes astronomy, mechanics, mathematics, and optics. These fields originated in antiquity; they matured as research endeavors in the ancient world; and they were, in fact, the sciences revolutionized in the Scientific Revolution. But even before their reformulations, these sciences were highly theoretical, with research aimed at solving specific problems. All things considered, the Classical sciences were not experimental in approach. Instead, they built on mathematical and theoretical foundations and were clearly the province of the schooled expert.

The other group of sciences, the "Baconian sciences," developed parallel to but largely separate from the Classical sciences during and after the Scientific Revolution. The name derives from the style of science advocated by Lord Bacon. The Baconian sciences—meaning primarily the systematic study of electricity, magnetism, and heat—were without roots as formal sciences in antiquity, but sprang into existence as domains of empirical investigation more or less as a result of the ferment surrounding the Scientific Revolution. That is, as the Classical sciences became transformed in the Scientific Revolution, the Baconian sciences took shape in the general intellectual excitement of the times. In contrast to the theory-dependent and more mathematical Classical sciences, the Baconian sciences were generally more qualitative in character and experimental in approach, and they therefore depended on instruments to a much greater degree than their Classical counterparts. The Baconian approach was only loosely guided by theory and typically involved the raw collection of data.

Newton's *Principia* provided the exemplar for the Classical sciences in the technical details of its physics, in its general approach, and in dictating problems that the community of mathematical scientists worked on over the course of the eighteenth century. The predicted return of Halley's comet, for example, and its appearance in due course in 1758–59 showed the awesome power of Newtonian mathematical science. In another confirmation and extension of Newtonian physics, other scientists measured the earth's curvature. In 1761 and again in 1769 international teams of observers clocked the rare crossing of Venus across the face of the sun and calculated the earth-sun distance reasonably well for the first time. On the Continent French and Swiss mathematicians extended research in theoretical mechanics into highly technical fields like hydrodynamics, the mathematics of vibrating strings, and elastic deformation.

Technical work of this sort continued into the nineteenth century. A famous example is the discovery of the planet Neptune in 1846. Based on observed irregularities in the orbit of the planet Uranus—itself discovered empirically in 1781 by William Herschel—British and French astronomical theorists predicted the existence of Neptune, and German astronomers in Berlin duly observed the planet the night after they received the prediction. This tradition of the Classical sciences may be fairly said to have culminated, at least conceptually, in the work *Celestial Mechanics* by P. S. Laplace (1749–1837). Newton's *Principia* with its abstruse geometrical diagrams seems quaint and antiquated in comparison to Laplace's magisterial work (five volumes, 1799–1825) written wholly in the language of the calculus. And where Newton saw God's presence in his physics, Laplace saw His absence. In a famous exchange the emperor of France, Napoleon Bonaparte, remarked that he found no mention of God in Laplace's opus. "But, Sire," Laplace reportedly replied, "I have no need of that hypothesis." The Classical sciences had progressed so far that Laplace could formulate a mathematically complete and ordered universe based on the fundamental laws of mechanics established by Newton and elaborated by his successors.

Newton's *Opticks* (1704) provided the conceptual umbrella beneath which the Baconian sciences developed in the eighteenth century. In a set of research questions that Newton appended to the *Opticks* the great man at one point posited a series of superfine, self-repulsive substances to account for phenomena. How can the heat of a warm room penetrate the glass of an evacuated receiver, Newton asked, except by means of such an ether? Similarly, Newton invoked various imponderable ethers to explain electrical, magnetic, certain optical, and even physiological phenomena.

Developments in eighteenth-century electricity exemplify these trends. Static electricity, of course, was known at least since antiquity. (Before the invention of the battery current electricity did not yet exist in the world of science.) The investigation of static electricity took off in the eighteenth century, as new instruments were developed to generate and store static electricity and as scientists applied themselves to investigating a host of new facts relative to electrical conduction, insulation, attraction, and repulsion. Benjamin Franklin's kite experiment—first performed in 1752—that identified lightning as an electrical phenomenon—seems entirely typical of this manner of experimental, qualitative research. On the theoretical side, Franklin offered a single electrical ether to account, not entirely successfully, for the phenomena. Other theorists countered with a two-ether theory. Notably, there was no agreement. Much the same points emerge in considering the scientific study of several other fields, all of which took their lead from the guidelines offered in the *Opticks*. With regard to magnetism, for example, the German-Russian scientist F.U.T. Aepinus (1724–1802), working at the imperial science academy in Saint Petersburg, articulated an ether theory to account for magnetic attraction and repulsion. The English physiologist Stephen Hales (1677–1761) deployed a "vegetable" ether in experiments on plants. While Franz Anton Mesmer's work on "animal magnetism" and early hypnotism may strike readers today as beyond the pale of science, quite the contrary is the case, because Mesmer (1734–1815) operated with the full authority of Newton's *Opticks* and the tradition of ether-based research and scientific explanations. Mesmer's sin was not invoking a superfine, self-repulsive magnetic ether to explain the seemingly miraculous cures he produced in his medical patients; it was, rather, his refusal to share the secrets of his discoveries with other members of the scientific and medical communities which led to his downfall and the repudiation of his mesmerizing ether.

One may well extend the concept of the Baconian sciences to in-

Fig. 14.1a–b. Electrical equipment. The scientific investigation of static electricity blossomed in the eighteenth century aided in essential ways by the development of new experimental apparatus, such as glass or sulfur ball generators, shown here. These engravings are taken from the Abbé Nollet's *Lessons on Experimental Physics* (1765). They illustrate something of the parlor-game character that often accompanied demonstrations concerning static electricity in the eighteenth century.

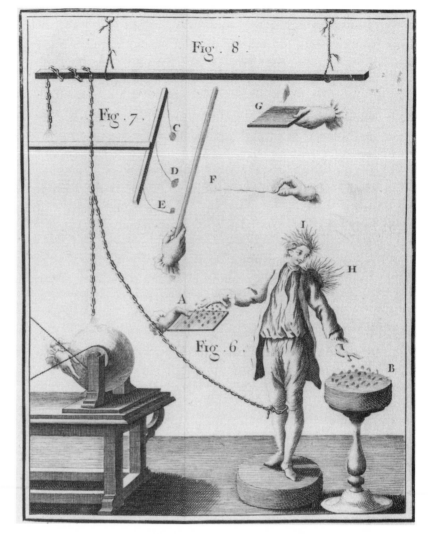

clude research in eighteenth-century meteorology, natural history, botany, and geology. Regarding meteorology, the scientific societies served as central depots and published reports sent in by individuals, and they independently sponsored several large-scale meteorological data-collecting projects. Obviously instruments (such as thermometers and barometers) were required, and collecting weather data allowed a provincial amateur, say, to feel as if he (or in a few cases, she) participated in the grand enterprise of European science in the eighteenth century. Analogous circumstances prevailed in botany and natural history, where the primary activity consisted in collecting specimens, often from far-flung corners of the world. Specimens found their way to central depots in London, Paris, or Uppsala, Sweden, where theorists such as the Count de Buffon (1707–88), Sir Joseph Banks (1743–1820), or Carolus Linnaeus (1707–78) attempted to develop rational systems of classification. "Botanizing," indeed, became something of a fad in the

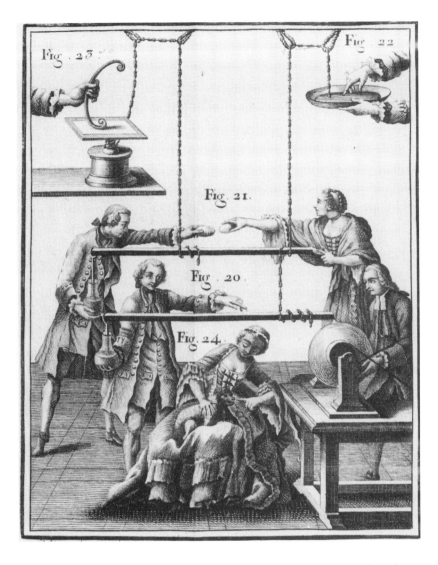

eighteenth century, where, armed with simple manuals for identifying plants (and perhaps a bottle of wine), individuals would pass the time in scientific communion with nature. Progress in geology in the eighteenth century likewise depended on systematic collections of data. In all these instances research proceeded without sophisticated theory and the other trappings characteristic of the *Principia*-based Classical sciences.

In considering the different traditions represented by the Classical and Baconian sciences in the eighteenth century, chemistry seems the odd science out. With deep roots in traditional alchemy, chemistry underwent no revolutionary restructuring in the Scientific Revolution of the sixteenth and seventeenth centuries, and chemistry in the eighteenth century does not fit well into either the empirical Baconian sciences or the more specifically problem-oriented research of the Classical sciences. Contemporary chemistry was highly experimental and instrument-dependent. In the early decades of the eighteenth century it developed an agreed-upon theoretical structure known as phlogiston chemistry, and at the end of the century chemistry underwent an independent conceptual revolution.

The history of the Chemical Revolution fits the pattern previously seen for scientific revolutions in general. The theoretical framework prevailing through the 1770s was that of phlogiston chemistry, phlogiston being the principle of combustion, vaguely akin to the ancient Greek concept of "fire" that is active and released in combustion. Thus, for example, according to the theory, a burning candle releases phlogiston; the candle goes out when covered by a jar because the air contained within becomes saturated with phlogiston, a circumstance that prevents further combustion (the very opposite of the view of combustion after the Chemical Revolution). Phlogiston theory, in giving a coherent account of a diverse range of phenomena—combustion, plant growth, digestion, respiration, smelting—provided a solid theoretical framework within which research in eighteenth-century chemistry unfolded.

Several factors brought about the downfall of phlogiston chemistry and its replacement by Lavoisier's oxygen theory of chemistry and combustion. The discovery of "fixed air" (what we know as carbon dioxide) by Joseph Black in 1756 and its identification as a distinct gas represents a major milestone. With its own specific properties, "fixed air" helped break down the traditional notion of "air" as a single element or entity, and with improved equipment chemists soon identified a series of other new "airs." Finally, a string of anomalies, which at first appeared minor, posed problems for phlogiston theory, notably that mercury seemed to *gain* weight in combustion (under certain conditions), whereas with the release of phlogiston, according to the theory, it ought to have *lost* weight. Such problems became more and more acute for theoretical chemists. Antoine Lavoisier (1743–94), a young chemist, knew phlogiston theory well, yet he began his theoretical researches with the radical notion that in combustion something was taken out of the atmosphere rather than released into it.

That "something" proved to be oxygen, although we hasten to add that even Lavoisier's mature views of oxygen were not the same as those taught in chemistry classes today. Nevertheless, with the discovery of oxygen gas and the identification of its role in combustion, Lavoisier effected a revolutionary reconceptualization of chemistry, aided in significant measure by his careful accounting for the inputs and outputs of the reactions he produced and reversed. Typical of revolutionary transformations in science, other chemists did not immediately subscribe to Lavoisier's radical new views. Indeed, led by the older English chemist Joseph Priestley (1733–1804), they modified phlogiston theory in ways that preserved satisfactory, rational accounts of chemical phenomena. Well into the 1780s chemists could still reasonably hold phlogistonic views, if only because the latter seemed more familiar. Indeed, Priestley never converted to the new chemistry and went to his grave in 1804 as virtually the last adherent of phlogiston chemistry.

If not through an irrefutable test or piece of evidence, why, then, did European chemists shift their allegiance to the new chemistry? Rhetoric or persuasion played a crucial role in the dynamic of the Chemical Revolution. Not only did Lavoisier and a band of like-minded colleagues make discoveries and publish experimental results, in 1787 they formulated an entirely new system of chemical nomenclature. In Lavoisier's new system, "inflammable air" became hydrogen, "sugar of Saturn" became lead acetate, "vitriol of Venus" became copper sulfate, and so on. The proponents of the new system wished to have language rationally reflect chemical realities. But, as a result, students schooled in the new chemistry could only speak the new language. This shift was compounded by a related step, the publication of Lavoisier's textbook, the *Elementary Treatise of Chemistry* in 1789, which taught only Lavoisier's chemistry. In it, phlogiston chemistry is banished. From being a central element of theory, phlogiston disappeared as an entity in the world.

One feature of Lavoisier's book deserves particular notice in the present context. In the opening section of his *Elementary Treatise* Lavoisier is careful to separate heat phenomena from truly chemical phenomena. Thus, for Lavoisier water remains chemically water even though it may change its physical state from ice to liquid water to water vapor. To account for changes in state and other thermal phenomena, Lavoisier introduced a new ether, caloric. Like the other ethers we have encountered caloric was a self-repulsive fluid-like material, much more fine than ordinary matter. Thus, caloric penetrated a block of ice, pushing its particles apart and melting the ice into water, the addition of more caloric transforming the water into water vapor. By introducing caloric Lavoisier normalized chemistry within the intellectual framework laid down by Newton in the *Opticks*.

The Second Scientific Revolution

A "second" Scientific Revolution began to unfold at the turn of the nineteenth century. Two essential trends characterize this pivotal historical transformation in the sciences: the *mathematization* of the previously more qualitative Baconian sciences and the theoretical and conceptual *unification* of the classical and Baconian sciences. That is, previously separate traditions became conjoined in a new scientific synthesis familiar to us today as "physics." As the Second Scientific Revolution unfolded and these processes of mathematization and unification proceeded, a single set of universal laws and a powerfully coherent scientific world picture began to emerge. That world picture, known as the Classical World View, seemed at once to integrate all the domains of the physical sciences and, by the last decades of the nineteenth century, to promise a complete understanding of the physical world and thereby the end of physics itself.

The pattern of mathematization and unification can be seen in many different specialties and areas of research in nineteenth-century science. Developments in electricity and their ramifications for magnetism and chemistry present a compelling example. Through the eighteenth century, the scientific study of electrical phenomena involved static electricity only. The accidental discovery of current electricity opened the door to a whole new area of research. In experimenting with frogs' legs in the 1780s, Italian scientist Luigi Galvani (1737–98) did not set out to extend the range of electrical science but, rather, he sought to investigate the etherial "animal electricity" that seemed to "flow" in an animal's body. His compatriot, Alessandro Volta (1745–1827), built on Galvani's work and in 1800 announced the invention of the pile or battery which could produce this flowing electricity. Volta's battery and the ever larger ones that soon followed manifested profound new connections between electricity and chemistry. The battery—layers of metals and cardboard in salt (later acid) baths—was itself a chemical-based instrument, and so the generation of current electricity was self-evidently associated in fundamental ways with chemistry. More than that, through electrolysis or using a battery to run electricity through chemical solutions, scientists, notably Humphry Davy (1778–1829), soon discovered new chemical elements, such as sodium and potassium, appearing at the poles of the battery. As a result, an electrical theory of chemical combination—that chemical elements were bound by electrical charges—predominated in chemistry during the first decades of the nineteenth century.

These discoveries in electrochemistry generally supported atomistic interpretations that had gained ground since the early nineteenth century. Following Boyle's lead, Lavoisier had been content to describe chemical elements as merely the last product of chemical analysis without saying anything about the constitution—atomic or otherwise—of these elements. In 1803, out of a background in meteorology and pneumatic chemistry, John Dalton (1766–1844) became the first modern scientist to propose chemical atoms—or true indivisibles—in place of the more vague concept of chemical elements. Atomism was not immediately accepted in all quarters, but by midcentury the doctrine had become a fundamental component of contemporary chemistry. In advocating chemical atomism Dalton and his successors established a link with "philosophical" atomism that had been so prominent a feature of the new science of the seventeenth century.

While scientists suspected some unity to electricity and magnetism, only in 1820 did the Danish professor of natural philosophy Hans Christian Oersted (1777–1851) accidentally demonstrate the sought-after link. Repositioning an electric circuit and a compass after a class-room lecture, Oersted discovered that opening and closing the circuit produced a magnetic effect if the wire stood parallel to the compass needle (rather than perpendicular to it, as might have been expected). By showing the magnetic effect of current electricity to be motion, Oersted unveiled the principle later applied to the electric motor. New discoveries followed, including the electromagnet and the attraction and repulsion of current-carrying wires.

This rash of work culminated in 1831 with the discovery of electromagnetic induction (or the creation of electricity by magnetism) by Michael Faraday (1791–1867), the self-educated experimentalist at the Royal Institution in England. Faraday generated an electrical current by plunging a magnet through a closed coil of wire. The significance of Faraday's discovery stemmed only in part from its technical potential in the form of the dynamo (or electric generator) or its being the sought-after analog to Oersted's production of magnetic effects through electricity. On a deeper philosophical level Faraday proved the interconnection of electricity, magnetism, and mechanical motion. After Faraday, given two of these three forces of nature, scientists could readily produce the third.

Faraday's explanations of electromagnetic phenomena, although initially idiosyncratic, proved highly influential in the long run. Lacking mathematical expertise, Faraday conceived the effects of electricity and magnetism visually as mechanical distortions in space. The case of iron filings distributing themselves around a magnet convinced Faraday of the reality of electromagnetic *fields* and "lines of force" emanating from magnets and moving currents. Faraday thus shifted attention from magnets and wires to the surrounding space and initiated field theory. Unmistakable in all this work is an incipient merger of scientific theory and technological applications, as the new science of electricity and novel electrical devices grew hand in hand, but about these points more later.

Complementary developments in optics constitute a central component of the Second Scientific Revolution. In the eighteenth century the authority of Newton was such that his particle theory of light predominated, even though scientists knew of an alternate wave theory espoused by Newton's contemporary, Huygens. In this case Newton's influence was oppressive and not a great deal of work in optics took place in the eighteenth century. That situation changed radically with the work of Thomas Young (1773–1829) and Augustin Fresnel (1788–1827) early in the nineteenth century. Dissatisfied with accounts of diffraction phenomena (or the slight bending of light around the edges of objects) Young proposed a wave interpretation in 1800. He envisioned light as a longitudinal pressure wave, something like sound. In France the young Fresnel upset the established scientific community by proposing that light consisted of transverse waves, like waves in the ocean. Fresnel's interpretation better accounted for the full range of optical

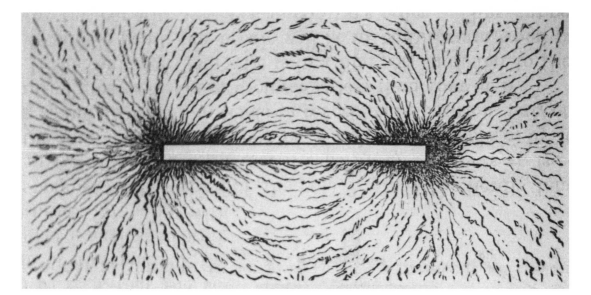

Fig. 14.2. Faraday's lines of force. Michael Faraday postulated that lines of force occur in a field surrounding a magnet. He illustrated this proposition by observing patterns that iron filings form in the space around a magnet.

phenomena, including interference phenomena where one set of waves interacts with a second set. A peculiar consequence of Fresnel's theory soon emerged, that under appropriate experimental conditions a white spot should appear at the center of the shadow cast by a circular disk. In a dramatic experiment Fresnel demonstrated that the predicted phenomenon in fact proved to be the case. He won the prize competition of the Paris Academy of Science in 1819.

With the gradual acceptance of the wave theory of light, old problems, such as polarization, had to be reevaluated, and new research areas opened up, such as determining wavelengths and analyzing the optical spectrum through spectroscopy. (The latter study, incidentally, showed unexpected ties between light and chemistry, with each chemical element emitting a distinctive light spectrum.) The wave theory of light also posed a major theoretical problem: what was the medium in which light waves propagated? The answer emerged that light consisted of waves in an all-pervasive cosmic ether. In the Second Scientific Revolution all the particular subtle fluids of the eighteenth-century Baconian tradition collapsed into this one unitary world ether.

The study of heat added to the conceptual innovations transforming the intellectual landscape of science in the nineteenth century. With his notion of caloric as a material substance Lavoisier had initiated a fruitful line of research in the measurement of heat. In his *Analytical Theory of Heat* (1822) Joseph Fourier (1768–1830) applied the calculus to the investigation of various modes of heat flow, but without pronouncing on the nature of heat. In 1824 the young French theorist Sadi Carnot (1796–1832) published his landmark tract, *Reflections on the Motive Power of Fire*. In this work Carnot analyzed the workings of the steam engine and elaborated what we know as the Carnot cycle, which describes what happens in the cylinders of all heat engines. For us his *Motive Power of Fire* has another telling importance. Carnot's became the first scientific investigation of the steam engine. By the time Carnot wrote, steam engines had been in use for more than 100 years, and the Industrial Revolution, propelled in large part by the application of steam power, was well under way in Europe. The case is precisely the opposite of the cliché of technology as applied science. Carnot's analysis of the steam engine provides the paradigm case, rather, of technology setting the agenda for scientific research.

The most remarkable development in heat studies—and, indeed, in all of the physical sciences in the nineteenth century—was the creation of an entirely new theoretical discipline, thermodynamics, which unified the sciences of heat and motion. In the years leading up to 1847, the recognition arose from a variety of sources that the forces of nature—heat, light, chemistry, electricity, and motion—might not simply interact with each other but might actually be mutually interconvertible. In the 1840s several individuals independently enunciated the first law of thermodynamics, the conservation of energy—the principle that the various forces of nature can change from one form to another and that an indestructible entity called energy is conserved in the transformation. In the steam locomotive, for example, chemical energy stored in coal is released, some of which becomes transformed into heat, light, and mechanical motion that drives the piston which in turn propels the train and heats the rails. What saves the first law of thermodynamics from being merely a metaphysical principle (although metaphysical principles concerning the unity of nature proved instrumental in its formulation) is that, quantitatively, the transmutation of energy from one form to another takes place according to fixed exchange rates. The English experimentalist James Prescott Joule (1818–89) worked out the mechanical equivalent of heat to a high degree of exactitude, with the fall of a standard weight exactly equal to a specific rise in temperature. Building on the work of the pioneers, in a series of fundamental papers in the 1850s and 1860s German physicist Rudolf Clausius (1822–88) formulated the second law of thermodynamics. This law concerns the behavior of energy over time; specifically, it postulates that in a closed, undisturbed system energy peaks and valleys will even out until no temperature differences exist in the system. The second law implies that energy, like water, naturally "runs downhill" and that, without additional work, reactions are not naturally reversible.

Thermodynamics was one of two entirely new scientific disciplines to emerge in the nineteenth century which fundamentally transformed our outlook on the natural world, the other being the theory of evolution (discussed in the next chapter). The concept of energy and the principles of thermodynamics united the physical sciences on a deeper and entirely unprecedented level and provided the basis for the conceptually unified worldview that coalesced at the end of the century.

For the physical sciences, at least, the Classical World View (or Classical Synthesis) of the latter half of the nineteenth century offered a comprehensive vision and a unified intellectual picture of the physical world unmatched historically since the Middle Ages and the heyday of the Aristotelian outlook. The unity of the Classical World View crystallized around the work of James Clerk Maxwell (1831–79). Maxwell mathematicized Faraday's more qualitative notions of the electromagnetic field and gave the world the elegant mathematical expressions that describe the electromagnetic field in the form of wave equations, known as Maxwell's equations. Two aspects of Maxwell's achievement proved instrumental in confirming the Classical World View. Electromagnetic waves possessed a finite velocity, and a constant (c) appearing in Maxwell's equations proved to be identical to the speed of light. When this understanding emerged it seemed to confirm a deep connection between electromagnetism (via Faraday-Maxwell) and optics (via Fresnel). Second, Maxwell's equations seem to imply that under appropriate conditions electromagnetic waves might be generated and transmitted. When Heinrich Hertz (1854–94) proved the existence of these electric waves in 1887–88—we know them as radio waves—Maxwell's equations and an integrated view of electricity, magnetism, light, and radiant heat seemed abundantly confirmed.

Taking its fundamental parameters from Newton and German philosopher Immanuel Kant (1724–1804), the Classical World View thus began with notions of absolute space and time—space being uniform and Euclidean and time flowing inexorably and constantly. The vision that converged around Maxwell's work then posited three entities in the world: matter, a universal ether, and energy. Matter consisted of chemical atoms that had no interior parts, that were individually identical and distinct from one another; in this view all oxygen atoms, for example, were identical and categorically dissimilar from all hydrogen atoms. The Russian chemist D. I. Mendeleev (1834–1907) arranged the elements into a chemical table of families, each atom with its own atomic number and weight. Although atoms could combine chemically, thus giving rise to the wealth of chemical substances evident in the world, the nature of the chemical bond remained obscure. Atoms, molecules, and larger bodies are endowed with mechanical energy and move; the dynamical motion of atoms and molecules determine their degree of heat; for gases, that motion came to be analyzed by the new discipline of statistical mechanics.

The material stuff of the world—its atoms and molecules—also possessed an intrinsic gravitational force that allowed particles to agglomerate into larger and larger bodies. The attracting force of gravity, therefore, provided a bridge between the invisible atomic world and the macroscopic world known to mechanics and astronomy. Scientists no more understood the nature of gravitational force at the end of the nineteenth century than they had in Newton's day, but all experience confirmed such a force. On the cosmic scale moving bodies like the earth, moon, planets, and comets obeyed the laws of classical physics, and in this way the Classical World View incorporated the tradition of the classical sciences initiated by Newton and perfected by two centuries of problem-solving research.

The world ether complemented ordinary matter. This universal ether, as we have seen, provided the substratum for radiation: light, radiant heat, and the electromagnetic field, all of which embodied energy. Regular matter, the ether, and energy were all interconnected and obeyed the laws of thermodynamics. Thus, mechanical, chemical, electrical, magnetic, and light energy can become transformed one into another. The second law of thermodynamics, in particular, by introducing what has been called the "arrow of time," provided a major pillar on which the Classical World View rested. For seventeenth-century mechanics, for example, the laws of impact, say, were perfectly reversible—theoretically billiard shots work the same in one direction or its reverse; in contrast, the second law of thermodynamics posited an irreversible direction to time and the behavior of energy. Although couched in abstract and highly mathematical terms, the second law envisioned the ultimate "heat death" of the universe with all energy equally diffused throughout the cosmos, with atoms and molecules ultimately vibrating with a uniform degree of heat just above absolute zero.

The Classical World View outlined here came together in the 1880s. It represented a powerful, coherent, and mathematically exact understanding of the physical aspects of the cosmos and the interconnectedness of natural phenomena. With the coalescence of the Classical World View, the enterprise of natural philosophy that had begun so long ago in ancient Greece seemed almost to draw to a close, at least as far as the physical sciences were concerned. Yet it would be a mistake to overemphasize the consensus that developed around the Classical World View or how much agreement the elaborate and esoteric claims of contemporary science actually attracted. In point of fact, serious and lively debates arose among scientists and philosophers as to whether the Classical World View represented the true and underlying reality of nature. Furthermore, a series of unexpected discoveries soon dispelled any overconfident sense of closure and set the stage for yet another revolution, the twentieth-century revolution in physics launched by Albert Einstein.

The focus on the physical sciences in the foregoing analysis should not obscure important developments to be noted in the history of the life sciences in the nineteenth century. That Jean-Baptiste Lamarck coined the term *biology* (or science of life) only in 1802 signals how much biology was a characteristically nineteenth-century field, particularly in the coming-into-being of laboratory-based and experimental approaches to investigating the chemistry and physiology of life. Although Robert Hooke had coined the term *cell* in the seventeenth century, cell theory did not emerge until the 1830s when German scientists M. J. Schleiden (1804–81) and Theodor Schwann (1810–82) looked through their microscopes and identified the cell as the basic unit of plant and animal tissue and metabolism. Claude Bernard's *Lessons in Experimental Physiology* (1855) and his *Introduction to the Study of Experimental Medicine* (1865) provided exemplars for the new style of research, out of which emerged the germ theory of disease, articulated by Robert Koch and Louis Pasteur in the 1870s. The germ theory eliminated once and for all the ancient view of humoral pathology as well as competing environmental explanations for the cause of disease. Contemporary experimental biology also took place under changed institutional and professional circumstances.

Science Reorganized, Again

The professionalization of science and careers in science represents a milestone along the route to today's science. Who is a scientist and how does one become a scientist? Historically, individuals who pursued investigations into nature have played many disparate social roles: the priests and anonymous scribes of the first civilizations, Greek natural philosophers, Arabic doctors and astronomers, Chinese mandarins and officials, medieval European university professors, Renaissance artists, engineers, and magicians, and Enlightenment academicians. What needs emphasis here is that the modern social role of the scientist first appeared in the nineteenth century coincident with the Second Scientific Revolution.

A major element involved in the creation of the modern scientist as a recognizable social type was the establishment of a new institutional base for science in the nineteenth century, a second "organizational revolution" comparable to the first that formed part of the initial Scientific Revolution. The mainstays of organized science in the eighteenth century—state-sponsored learned societies—continued in the nineteenth, but less as centers for original research and more as honorary organizations rewarding past scientific achievement. In their place a complementary set of more vital institutions emerged for the practice of science. Established in 1794 in revolutionary France, the École Polytechnique provided a key institutional setting where France's leading scientists taught advanced theory to an exceptional generation that made France the leading scientific nation through the 1830s. (It also became the model for West Point and polytechnic colleges in the United States and elsewhere.)

In England the Royal Institution, founded in 1799, provided a home for such luminaries as Davy and Faraday. A revealing institutional novelty and sign of the times was the creation of hundreds of Mechanics Institutes throughout Great Britain and North America which at their height in the nineteenth century provided instruction in the sciences to over 100,000 artisans and interested middle-class amateurs.

The reform of the German university system represents the most significant manifestation of this new organizational basis for nineteenth-century science. Beginning with the creation of the University of Berlin in 1810, the natural sciences gradually gained a powerful new position within the network of universities in the German-speaking states. Nineteenth-century German universities became secular state institutions, and science instruction fulfilled a service function for the state in helping to train secondary-school teachers, physicians, pharmacists, bureaucrats, and other professionals.

An unprecedented emphasis on scientific research distinguished science education in this new context. That is, the role of a science professor was not merely to transmit old knowledge to students, but to lead the way in the production and dissemination of new knowledge. Several new pedagogical modes emerged to facilitate the new research agenda within universities, including the now-familiar teaching laboratory (inaugurated in 1826 by Justus von Liebig's chemistry lab at Geissen), graduate-level scientific seminars and colloquia, and specialized institutes within universities equipped for advanced research. The formal textbook as a medium for instruction in the sciences emerged for the first time as part of these developments, and the Ph.D. became a requirement for a career in science. The decentralized nature of the German university system spurred competition among the separate German states for scientific talent and thereby raised the level of research in science. The advent of polytechnic schools in later nineteenth-century Germany—the Technische Hochschulen—strengthened these trends. And the importance of science and scientists within the context of German higher education grew further as connections to technology and industry developed in the second half of the nineteenth century, notably in the chemical industry, electrotechnology, and precision optics. The model of the research university soon spread outside of Germany, as the example of the Johns Hopkins University (1876) illustrates.

A characteristic feature of the first Scientific Revolution had been the social and intellectual turning away from the medieval university by the vanguard of science. After two centuries in the background the university once again became the leading institution for the natural sciences as part of the Second Scientific Revolution of the nineteenth century. Even in England, where the universities were slow to respond to the upsurge in scientific research, by the third quarter of the century the two oldest universities, Oxford and Cambridge, along with newly founded ones in London (1826) and elsewhere in Great Britain, established new scientific professorships and sponsored research, some of it in fields bordering on technology and industry. None of these developments, however, changed the essentially male-dominated character of nineteenth-century science; as in the past, a mere handful of women became directly involved in science, usually exercising subsidiary functions. Some nontrivial exceptions include the American astronomer Maria Mitchell (1818–89), and the Russian mathematician Sonya Kovalevsky (1850–91); the latter received a Ph.D. from the University of Göttingen in 1874.

The professionalization of science also came to entail specialized institutions. While the traditional learned societies generally represented all of the sciences, in the nineteenth century new organizations devoted to single disciplines gradually supplanted the former as the locus of expert interests. England led the way in new institutions of this sort with the Linnaean Society (1788), the Geological Society of London (1807), the Zoological Society of London (1826), the Royal Astronomical Society (1831), and the Chemical Society of London (1841). New publication patterns also emerged, wherein specialized journals competed with the general journals of the traditional scientific societies as the loci of original publication in the sciences. Among the noted early specialized journals one might mention Lorenz Crell's *Chemische Journal* (1778), *Curtis's Botanical Magazine* (1787), the *Annales de Chemie* (1789), and the *Annalen der Physik* (1790), with similar specialized journals proliferating as the nineteenth century progressed. Finally in this connection, the nineteenth century saw the appearance of truly professional societies representing the interests of scientists, including the Association of German Researchers (the Deutsche Naturforscher Versammlung from 1822), the British Association for the Advancement of Science (from 1831), and the American Association for the Advancement of Science (1847).

That the English word *scientist* was coined in 1840 is powerful testimony to the profound social changes surrounding science and scientific investigators at that time. Obviously, science and its practitioners had been part of the learned world at least since the dawn of civilization in ancient Mesopotamia. Yet it is telling of the changed circumstances of organized science in the nineteenth century that only then did the "scientist" emerge full blown as a social and professional entity.

Applying Science in Industry

Science and industry and the cultures of science and technology generally began their historical merger in the nineteenth century. The main thesis of this book has concerned the historically limited degree of applied science prior to the nineteenth century. In the earliest civilizations and state-level societies thereafter, governments patronized useful knowledge and applied science in the service of administration. In Europe, state support for what were deemed useful sciences appeared slowly after the Middle Ages, as noted in the case of cartography, for example, or somewhat later as an outgrowth of the Scientific Revolution in the creation of state scientific societies. The conviction that natural philosophy ought to be turned to public utility became an ideological commonplace in the seventeenth century. More readily apparent, however, are the intellectual and sociological disjunctions between the sciences and the vast body of technology as these have developed throughout history. In the Industrial Revolution in eighteenth-century England the worlds of science and technology drew closer together, but we were hard-pressed to find historical evidence to support the view of contemporary technology as applied science. In the nineteenth century, however, several important novelties appeared that began to recast the age-old separation of science and technology that originated in Hellenic Greece. Firm connections came to link theoretical science and industry. To be sure, much of science and technology remained separate, but the new dimensions of applied science that arose in the nineteenth century in the context of industrialization represent historical departures of great consequence that solidified in the twentieth century on a global scale.

The new nineteenth-century science of current electricity spawned several new applied-science industries, of which the telegraph represents a prime example. Following the discovery of electromagnetic induction by Michael Faraday in 1831, the scientist Charles Wheatstone and a collaborator invented the first electric telegraph in 1837. Wheatstone and other European and American scientists and inventors worked to create a telegraphic industry, spurred in part by the utility of the telegraph as an adjunct to railroad development. These efforts quickly culminated in the system patented by Samuel F. B. Morse in 1837 and field-tested in 1844, incorporating Morse's renowned alphabetic code using dots and dashes. London and Paris became connected by telegraph in 1854, the first trans-Atlantic telegraph cable was laid in 1857–58, the first transcontinental telegraph in North America linked New York and San Francisco in 1861, and the telegraph and the railroad spread together worldwide thereafter. The upshot was something of a communications revolution.

Even though the telegraph tapped a body of preexisting scientific knowledge, the development of the new technology of telegraphy involved the solution of a myriad of problems—technical, commercial, and social—that had little or nothing to do with contemporary scientific research or theory. In other words, the coming into being of a science-based technology usually involves the creation of a complex technological system to the point where it is misleading to think of such systems as merely "applied science."

Much the same conclusion is evident in considering the electric lighting industry that arose in the last quarter of the nineteenth century. Such an industry clearly derived from prior work in the new science of electricity. As celebrated in traditional biographies of great inventors, Thomas Alva Edison (1847–1931) in New Jersey and Joseph Swan (1828–1914) in England independently created the incandescent light bulb through elaborate empirical trials in 1879. By the 1880s and 1890s the science involved in the developing electric lighting industry was hardly new. Some of it, with regard to insulators, for example, harked back to work in the eighteenth century. This example brings home that in considering applied science it can be analytically fruitful to distinguish between whether the science involved is, say, "boiled down" science or whether it represents the application of more up-to-date and cutting-edge theory. Furthermore, the light bulb itself hardly constitutes the establishment of a practical electric lighting industry. Again, a large and complex technological system had to be brought into being before an electric lighting industry could be said to have existed, a system involving generators, power lines to distribute electricity, appliances, a device to measure consumption, and methods of billing customers, to name only a few of its many elements.

Another of the early instances where science and up-to-date scientific theory became applied to technology and industry was the case of radio communications, where practical application followed closely on the heels of theoretical innovation. Seeking to confirm Maxwell's theory of electromagnetism, Heinrich Hertz demonstrated the reality of electromagnetic waves in 1887. Hertz worked exclusively within the abstract tradition of nineteenth-century theoretical and experimental physics, but when the young Italian Guglielmo Marconi (1874–1937) first learned of Hertzian waves in 1894, he immediately began to exploit them for a practical wireless telegraphy, and by the following year he produced a technology that could communicate over a mile in distance. Marconi, who went on to build larger and more powerful systems, received his first patent in England in 1896 and formed a company to exploit his inventions commercially. In 1899 he sent his first signal across the English Channel, and in a historic demonstration in 1901 he succeeded with the first radio transmission across the Atlantic. The creation of this new technology involved much more than the application of scientific theory, however direct, and, although Marconi's contribution was essentially technical, in this instance the line between science and technology became so blurred that in 1909 Marconi received a Nobel prize in *physics* for his work on telegraphy. The case is also noteworthy because it illustrates that the outcome of scientific research and technological change cannot be foreseen. What drove Marconi and his research was the dream of ship-to-shore communications. He had no prior notion of what we know as radio or the incredible social ramifications that followed the first commercial radio broadcasts in the 1920s.

The growth of applied science in the nineteenth century was not limited to physics or industries connected solely to the physical sciences. The emerging germ theory of disease, for example, and ideas about *microbes* in the 1850s led the great French scientist Louis Pasteur (1822–95) to his studies of fermentation. The resulting process of pasteurization produced practical and economically important conse-

Fig. 14.3. Invention without theory. Thomas Edison was a prolific inventor (with more than 1,000 patents) who had little education or theoretical knowledge. His ability to tap "boiled down" science and his organizational skills in institutionalizing invention in his laboratory in Menlo Park, New Jersey, proved essential to his success.

quences for a variety of industries, including dairy, wine, vinegar, and beer production. Related work on silkworm diseases produced like effects for the silk industry, and Pasteur's later medical experiments to develop inoculations against anthrax, rabies, and other diseases represent the advent of a truly scientific medicine.

Chemistry proved yet another domain where important practical applications in nineteenth-century industry were forthcoming from science. Through the middle of the century the dye industry in Europe remained a traditional craft activity with no contact whatsoever with the world of science. Then, in 1856, out of the background of German advances in organic chemistry, the English chemist William Perkin discovered an artificial dye producing a purple color. The economic value of bright, synthetic dyes became immediately apparent, and mastery of the chemistry of dyes and dyestuffs derived from coal tar became essential to the textile industries. Until the various German states adopted a uniform patent code in 1876 competition between firms amounted largely to raiding one another's expert chemists. After 1876, however, with patent rights assured, the emphasis shifted to research and development of new dyes. A new institution uniting science and technology—the industrial research laboratory—emerged as a result. The Friedrich Bayer Company created a research division and hired its first Ph.D. chemist in 1874. In 1896 the number of its salaried staff scientists reached 104.

The style of applied research undertaken at the Bayer research laboratory deserves emphasis. The story hinges on the fact that the German chemical industry established close contacts with research universities. Industry supplied universities not only with materials and equipment for advanced studies in chemistry, but also with students and with opportunities for consulting. Universities reciprocally offered industry trained graduates and prospects for scientific cooperation. In a further division of labor, fundamental research became the province of the universities, while industry undertook mainly empirical research and routine experiments to test dyes and their fastness on different materials in the hope of developing useful products. For example, in 1896 the Bayer Company subjected 2,378 colors to an assortment of tests but marketed only 37. As this case indicates, even where theory was applicable, "research" often still took the form of trial and error, albeit conducted by scientists in a research environment. The reality of applied science was (and is) often far removed from the assumption that technology is merely the translation of scientific theory into practice.

The model of the research laboratory spread widely in late nineteenth- and early twentieth-century industry. Thomas Edison's laboratory at Menlo Park, New Jersey, established in 1876, represents an early example. Others include Standard Oil (1880), General Electric (1901), DuPont (1902), Parke-Davis (1902), Corning Glass (1908), Bell Labs (1911), Eastman Kodak (1913), and General Motors (1919). Today thousands of such labs operate in the United States alone. The advent of industrial research has been hailed as the "invention of invention," although this view is somewhat misleading in that research labs are generally not innovators of new technologies. For the most part research labs concern themselves with the development and extension of existing technologies, and they often function as part of a business strategy to develop and control patents to ward off competitors. Even today, as we will see further, pathbreaking inventions, like xerography and the personal computer, may still be the work of independent inventors rather than established scientists or engineers working in industry.

CHAPTER 15

Life Itself

Science usually changes incrementally, and the increments are usually small. A new chemical element might be discovered now and then; a new fossil might be found, or a new star. But these incremental discoveries leave the theoretical framework of a science undamaged. Indeed, they generally strengthen it. But occasionally a major upheaval occurs, and when the dust settles a new theoretical framework has arisen in place of the old. These upheavals—scientific revolutions—not only replace old and long-held ideas with sharply revised concepts, they also modify research boundaries and create new problems for research which the old science could not even formulate. The Copernican Revolution was one such revolution that unfolded in the sixteenth and seventeenth centuries; in the nineteenth and twentieth the Darwinian Revolution similarly refashioned the intellectual landscape of science.

Charles Darwin published *The Origin of Species* in 1859. The year and the book represent another watershed in the history of European science. On the one side of the Darwinian divide stood what might be labeled the traditional Christian worldview. This view, sanctioned by biblical authority and apparently confirmed by both folk wisdom and scientific observations, held that the species of plants and animals were created separately and remain fixed—like producing like. This worldview was largely static, not admitting of significant change. It included the notions that each distinct species was separately created at a not-very-distant point in time, possibly only 6,000 years ago; that various catastrophes, notably Noah's flood, explain our observed geological and biological surroundings; and that humans occupy a special place in the cosmos, one created by God who is the architect of all the apparent design in the world and who plays an active, divine role in its historical unfolding.

On the other side of the landmark year of 1859 radically contrary views, springing from Darwin's seminal ideas, gained ground to the effect that the species are not fixed, that there were no separate creations, that the life forms we observe around us evolved through the process of natural selection, that biological evolution and geological change have been unfolding gradually over aeons of time, that humankind is no more than an artifact of the history of nature, and that the study of nature offers no evidence of miracles or any divine plan.

The Copernican and Darwinian Revolutions display parallel characteristics. The Copernican Revolution departed from a set of astronomical beliefs that in their essential form had been held for 2,000 years, and because of their seemingly self-evident truth—the stability of the earth and the motion of the sun—had been taken for granted in both astronomy and religion. The Darwinian Revolution departed from the age-old belief, also preserved in the biblical tradition, of the fixity of the species. Both Copernicus and Darwin held back publication of their new ideas, not out of fear of religious or political authorities, but rather out of fear of ridicule for far-fetched theories that in their day they could not prove. And both of them were vindicated by subsequent research. The result of these two revolutions is the scientific worldview in which heaven and earth obey the same physical laws while man and beast share the same biological origins.

Natural Theology and the Background to Darwin

Consistent with Galileo's advice, a consensus now reigns in scientific and most theological circles that in the study of nature biblical authority must defer to scientific investigation. Historians of science long ago dismissed the idea of an innate and abiding conflict between science and religion. Indeed, one outcome of the Scientific Revolution of the seventeenth century was precisely to strengthen ties between science and the traditional Christian worldview. Natural theology or the idea that one can gain insight into the divine plan by examining God's handiwork in nature gained significant ground, particularly in England. The conviction was that, by investigating apparent design in nature, we could better understand the Great Designer and his providential provisions for humans and their needs. In other words, as evidenced by such tomes as John Ray's natural history, *Wisdom of God in the Creation,* and Thomas Burnet's geology, *Sacred History of the Earth* (both 1691), seventeenth-century religious sensibilities stimulated scientific research toward a harmonious reconciliation of genesis and geology.

Empirical research in botany, natural history, and geology exploded in the eighteenth century, and by 1800 scientists knew much more about the world than a century earlier. Yet the strand of natural theology that sought to find God in His handiwork remained no less an element, particularly of English science at the turn of the nineteenth century, than it had been in the seventeenth. The first edition of William Paley's argument that design requires a Designer, *Natural Theology, or*

Evidences of the Existence and Attributes of the Deity Collected from the Appearances of Nature, appeared in 1802, and it brought up to date for a new generation of Englishmen, including the young Charles Darwin, the idea that the natural sciences and received religion were two faces of the same coin. The argument, embraced by the young Darwin, was that just as a pocketwatch found by the roadside implies the existence of a watchmaker, a beetle or a butterfly, which are infinitely more complex and purposeful, imply a Creator. The tradition of natural theology continued strongly in England into the 1830s when the earl of Bridgewater commissioned the so-called Bridgewater Treatises, eight scientifically based studies designed to show the "power, wisdom and goodness of God in the works of creation."

In the eighteenth century, Swedish botanist Carolus Linnaeus (1707–78) had brought order to the explosion of knowledge in the organic realm—what he thought of as the plan of God—by arranging plants and animals in the "binomial" system of classification still in use today. (Each plant or animal receives two Latin names, the first, the *genus*, places it in a general category, while the second, the *species*, refers to its specific characteristics.) Although his strict classifications seemed to confirm the fixity of species, late in life Linnaeus began to wonder about the distinction between species and varieties within species and hinted vaguely that both might be, in his words, "daughters of time."

Just as suggestions that the planetary system may be sun-centered predated Copernicus, a series of precursors to Darwin circled around the principle of the transformation of species. French naturalist and superintendent of the royal gardens in Paris, the Count de Buffon (1707–88) held that the species evolved. Buffon believed not in progressive evolution but that the plants and animals we observe around us today devolved or degenerated from more robust ancestors that existed at earlier eras in time. He never provided a mechanism to explain how such transformations were possible, however, and he later recanted his belief when confronted with religious criticism as to the antiquity of the earth.

Another French naturalist, Jean-Baptiste Lamarck (1744–1829), went further. He postulated a *mechanism* whereby evolution could occur—what has become known as the inheritance of acquired characteristics. Essentially, Lamarck proposed that an individual organism modifies itself by striving to adapt to its changing environment. Moreover, the modifications are supposedly inherited by the organism's offspring, establishing a line of descent that from time to time produces new species. It was an appealing and influential idea, and Darwin himself found refuge in it when attempting to account for the source of variation within species. Lamarckian inheritance of acquired characteristics became empirically discredited but survived into the twentieth century where it intermittently served as an anchor for those who cling to a hope that evolution contains an element of guidance or purpose, or at least of direct environmental influence and is not merely the result of random events.

Around the time that Lamarck formulated his ideas, Charles Darwin's own grandfather, Erasmus Darwin (1731–1802), put forth similar notions in a series of scientific poems. The elder Darwin believed that useful traits are passed on through biological inheritance and their slow accumulation results in diverse forms. To start the process life emerged from inanimate matter:

Hence without parents, by spontaneous birth,
Rise the first specks of animated earth.

Some of these new ideas were consistent with old ones. It had long been assumed that the natural world formed what has been termed the Great Chain of Being from the simplest globule of life to the highest and most divine form, all in accordance with a law of continuity that stated that the links in the chain are infinitely small and there are no jumps or gaps between them. Both the prevailing theory of separate creation and the new theories of evolution embraced the idea of an organic continuum unbroken by any missing links. But, apart from a few such points of intersection, the two approaches necessarily diverged in the face of the evidence of natural history. With regard to another telling point—the age of the earth—traditional belief was content to accept either the biblical chronology or at most a modest extension backward in time. But the new view seemed to call for a much greater antiquity of the earth since the postulated evolutionary changes were necessarily small and incremental and required long stretches of time to produce the diversity that the natural world displays. Only slowly did some naturalists by the end of the eighteenth century become convinced that the earth is substantially older than the few thousand years granted by biblical authority and the story of Noah's ark and the flood.

On the other hand, the accumulating evidence from the fossil record and geological strata proved inconsistent with traditional beliefs about the age of the earth. Some fossil impressions were embedded in what appeared to be extremely old rocks. And some were found in puzzling strata; for example, marine fossils found at high elevations away from any body of water suggested great upheavals, evidently over long periods of time, in the earth's surface. Traditionally fossils were thought to be geological accidents or "sports of nature" and not the truly fossilized remains of organic beings. Around 1800, however, the discovery and recognition of large vertebrate fossils ("elephant bones") dramatically changed the debate. Based on the science of comparative anatomy of living creatures, experts could reassemble newly discovered fossil remains. The stunning revelation of the reality of biological extinction and a lost world of prehistory emerged with shocking clarity. By 1830 there was no denying that strange large creatures, now extinct, once roamed the earth.

Squaring the mounting new evidence with biblical convictions and traditional chronology tested the imaginations of many naturalists. A conservative theoretical synthesis known as *catastrophism* coalesced around the work of the eminent French naturalist, Baron Georges Cuvier (1769–1832), the successor to Buffon at the Museum of Natural History in Paris. Cuvier concluded that seemingly great changes occurred through a few catastrophic events spaced over a relatively brief past. The apparently sudden breaks observed in the geological and fossil records could then be accounted for by this series of catastrophes (floods, fires, volcanoes). Furthermore, the apparent sequence of life forms shown in the fossil record (of reptiles and fish followed by birds and mammals, of extinct forms preceding living ones) would seem naturally to point to the coming of Man. In this manner the Cuvierian synthesis preserved the (relatively) brief history of the world required in the traditional view. It also admitted the reality of extinction and of progressive biological change over time, but without the necessity of any organic transformations or violation of the principle of the fixity of species.

At the end of the eighteenth and beginning of the nineteenth centuries, "catastrophist" doctrines in geology were challenged by what became known as *uniformitarianism*—the claim that the physical processes now operating on the surface of the earth have been operating uniformly, rather than catastrophically, over long periods of time and have slowly produced the changes that the geological record reveals. In 1795 Scottish geologist James Hutton (1726–97) published his landmark work, *Theory of the Earth,* in which he attributed the geological features of the earth to the combined actions of two opposing forces— a leveling tendency induced by gravitation and a lifting tendency produced by the heat within the earth. These forces, which as we observe them at present produce their effects slowly, would have required immense stretches of time to account for present geological conditions. They would have acted uniformly, as they evidently do today, and would have produced a world with, as Hutton put it in his majestic aphorism, "no vestige of a beginning, no prospect of an end." No rational hope existed for reconciling the limitless time span of uniformitarian geology with the traditional biblical account.

In his *Principles of Geology* (three volumes, 1830–33), Charles Lyell (1797–1875) revived and strengthened Hutton's uniformitarian argument that the physical features of the earth result from processes we observe today working slowly and continuously over long periods of time, in opposition to catastrophism which granted little time to geological processes even though it recognized extinction and biological change. While uniformitarian geology admitted infinities of time, it

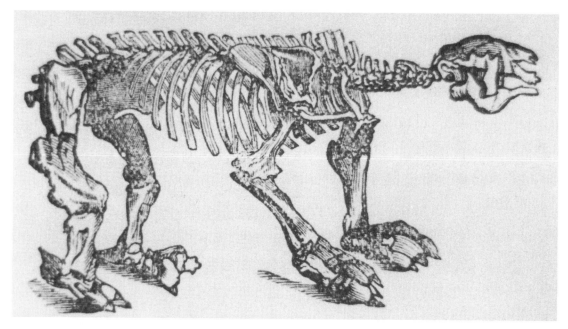

Fig. 15.1. The Megatherium. In the first decades of the nineteenth century the discovery and reconstruction of the fossils of extinct large animals revealed a "lost world of prehistory" and heightened the problem of how to account for apparent biological change over time.

hesitated over the issue of biological change. Lyell himself, the most eminent geologist in Britain and later a partial convert and close friend of Darwin's, remained adamant in denying the possibility of the transformation of species: "There are fixed limits beyond which the descendants from common parents can never deviate from a certain type." In this intellectual environment the question of how to account for biological change on uniformitarian principles awaited its prophet.

Darwin

Charles Darwin (1809–82) was born into a well-to-do family of rural English gentility. His father, like his grandfather Erasmus, was a successful country physician. His mother was a sister of Josiah Wedgwood, the potter-entrepreneur, but she died when Charles was only eight years old, and Darwin's older sisters largely saw to his upbringing. In his privileged youth Darwin showed little passion for scholarship, although something of the future naturalist might be read into his love of the outdoors, of dogs, and especially in his fascination with beetles. His father demeaned the young Darwin, saying, "You care for nothing but shooting, dogs, & rat catching, & you will be a disgrace to yourself and all your family."

Darwin entered the University of Edinburgh to study medicine but, sickened by the sight of blood, he soon disappointed his father by dropping out. He then transferred to Christ's College of the University of Cambridge with the intention of studying to become an Anglican clergyman. Darwin was a middling student who enjoyed cards and hunting, but his naturalist's interests led him to make contact with the eminent science professors at Cambridge, the botanist John Henslow (1796–1861) and the geologist Adam Sedgwick (1785–1873). Darwin received his degree in 1831 at age 22, still with a vague intention to become a clergyman. Indeed, that lack of firm commitment settled his fate, for shortly after graduation he was presented with the opportunity to be the naturalist on board the HMS *Beagle* on a projected two-year mapping project in South America.

In the end, the voyage lasted five years from September 1831 to October 1836, and it transformed both Darwin and ultimately the world of science. Darwin proved to be an acute observer, an avid collector of plant and animal specimens, and a proficient writer of letters to his former teacher, John Henslow, who transmitted them to the larger community of English naturalists.

Darwin's mind was prepared for his conversion by the doctrine of uniformitarianism which he received from Lyell's *Principles of Geology*, the first volume of which he carried on the *Beagle* and later volumes of which he received en route. In the face of his Christian roots and his respect for Lyell's teaching, only with the greatest intellectual effort did Darwin eventually arrive at his belief that as the physical features of the earth changed, plants and animals well-adapted at one point in time became ill-adapted—unless they, too, became modified in step with the geological sequence.

Everywhere the *Beagle* landed Darwin saw evidence of change. Traveling along the coast of South America and across the pampas, he noted biological change over space as related species of birds abruptly gave way to others. Discovering the fossils of extinct giant armadillos where their miniature descendants scurried about in his day, he confronted the question of biological change over time. But nowhere was the puzzle of biological diversity and change more pronounced for Darwin than on oceanic islands. Two weeks after leaving England the *Beagle* arrived at the Cape Verde Islands off the west coast of Africa where Darwin observed animals similar but not identical to animals on the nearby mainland. And later, when he explored the Galapagos Islands off the west coast of South America he recorded a strikingly similar pattern—animals similar to those on the coast of Ecuador but not quite the same. Since the Cape Verde and Galapagos Islands seem environmentally similar, why were their animal populations similar to their respective mainlands rather than to each other? He also observed a variety of birds on the Galapagos Islands—since known as "Darwin's finches"—and he noted their different features island to island. But Darwin was not yet a Darwinist. What he later recognized as separate species, on the voyage of the *Beagle* he generally took to be merely varieties. Only once at this stage did he speculate that if varieties diverged sufficiently it "would undermine the stability of Species." And only after he returned to England and consulted with zoologists about the specimens that he had collected did it dawn on Darwin that these pat-

GENERAL CHART shewing the PRINCIPAL TRACKS of H M S BEAGLE _ 1831-6.

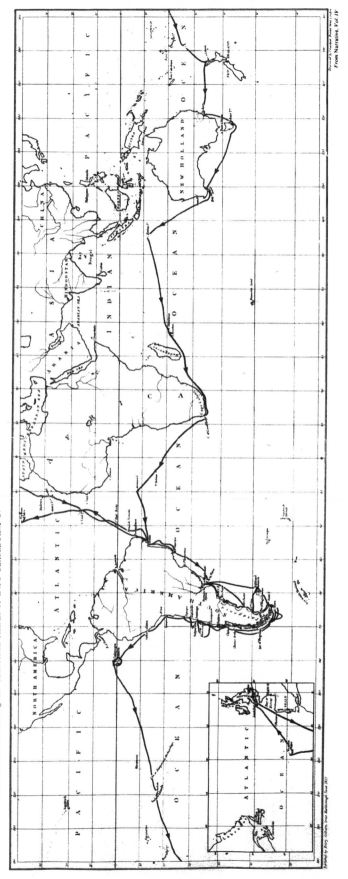

terns he observed on the Galapagos of minor variations in restricted geographical locales could not be explained by the doctrine of divine creation, but rather could only be accounted for through some sort of evolutionary process.

Poor Darwin was seasick the whole time the *Beagle* was at sea. After his return he did not travel much and never left England again. In 1837, a few months after the *Beagle*'s circumnavigation of the earth, Darwin began a notebook on "the species question." He had become convinced of the fact of biological evolution, but he still groped for a principle that would drive the process. He found it in the writing of an English clergyman.

In his *Essay on the Principle of Population* (1798/1803) the Reverend Thomas Robert Malthus reached the conclusion that "population, when unchecked, increases in a geometrical ratio. Subsistence increases only in arithmetical ratio. A slight acquaintance with numbers will shew the immensity of the first power in comparison with the second." In other words, as population necessarily outpaces food supplies, competition for resources becomes increasingly fierce, a competition in which the swifter, stronger, hardier, and more cunning are most likely to survive and reproduce. That insight into the pressure of population against resources, resulting in the "warring of species," gave Darwin the decisive piece of the puzzle. Malthus had applied it to social change; Darwin redirected it to plants and animals. He now possessed, in 1838, the theory of evolution by natural selection: despite high rates of reproduction, populations of species are kept in check by high mortality as individuals compete for resources, and those best adapted to the habitat are more likely to survive and produce offspring. That process, he believed, would account for the patterns of variation he had observed on the voyage of the *Beagle* and which he was already convinced could only be explained by descent from a common ancestor.

Darwin was fully aware of the importance of his discovery, and like all scientists he had every incentive to publish in order to secure his priority. In 1842 he wrote a short sketch of his theory, which he expanded in 1844 into a substantial manuscript of 231 folio pages. He left a letter with instructions that the manuscript should be published if he died prematurely. But he still held back from announcing the great work publicly. Although he was convinced in his own mind, he knew that he could not prove definitively that one species can be transformed into another in accordance with his theory. From his original conception in 1838 it would be more than 20 years before Darwin published his thoughts. Just as Copernicus had refrained from publishing out of fear of being "hissed off the stage" for a theory that he could not prove, Darwin remained reluctant to publicize a theory that was at least as unsubstantiated and even more provocative.

Over the next years Darwin collected his thoughts and planned the great work that would substantiate his theory. He settled first in Lon-

Map 15.1. The voyage of the *Beagle*. Charles Darwin sailed as naturalist aboard the HMS *Beagle* on its five-year voyage to map the coasts of South America (1831–36). A keen observer, Darwin became exquisitely sensitive to the distribution of plants and animals, particularly on the Galapagos Islands off the coast of Ecuador which he visited toward the end of his voyage. He returned to England convinced of the fact of evolution but struggled to articulate a theory or mechanism to account for evolutionary change. *(opposite)*

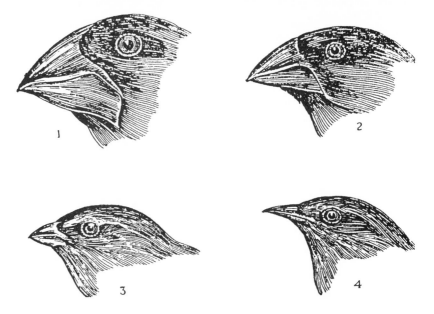

Fig. 15.2. Darwin's finches. The variations in beak formation of the birds that Darwin collected on the Galapagos Islands were later interpreted in terms of evolution from an ancestral stock that arrived on the islands from the west coast of South America. Darwin saw that different finches with different styles of beak survived better in different ecological niches.

don and began to make his way around the Linnaean Society and the Geological Society of London as a junior member of the growing British scientific establishment. In 1839 he married his cousin, Emma Wedgwood Darwin, with whom he was to have ten children. He hated London, calling it an "odious dirty smokey town," and in 1842 he purchased an 18-acre estate in Down, Kent, a convenient 16 miles from London. There, with his independent means, he settled into the life of a Victorian country squire. Gradually over the succeeding years, Darwin went from a diffident Christian to an outright scientific materialist who attributed all being and all changes in the natural world to the action of natural processes.

Soon after his return to England Darwin began to suffer from the debilitating ailments that plagued him the rest of his life. Symptoms included chronic fatigue, violent gastrointestinal disruptions, and assorted dermatological ailments. Scholars still debate the causes of his medical condition: perhaps he caught a blood disease in South America; very likely he became psychologically crippled by panic attacks caused by the knowledge that he alone possessed the secret of organic evolution, a secret so much at odds with the prevailing Victorian culture he represented. In any event, Darwin remained frequently "unwell" for days, and when he did work at his science at Down it was only for a few hours in the morning.

Nevertheless, over a span of more than 40 years, undistracted by any teaching duties, his scientific output was prodigious. During the 1840s and early 1850s Darwin published more than half a dozen books and became recognized as one of Britain's most eminent naturalists. He spent years in painstaking research on barnacles, in large measure to establish his scientific credentials as a rigorous systematist and not merely a field naturalist.

Darwin received a cautionary lesson in the highly negative reception of Robert Chambers's *Vestiges of Creation*, which first appeared in 1844. Published anonymously it popularized the idea of transmutation of species as an innate part of design and God's plan, what Chambers called the "law of development." The book was very poorly received, and was severely criticized on scientific, philosophical, and religious grounds. The continuing debate brought about by Chambers's replies in subsequent editions of the *Vestiges* did little to promote serious scientific consideration of evolution and the species question. It also did nothing to expedite Darwin's own publication. He continued to mull over his "discovery," and he hoped eventually to produce a massive work that would carry the argument by the sheer weight of factual evidence. But in 1858 Darwin's reticence was cut short by an event every scientist dreads: he had been scooped. A letter arrived from the South Seas, the celebrated letter from Wallace, showing that Darwin's ideas had been arrived at independently.

Alfred Russel Wallace (1823–1913) was, like Darwin, a scientific maverick. He received only an elementary education and became a naturalist through self-study, eventually producing a body of research and publications which, even without his independent discovery of the principle of natural selection, would have established him as one of the most eminent naturalists of the nineteenth century. His spectacular collections of specimens were matched, almost from the beginning of his investigations, by a theoretical curiosity about the origin of species. By the 1840s, when he traveled to South America to explore regions of the Amazon basin, he was already convinced that species develop naturally and not by the intervention of a divine power, although he, like Darwin, did not initially possess any mechanism for evolutionary change. Ten years later he explored the Malay Archipelago, and there he arrived independently at the principle of natural selection as the driving force of evolution.

Even before he shocked Darwin with his "bombshell" of 1858 Wallace had begun to publish his views on evolution, and Darwin, in response, began a long version of what became *The Origin of Species*. When Wallace's letter describing his thoughts on evolution arrived in June of 1858 Darwin feared that he might lose credit for the theory of evolution, and he allowed his friends Charles Lyell and Joseph Hooker (1817–1911) to arrange a simultaneous announcement at the Linnaean Society through which Darwin and Wallace were recognized as codiscoverers of the theory. Yet, perhaps because of Darwin's superior social and scientific status, Wallace still receives only token credit for his independent discovery of natural selection. Both men remained on amicable terms until Darwin's death and often exchanged views on evolutionary theory. While Darwin remained a strict selectionist, Wallace retreated on the question of human origins and adopted a spiritualist account in which a divine power, a "Higher Intelligence," played a role.

The Origin of Species

Darwin could no longer linger over a definitive work that would clinch the argument for evolution by natural selection. Within 18 months he produced an "abstract" which he titled *On the Origin of Species by Means of Natural Selection, or the Preservation of Favoured Races in the Struggle for Life*. The *Origin* is, as Darwin put it, "one long argument," and he marshaled an impressive mass of evidence to demonstrate the greater plausibility of evolution over special creation. In the fourteen chapters of the original edition Darwin's argument unfolds in three phases, something like three acts of a highly dramatic play.

In the first act, he outlined his model for evolution. Since he could provide no direct evidence for what he knew was a highly controversial theory, he began by introducing a grand analogy in discussing the achievements of plant and animal breeders in producing domesticated varieties by means of artificial selection. He wanted to suggest that the great variety among domesticated dogs and pigeons, for example, warrants the view that natural varieties may, over much longer periods, become sufficiently extreme to result in separate species and not merely races of a single species.

Darwin then presented the fundamental elements of his own model: variation, struggle for existence, and natural selection leading to species change. His guiding principles were that individuals of each species will display variations for each characteristic, that Malthusian population pressures inevitably lead to competition and a struggle for existence, and that through a process of "natural selection" the best-equipped individuals will reproduce at a higher rate, thus diverging from the ancestral stock and in the long run producing species change. Compared to what humans have wrought by artificial breeding over a comparatively short period of time, how much greater will have been the effects of the "struggle for existence" acting through natural selection over "whole geological periods"?

In the middle chapters of the *Origin*—the second act of this great intellectual drama—Darwin raised what he called "difficulties" with the theory. He had become his own best critic, and he set out in advance to explain away what seemed to be the gravest problems besetting his views. Through small incremental changes, how could he account for organs of extreme perfection, such as the eye of an eagle, without the notions of design or Providence? How can natural selection explain the existence of animals with extraordinary habits or instincts, such as the cuckoo bird, which lays its eggs in other birds' nests, the honey bee with its exquisite hive-making abilities, ant colonies with sterile castes, or ant species that capture slaves? In each instance Darwin strained to show that the incremental addition of small variations could ultimately account for the extraordinary behaviors and traits observed in nature. With regard to the geological and fossil records, which in Darwin's day showed little evidence of the transitional forms between species that the theory required, Darwin reasoned that the geological record was simply imperfect and preserved only random samples from the vast history of life on Earth.

In the final chapters—the last act of the play—Darwin turned the tables and raised issues explained only with great difficulty by special creation yet readily accounted for by his theory of evolution. Why does the fossil record, even though imperfect, provide evidence of extinction and species change over time? How can we most reasonably account for the geographical distribution of plants and animals? How, Darwin asks, should taxonomists classify the obviously related varieties, species, and orders of living forms observed in nature, except through notions of descent with modification? How are the striking similarities of embryos of wildly divergent species to be explained except by evolutionary principles? Darwin also cited the many known cases of useless organs—"rudimentary or atrophied organs . . . imperfect and useless"—such as "the stump of a tail in tailless breeds." For the creationist doctrine rudimentary traits present an unbending difficulty. What deity would produce a kingdom of animals with sets of useless organs? For the evolutionist such traits actually strengthen the hypothesis—they represent "records of a former state of things" as organisms become transformed from one species to another:

> On the view of descent with modification, we may conclude that the existence of organs in a rudimentary, imperfect, and useless condition, or quite aborted, far from presenting a strange difficulty, as they assuredly do on the old doctrine of creation, might even have been anticipated in accordance with the views here explained.

For nearly 500 pages Darwin piled detail upon detail and argument upon argument, creating a preponderance of evidence for his theory of evolution. *The Origin of Species* was an immediate bestseller among scientists and educated amateurs alike, and, in the end, Darwin left his readers with a haunting sense that the ground had shifted, that the biblical account of the origin of species was implausible and would be replaced by a naturalistic account.

Through almost the entire book Darwin refrained from implicating humankind in his great revision, but in the third paragraph from the end of his long monograph he opened the debate on humanity's place in nature with perhaps the most fateful statement in the history of science: "Light will be thrown on the origin of man and his history." Darwin hoped that others would develop this line of inquiry, but, dissatisfied with what he saw as the tendency, even among his supporters, to envision humankind in a different light, Darwin himself took up the cudgels in his *The Descent of Man* (1871). In this work he made explicit the human connection to an ape-like ancestor, even insisting that evolution had not only produced humankind's physical features but also played a role in the development of instinct, behavior, intellect, emotions, and morality.

The degree to which Darwin's theory has been confirmed in our day testifies to the power of his theoretical sagacity. At the time he wrote nothing was known about genetics, or, in general, about the biochemistry of inheritance. No convincing evidence existed that the earth was more than a few hundred thousand years old. And, since no human-like fossils had yet been found (the bones of Neanderthal man were at first assumed to be those of a modern European) there seemed to be a yawning gulf between humans and apes. Over the next century the pieces fell into place with a regularity and precision that transformed theory into accepted fact. In Darwin's lifetime, however, his theory did not win universal acceptance in the scientific community.

Darwin's understandings of how inheritance works and how variations arise and become stabilized in a breeding population—"the various, quite unknown, or dimly seen laws of variation," as he put it, remained weak links in his argument. Along with his contemporaries he assumed that inheritance embodies a blending process akin to mixing two pigments and producing an intermediate color. But if inheritance occurs through blending how can a more fit individual spread its fitness to the entire population? On the contrary, the improved trait will, in a few generations, be first diluted and then completely wash out through successive blending. At one point Darwin had thought that species originate on islands or in other small and isolated groups where the effects of blending would be minimized and, by the laws of chance, an improved variety might then spread over the whole population. But even in small breeding groups, it was shown, blending would prevent the formation of new species. Based on the same reasoning and with equally compelling force, critics likewise posed grave difficulties concerning the evolution of mimicry in nature—or how two different species of butterfly, for example, can come to have the same outward appearance.

Another devastating source of criticism sprang from an unexpected and authoritative quarter: the community of physicists. That evolution requires long periods of time stood as a fundamental tenet of Darwin's theory. In the first edition of *The Origin of Species* Darwin mentioned a figure of 300 million years for a single geological formation, the implication being that life probably evolved over hundreds of millions or even billions of years. But just as Darwin's views became known, the discipline of physics, based in particular on the first and second laws of thermodynamics, began to assume an unprecedented intellectual coherence and mantle of authority. Given what was known of the radiational cooling of the earth and the presumed combustion of the sun (thought to be something like a burning lump of coal), physics—notably in the august personage of William Thomson, Lord Kelvin—

pronounced against the extraordinary amounts of time required by Darwin's entirely qualitative theory.

Over the next two decades Darwin responded to his critics. In the process he ceased to be a true Darwinist and acquired the characteristics of a Lamarckian. He maintained an emphasis on his beloved natural selection, but he now admitted that natural selection was perhaps not the exclusive mechanism for species change. But he refused to capitulate to the authority of the physicists regarding the span of time available for evolution, saying that we do not "know enough of the constitution of the universe and of the interior of our globe to speculate with safety on its past duration," and he suggested that perhaps the *rate* of evolutionary change was more rapid at an earlier point in time. He even allowed for the possibility that environmental factors may affect whole groups of plants and animals in a Lamarckian fashion, inducing wholesale changes and thereby minimizing the significance of individual variations. In a special section of his 1867 book on *Variation of Animals and Plants under Domestication* Darwin outlined a possible mechanism he called "pangenesis" for the inheritance of acquired characteristics. In Darwin's case the first edition of his great work gives clearer insight into his doctrines than the intellectually troubled later versions.

The Origin of Species provoked immediate and strong reactions from a conservative Victorian world forced to confront Darwin's views on evolution. The tone was set in 1860, just months after the publication of the *Origin* at the famous meeting of the British Association for the Advancement of Science at Oxford; there, in a staged debate the Anglican bishop of Oxford, Samuel Wilberforce, denounced Darwin and evolution for equating humans and apes; T. H. Huxley, Darwin's ardent supporter who earned the nickname of "Darwin's Bulldog," retaliated with devastating sarcasm.

While his minions pushed the fight in the public arena, the retiring Darwin continued to work and to tend his gardens and family at Down. As the years unfolded he published substantial volumes on the emotions of animals, fertilization and cross-fertilization in plants, insectivorous plants, climbing plants, and the actions of worms. All of these works embodied evolutionary perspectives and were framed by the principle of natural selection, but none carried the day in fully persuading the contemporary world of science that the history of life was to be understood in evolutionary terms. Darwin died in 1882. Even though his ideas threatened the foundations of the Victorian social order and he never became Sir Charles, the kind old man himself clearly stood with the greats of British science and culture. He was buried in Westminster Abbey alongside Sir Isaac Newton.

The Neo-Darwinian Synthesis

Ironically, one of Darwin's main problems was solved only six years after the publication of the *Origin* when Gregor Mendel (1822–84), an Austrian monk, described his experiments on plant hybridization which showed that heredity is not a blending process but instead maintains traits in discrete units. These units—later designated genes—pass from generation to generation, and since they are never blended out they may preserve a useful trait and eventually spread it through the breeding pool. Mendel's publication was almost wholly neglected, however, and by 1900 Darwin's theory of evolution through natural selection remained far from established in the scientific community. Indeed, at the turn of the twentieth century it would be hard to speak of a Darwinian revolution.

Several scientific trends combined after 1900 to produce a refurbished evolutionary theory known as the Neo-Darwinian synthesis. The discovery of radioactivity and Einstein's 1905 equation of matter and energy justified Darwin's skepticism about Lord Kelvin's estimate of the age of the earth. During the first half of the twentieth century estimates of the age of the solar system progressively increased first to millions and then to billions of years, and evolutionary theory ended up with all the time it needed.

Several researchers simultaneously rediscovered the Mendelian principle of particulate inheritance at the beginning of the twentieth century. In 1900 Hugo de Vries introduced a theory of *mutations* which provided evolution with another mechanism of variation and also, by accelerating the process of evolutionary change, added to the plausibility of accounting for the origin of species within a limited time frame. Ironically, Mendel's laws of inheritance along with the mutation theory suggested to some scientists that these principles alone, without the operation of Darwinian natural selection, could account for evolution. Darwinism was finally adopted as the new paradigm of biological diversity during the 1940s with the recognition that genes, those discrete units of heredity, combine in extraordinarily complex patterns to produce the minute variations on which natural selection operates.

Prior to the turn of the twentieth century, improved staining techniques had already led to the discovery of chromosomes—thread-like structures in the nucleus of the cell—and in the 1880s August Weismann speculated that they contained the units of heredity, a speculation soon confirmed. In the 1910s and 1920s, working in the famed "fly room" at Columbia University, Thomas Hunt Morgan (1866–1945) and his colleagues experimented with the fruit fly, *Drosophila melanogaster*, confirming Mendel and the chromosomal basis of inheritance. In England R. A. Fisher and other specialists in population genetics demonstrated mathematically that a single variation conferring an advantage on an individual in the struggle for existence could in the fullness of time come to characterize an entire population. These trends converged in a series of scientific works appearing in the late 1930s and early 1940s, notably *Genetics and the Origin of Species* (1937) by Theodosius Dobzhansky, *Evolution: The Modern Synthesis* (1942) by Julian Huxley, and *Systematics and the Origin of Species* (1942) by Ernst Mayr. These and related works provided a full and compelling theory of evolution in line with the original doctrine outlined by Darwin in 1859.

But the material, biochemical basis of inheritance still eluded scientists. In the 1940s the key nucleic component seemed to be an organic acid, DNA—deoxyribonucleic acid, a white powder composed of carbon, hydrogen, nitrogen, oxygen, and phosphorus. In 1953 a maverick young post-doc and an older graduate student, James Watson and Francis Crick, decoded the molecular structure of DNA. They showed the molecule to be in the form of a double helix, an elegant structure that provided a mechanism for inheritance and the appearance of new variations, a major discovery for which they received a Nobel Prize in 1962. It represents the crowning achievement of a century of research—the code of life had become reduced to chemistry.

Social Struggles for Existence

Practitioners in the biological sciences and a variety of related fields now accept Darwinian evolution, and its principles now guide research and scientific understanding. Still, in other sciences some of the implications of evolutionary theory have been and remain controversial.

From the very outset, Darwinism, through its substitution of natural for divine processes, courted a duel with the religious mentality and biblical authority. The Tennessee "monkey trial" of 1925, where public-school teacher John Scopes was convicted of violating a state law against teaching Darwinism, remains a notorious example. Controversies continue to flare up today, not in the scientific world over evolution itself, but rather over the teaching of biological sciences versus "scientific creationism" in public schools, and textbook publishers have, to their shame, simply removed most evolutionary references from their works. In the realm of theology, fundamentalism excepted, major religious authorities soon accommodated themselves to the general principle of evolution by adopting, *pace* Galileo, a nonliteral interpretation of scripture and allowing divine intervention into the evolutionary process only to infuse a nonmaterial soul into humankind at some moment during its physical evolution.

In the social sciences and philosophy, fierce conflicts have arisen over the application of Darwinian concepts to explanations of human nature. The first such conflict was provoked by what became known as Social Darwinism—the claim that the organization of society, and specifically the stratification into classes—can be accounted for bio-

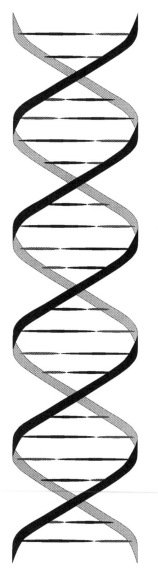

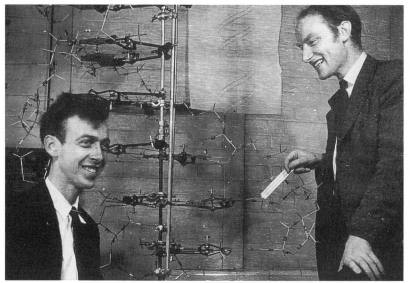

Fig. 15.3 a–b. The double helix. The discovery of the double helical structure of DNA by James Watson and Francis Crick in 1953 demonstrated the material basis of inheritance and the substrate for how new variations can lead to species change. Their discovery marked the beginning of a new era in genetics and biochemistry.

logically by the struggle for survival. In the United States Social Darwinism tended to color social and psychological analyses by suggesting that race as well as class determined success in life. The general implication was that the status quo with all of its inequities is determined by hereditary biological characteristics and cannot readily be improved by social reform.

Various strains of Social Darwinism provided ideological support for an array of political, economic, and social positions. For example, in his book *Gospel of Wealth* (1900), the American industrialist Andrew Carnegie justified cutthroat laissez-faire capitalism, the sacredness of property and profit, and scientific opposition to socialism and communism on social Darwinist grounds. Other apologists have put forward nominally similar, and equally specious, arguments in favor of Nazi racial claims, for class struggle, for the eugenics movement and the sterilization of "inferior" persons, and for the superiority of men over women. However, the mainstream of American social scientists repudiated such applications of Darwinian thought.

How to think scientifically about human evolution likewise turned into an arena of dispute. Fossils have now been found that represent at least three species of australopithecines—an extinct classification of bipeds that link humans with an ancestral ape living 5 million years ago—and at least two species of ancestral human beings who successively followed the australopithecines before the evolution of our species, *Homo sapiens*. While these discoveries can be seen as confirmations of Darwin's theory of evolution, there has been consensus only on the evolution of human physical traits. Mental and behavioral traits remained the province of the social sciences, where a medley of doctrines defended the sanctity of consciousness against the encroachment of the gene and natural selection. In reaction to Social Darwinism sociologist

Emile Durkheim enunciated the principle that social conditions must be explained by strictly social events, thereby erecting a barrier between the social and biological sciences. Since the scientific basis of Social Darwinism remained insecure and since most prominent social scientists had liberal leanings Durkheim's dictum held sway during the first half of the twentieth century, and as a result biological approaches to the social and psychological sciences were treated with disdain.

The situation began to change during the 1930s along with the clarification of Darwinism itself. Austrian zoologist and Nobel Prize–winner Konrad Lorenz (1903–89) inaugurated the discipline of ethology, or the scientific study of animal behavior under natural conditions, and he and his followers drew attention to the genetic basis of mental and behavioral traits. Inevitably, some natural scientists focused their attention on the evolution of human behavior, a trend that was reinforced in 1975 by Edward O. Wilson's book, *Sociobiology*, a landmark work that included a suggestion that human social life can be studied in terms of Darwinian evolution. It provoked ideological objections on the grounds that human sociobiology constitutes "biological determinism" lending itself to justifications of some of the least appealing features of the status quo. However objectionable sociobiology may seem to some, its advocates are innocent of the scathing political charges that were once leveled against them. Since the 1970s there has been an explosion of research in sociobiology, mainly directed at the social behavior of nonhuman animals.

In the course of its final triumph Darwinism had to vanquish still another doctrine—Lamarckism, reduced to its essential claim that acquired characteristics can be inherited. The doctrine's appeal is based on both its postulate of a direct influence of the environment on the process of evolution and the rapidity of the process in producing new forms. Darwin himself, as we saw, accepted the principle while casting about for an explanation of variation, and Lamarckism continued to find favor in a few circles well into the twentieth century. The eminent psychologist Jean Piaget, for example, adopted it at one point, and for a few years during the 1940s the government of the Soviet Union sanctioned it in the forlorn hope that plant breeders could improve agriculture rapidly. Since then Lamarckism has found no support and is generally repudiated. What is sometimes referred to as the "Central Dogma of Genetics" asserts that while DNA carries the codes for all hereditary traits no mechanism exists whereby the modification of a trait can effect its determining DNA. In the end, the chemistry of the double helix sounded the death knell of Lamarck's doctrine.

Darwinism today provides the paradigm for many fields of research. While technical differences of opinion divide scientists over precisely how evolution operates—whether it proceeds gradually or in spurts, for example, or the extent to which hereditary traits are adaptive— virtually universal consensus has crystallized around its fundamental tenets that species are the fruits of evolution and that natural selection represents its main method. In the past century and a half the world of the biological sciences has largely recapitulated the trajectory of Darwin's biography—from a divinity student at Christ's College to the recognition that the diversity of life is an artifact of nature.

CHAPTER 16

Toolmakers Take Command

The Industrial Revolution first transformed the economy and society of England in the later eighteenth and early nineteenth centuries. From a global perspective, however, the process of industrialization did not end in the nineteenth century. Indeed, the shift from agrarian to industrial economies accelerated as the century progressed, and it is a process that continues today. Furthermore, the ultimate outcome of industrialization is still not clear, and in the balance hangs the fate of humanity on planet Earth.

The Spread of Industrial Civilization

Global industrialization took off in the period 1820–40, spreading in waves outward from England. Belgium, Germany, France, and the United States were the first countries to be affected. Germany reached 50 percent urban population in 1900, France and the United States by 1920. While the Netherlands, Scandinavia, parts of Spain, and northern Italy joined the core area of industrial civilization, the spread of industrialization over the European continent and North America was decidedly uneven. Most of Eastern Europe, for example, and pockets within industrialized countries (such as Ireland and the American South and Southwest) remained overwhelmingly agrarian and pastoralist well into the twentieth century, as did much of the rest of the world.

Although notable national and regional differences characterize the process, industrialization in each case centered primarily on the iron industry, textiles, the railroad, and, later, electrification. As industrial intensification continued, new areas of activity arose in construction, food processing, farming, and housework. As discussed in a previous chapter, whole new science-based industries emerged coincident with advances in electricity and chemistry. An expanded service sector complemented the core industries, often employing women as clerks, schoolteachers, nurses, secretaries, and, following the invention of the telephone by Alexander Graham Bell in 1876, telephone operators. And wherever industrial development took root, higher education eventually reflected its progress. Engineering, nursing, teaching, and architecture became professions in the nineteenth century through the establishment of university programs. In the case of engineering such programs leaned heavily (and increasingly) on the basic sciences, even when they were not fully applicable.

By the 1870s these developments transformed Europe and the United States into the dominant powers in the world. A new era of colonial development unfolded, and the so-called new imperialism of the nineteenth century ensued with the solidifying of English rule in India, the creation of a new French colonial empire in Southeast Asia and Africa, the spread of Russian power eastward across Asia, Western incursions into China, the forcible opening of Japan to the United States in 1853–54, and the "scramble for Africa" by European powers after 1870. The point that deserves emphasis here is that Europe and America could dominate less-developed countries because they monopolized the process of industrial intensification. Other countries did not possess the resources or technical capabilities to match the nations of Europe or America in output of armaments, railroad track, electricity, or shipping. By 1914 the global empires of the Western powers enveloped 84 percent of the world.

The spread of industrialization is intimately linked to the history of European colonialism, of which India provides a relevant example. Prior to the Industrial Revolution in Europe and the formal takeover of India by the British government in 1858, India ranked as one of the most technically advanced regions of the world. With the loss of its political independence, what followed amounted to the deindustrialization of a flourishing traditional economy. Railroads, of course, proved essential to effective British control over India and spread rapidly. The first railroad began operations in India in 1853. By 1870 engineers had laid over 4,500 miles of track; by 1936 mileage had risen to 43,000 miles, creating the fourth largest railroad system in the world. Hardly a decade after Morse's first telegraph line in America, a telegraph system arose in India literally alongside the railroad. By 1857 4,500 miles of wire had been strung, and in 1865 India became connected to Britain by telegraph cable. But notably, these technologies did not spur industrial development in India. Rather, they served as instruments of colonial control and as transportation and communications conduits to extract raw materials and commodity products and to bring British manufactured goods to Indian markets. Thus, for example, India went from being the world's largest exporter of textiles to a net importer. The traditional shipbuilding industries that flourished in India through the nineteenth century—often fulfilling European contracts—became completely outmoded and outdated once the shift to steamships began. Although indigenous industrial activity developed in India later in the nineteenth and twentieth centuries, the initial effect of contact with industrial Europe resulted in increasing poverty through technological displacement, unemployment, and neglect of the traditional agricultural infrastructure. The pattern of colonial exploitation observable in India repeated itself in virtually every other European encounter with the nonindustrialized world. In a word, the success and power of the Western industrial economies virtually precluded the independent development of industrialization elsewhere. Poorer and less-developed nations stood hardly a chance against the technical and economic power of the European onslaught.

A constellation of new technological practices provided the basis for further European dominance in the nineteenth and twentieth centuries and the imposition of European imperialism on an even greater worldwide scale, particularly in Africa and Asia. At the heart of the matter is what has been labeled the "industrialization of war," that is, the application of industrial methods to the production of war materiel. Consider, for example, the steamship. Developed in the first decade of the nineteenth century, the steamship quickly and reliably moved goods and people and furthered the penetration of river systems around the world. As a military vessel—the flat-bottomed steamship was introduced into the Royal Navy in 1823—it secured British victories as early as the 1840s in the opium wars in China. Development of the steamship culminated in the steel-hulled coal-fired ship armed with breech-loading rifled artillery firing high-explosive shells. Floating citadels, like the massive HMS *Dreadnought* launched in 1906, allowed the projection of Western power to virtually any coastal area of the world. Ultimately, with 15-inch guns and elaborate sighting, loading, and recoil systems, they could strike targets within a 20-mile range. Explosive shells, in place of cannonballs, spelled the end of the wooden warship. Only those nations with the resources and know-how to produce steel could play the deadly new game.

The arms race that developed in the second half of the nineteenth century in Europe among England, Germany, and France escalated these military developments and led to the creation of machine guns, rapid-firing bolt-action rifles, new types of bullets, submarines, self-propelled torpedoes, destroyer battle groups, and the full armamentarium of modern naval warfare and "gunboat diplomacy" so characteristic of Western relations with the non-Western world. The American Civil War, mainly through the use of railroads to move military units, presaged the effects of the industrialization of land war in the 1860s, and the victory of the United States over Spain in the Spanish-American War in 1898 signaled the effectiveness of the new style of gunboat and the beginnings of extraterritorial American imperialism. The industrialization of war expanded in World War I, with combatants dependent on heavy industry, submarines, tanks, railroads, nascent air power, and poison gas, the last clearly indicating the participation of scientists, science, and the chemical industry in the war effort. Indeed, World War I has been called the chemists' war. And, finally along these lines, in the interwar period the rapid development of the warplane and motorized military vehicles gave European imperialism and colonial powers new weapons to enforce their rule. Military technologies and industrial development stimulated each other at the end of the nineteenth century, in a manner akin to the effects of the steam engine and railroad-building over the previous century.

The industrialization of Russia extended industrial civilization eastward from its European base. Only partly a European nation, nineteenth-century Russia preserved a traditional, agrarian economy. The Tsarist government prevented the outright colonization of the country by foreign powers, but industrialization depended on outside expertise and capital (mostly from England). At the same time, railroad building provided a mechanism for Russia's own imperialist expansion across Asia. The Moscow–Saint Petersburg line opened in 1851, and the number of railroad miles mushroomed from 700 in 1860, to 12,500 in 1878, to 21,000 in 1894, and over 36,000 in 1900. The first leg of the Trans-Siberian Railroad opened in 1903. By 1913, with over 43,000 miles of railroad, Russia possessed the fifth largest industrial economy in the world.

Following the Russian Revolution of 1917, the Soviet government speeded the industrialization of the nation, achieving during the 1930s the most rapid rate of industrial development in history. By the late 1940s, despite the heavy toll of World War II, the Soviet Union emerged as the second largest manufacturing economy in the world (the United States was first), with leading heavy industries in coal, iron, and steel production, chemicals, and electric power generation, and with increased urbanization. Officials achieved the transformation of the Soviet economy through heavy-handed state planning, extraordinary inefficiencies, and at great human cost, but also with more attention to worker issues, including concern for education, medical and child care, and recreation. From the point of view of the history of industrialization, the case of the Soviet Union stands out, especially insofar as developments there took place largely independently of industrialization elsewhere and in the face of hostile reactions in the West, including an American embargo on trade that lasted, at least officially, until 1933. The future of industrial development in the former Soviet Union is a question that historians will be keen to judge.

Japan presents another historically notable case in the elaboration of the world industrial economy. Through 1870 industrial intensification remained confined to the West, and Japan became the first to break the European stranglehold. Forcibly opened to outside trade in 1854 Japan, unlike India, managed to maintain its political independence and never became subjected to outright foreign control. The Meiji Restoration of 1868 signaled the end of feudal government in Japan, and a combination of traditional merchant classes and progressive government civil servants began to promote industrial development. Created in 1870, a Ministry of Industry set an enduring pattern for state planning and the financing of Japanese industry. Initial investments went to the railroad, and the first line linking Tokyo and Yokohama opened in 1872. Japanese shipbuilding also received early support to promote trade and in order to compensate for the lack of raw materials in Japan itself. The mechanization of silk production was another arena of initial industrial policy. A spurt in population from 30 million in 1868 to 45 million in 1900 (and to 73 million in 1940) provided a population base for an industrial working class. In stark contrast to Europe, a large proportion of women in the workforce (above 50 percent) was a unique feature of early industrialization in Japan. Paternalism and the group-identity characteristic of Japanese culture generally helped ease the transition to an industrial economy, with less social and political strife than seen in the West. In the twentieth century Japan itself became an imperial power, with the victory of the Japanese in their war with the Russians in 1904–5 and expansion in East Asia and the Pacific in the 1930s and 1940s. Regarding the scientific underpinnings of Japanese industrial development, the first university in Japan was founded only in the 1880s.

"See the USA in Your Chevrolet"

The United States became the world's leading industrial power in the twentieth century. The explanation why a predominately rural and agrarian society could become transformed into a manufacturing giant remains complex. Its enormous agricultural potential, along with ample resources of raw materials, were surely major influences, but the development of the American automobile industry and the techniques of mass production pioneered by Henry Ford rank high among the factors propelling America to the forefront of industrial civilization.

The steam engine remained the prime mover in industry and in transportation systems—railroads and steamships—to the end of the nineteenth century. With the development of effective gasoline and then diesel internal-combustion engines by German engineers in the 1880s a new prime mover appeared that, mated to the wagon and the plow to create the automobile and the tractor, came to supplant the omnipresent horse. Self-propelled cars powered by internal-combustion engines were first developed in the 1880s, and the "horseless carriage" began to make a public impact in the 1890s. Total auto sales in the United States in 1900 topped 4,000. By 1911 600,000 automobiles puttered along on U.S. roads; by 1915 that number had risen to 895,000, and by 1927 it had soared to 3.7 million. Some of the early automobiles were electric or steam-driven, but the internal-combustion engine eventually won out as the most successful power source.

In more heroic versions of the history of American technology, Henry Ford (1863–1947) stands as a lone visionary personally responsible for the creation of the automobile industry in America. A self-educated mechanic with a lifelong disdain of experts with university degrees, Ford built his first automobile in 1893; a decade later he founded the Ford Motor Company. Ford intended to produce "the car for the great multitude," and to do so he had to gear mass production for mass consumption. He was not the first manufacturer to use interchangeable parts or to run an assembly line, but in his quest to produce an inexpensive and standardized product Ford perfected assembly-line production techniques, and the results proved dramatic. In 1908, before he introduced the assembly line, Ford made 10,607 Model Ts—the "Tin Lizzie"—which he sold for $850 each. He shifted to an assembly line in 1913, and production quickly rose to 300,000 cars a year. In 1916 he sold 730,041 Model Ts for $360 apiece, and in 1924 he produced 2 million of the cars retailing at $290 each. A total of 15 million Model Ts rolled out of Ford plants before production ceased in 1927. Ford achieved these unprecedented feats of production by making his factories incredibly efficient. Prior to Ford, it took over 12 hours to assemble a car; his first assembly line turned out a Model T every 93 minutes, and by 1927 Ford made a Model T every 24 seconds! The Ford Motor Company became not just the world's largest automobile manufacturer, but the world's largest industrial enterprise.

Ford indeed created a car for the masses. By the 1920s the automobile was no longer an esoteric toy for enthusiasts or the rich and idle, but was well on its way to becoming a necessity of modern life and a mainstay of the modern global economy. By 1980, for example, 300 million cars and 85 million trucks moved people and things around in the world and spewed tons of pollutants in the process. In 1992 automobile companies in the United States employed 212,000 workers and produced $141 billion worth of motor vehicles. The model of mass production and mass consumption originated by Ford with the Model T would come to be applied to many other "necessities" of modern life, one of the many cases where invention proved the mother of necessity.

It is too easy to see Ford as the heroic inventor single-handedly changing the landscape of American industry and culture. Our understanding of his accomplishment is enriched if we think of him, rather, as a system builder who orchestrated the efforts of thousands of others and who oversaw the coming-into-being of a diverse and self-sustaining technological enterprise—modern automobile manufacturing. Within his own company Ford headed a sizable team of talented and enthusiastic young engineers, foundrymen, and toolmakers to bring his first assembly line together, and other companies quickly hired some of them away. If anything, Ford's personal character and rigid and controlling management style hurt the long-term growth of Ford Motors in that

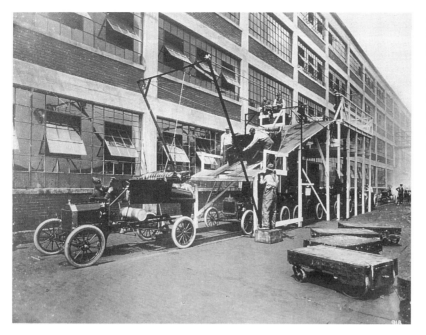

Fig. 16.1. The assembly line. Improved by Henry Ford, the assembly line greatly increased production. The case of the automobile illustrates well the nature of a technological system that must combine many subsystems.

he refused to introduce improvements and stuck with the Model T well past its prime, thereby allowing competitors such as General Motors and Chrysler to take a larger market share for cars. Ford stands out as a key figure in the history of technology, to be sure, but the history of the automobile industry in America is hardly encapsulated in one individual.

Ford created an elaborate organizational structure to secure materials, make cars, and then market them, and the automobile industry provides a clear example of what historians of technology mean when they speak of a technological system. A piece of technology does not exist in a social vacuum but is connected with makers, users, and other technologies in often complex ways. The automobile itself can be thought of as a technological system, one composed of thousands of parts that integrate fuel, carburetors, an engine, transmission and power train, brakes, suspension, and electrical subsystems, to name only the most major. The invention of the electric starter in 1912 and the introduction of balloon tires in 1921 were innovations that significantly improved the automobile considered as a collection of subsystems. The starter replaced the hand crank and, incidentally, brought women into the automobile market.

Another aspect of the automobile-as-technological-system emerges in focusing on the manufacture of automobiles. Multifaceted production mechanisms had to come into place before the automobile could become a significant piece of technology in American life. The Ford factories at Highland Park and River Rouge, for example, became vast industrial sites that grouped together several subsidiary plants to service the main assembly lines, including a coking plant, a foundry, a steel plant, and a cement plant. Ford had his own coal mines, glass plants, rubber plantations, ships, and railroads. Labor comprised a key feature of this intricate manufacturing operation, and Ford gained fame for dramatically increasing the wages of his factory workers. While the prevailing rate was $11 a week, Ford announced that he would pay workers $5 a day for an eight-hour day. Otherwise no friend of organized labor, Ford took this bold step primarily in an effort to create a stable workforce, but it had the effect that Ford factory workers could now be consumers of the very products they made.

Cars and car plants themselves form part of an even larger network of related technologies and social practices required for the car to be an effective technological system. For example, cars run on gasoline; the explosive growth of the automobile industry both occasioned and would have been impossible without an equally massive expansion of the oil industry and improved techniques of "cracking" crude oil to make gasoline. The local gas station thus became an essential element in the larger technological system, as did garages, repair shops, and replacement auto parts. The same point can be made about the necessity of reliable roads and road systems, and about traffic signs, driving conventions, auto insurance, and the bureaucracies of state motor vehicle departments. Similarly, without auto dealerships and advertising the automobile industry would be very different, to say the least. Along these lines, the car loan, introduced in 1915, and the trade-in were marketing innovations that continue to play a large role in keeping the system going.

More than any other technology, mass-produced automobiles defined the culture of the twentieth century. Automobile manufacture has spurred industrialization around the world, and from creating suburbia to transforming the sex lives of adolescent drivers, the social repercussions of the technological system of the automobile have been immense.

Rich and Poor in the Industrial World

Technical innovations further changed conditions for industrial civilization in the twentieth century. The widespread use of electricity as an especially versatile energy source in manufacturing and domestic use represents one key innovation. As a result of these and like changes, world production increased from $1 trillion in 1900, to $4 trillion in 1950, and $14 trillion in 1973, while urbanization on a world level has jumped from 15 percent in 1900 to 45 percent in 1990 to a projected 50 percent by the year 2000. These developments have been accompanied by an expanded middle class and higher standards of living, but also increasing stratification of rich and poor and a more hectic pace of life.

Industrial civilization spread unevenly in the twentieth century, and

Fig. 16.2. Industrial pollution. In developed societies today scenes such as this one from the 1920s are less and less common, as industries and nations begin to take care to control their emissions and preserve a healthy environment. Such concerns are less strongly felt in developing regions where the expense of pollution controls is less easily met. It remains to be seen if industrial civilization can achieve a sustainable balance with the earth's ecology.

at the end of the century the Western industrial societies and Japan remain the dominant industrialized powers. By 1950 most of the world's population still did not live in industrialized countries. Indeed, a striking feature of industrialization in the twentieth century has been an increasing divergence of wealth and poverty on a global scale. In 1990, for example, the West, with 20 percent of the world's population, consumed 70 percent of commercial energy resources. The top 10 percent of the world's population consumed nearly 40 percent of the world's energy.

The less-developed countries remain subservient to the industrialized nations. Decolonization after World War II brought political independence for many former colonies, but in most cases their dependent economic status remained unchanged. Nevertheless, new industrial or industrializing nations have emerged since the Second World War, notably the Pacific rim countries—Japan, Hong Kong, South Korea, and Taiwan. A series of "emerging markets" now complement the more established industrial societies, including Malaysia, Singapore, Thailand, the Philippines, India, China, and countries of Latin America. In some cases, such as the development of the oil industry in the Persian Gulf region or diamond and gold mining in South Africa, only certain sectors of a nation's economy have become industrialized. In all cases state planning seems to be a key to success. The current industrializations of China and to a lesser extent India, with their combined populations of over 2 billion of the world's nearly 6 billion people, are now at the forefront of change in the ongoing history of industrial civilization.

At the same time several factors combined to transform the more mature industrial economies, notably an increasing emphasis on service industries, the management of information, electronic industries, and biotechnologies. The entry of substantial numbers of women into the workforce in the West has effected great social and economic change. The emergence of the multinational corporation has helped forge an interdependent global economic system. Many multinational corporations are economically more powerful than many whole countries, and to a considerable extent they rival the importance of many nations and national economic policies.

The processes set loose in the Industrial Revolution continue to unfold across the globe. For a very large number of people the results have been historically unprecedented lives with good health, comfort, and technological trinkets of amazing variety. By the same token, the material progress that many enjoy has not been achieved without heavy costs. Recent years have seen a drop in real wages in advanced countries. Consumerism now represents the dominant values in many parts of the world. Current environmental problems with pollution, oil spills, acid rain, the depletion of the ozone layer, waste disposal, loss of biodiversity, and like concerns reflect the tremendous and likely irreversible ecological degradation accompanying industrialization. The ultimate outcome of events that began with the Industrial Revolution in England is not clear, but it seems unlikely that the world can long sustain further industrial intensification. In the meantime, thinkers pondering nature continue to elaborate a great tale.

CHAPTER 17

The New Aristotelians

The Hellenic tradition of natural philosophy—the disinterested inquiry into nature that sprang to life in ancient Greece—continued to form a defining element of science in the twentieth century. As a result, a host of theoretical novelties has emerged across a broad range of disciplines to transform our contemporary vision of the world. The significance of science in the twentieth century may ultimately stem less from its natural philosophical import than from the great practical impact it is having on society at large. Nevertheless, the tradition of pure science continues to shape our intellectual culture.

Einstein, Relativity, and Quantum Physics

One of the most notable achievements of scientific thought in the twentieth century was the overthrow of the Classical World View of nineteenth-century physics and the substitution of revolutionary new parameters for understanding the physical world. This revolution is ofttimes labeled the Einsteinian Revolution, and it played a decisive role in shaping how we now think about nature.

A previous chapter outlined the coming together of the grand intellectual synthesis known as the Classical World View. The reader will recall its major tenets: a Newtonian framework of absolute space and time; a world of immutable and indivisible atoms; an etherial substrate that provided the basis for the electromagnetic field and for the undulatory propagation of light and radiant heat. The principles that matter and the ether mutually interacted and that the medium of energy obeyed the strict mathematical laws of thermodynamics added great unity, simplicity, and harmony to the worldview fabricated by physicists in the second half of the nineteenth century.

Unlike Aristotle's scientific vision, which endured for 2,000 years, the Classical World View fell apart almost as quickly as it came together. In the last decade of the nineteenth century a nagging series of problems began to undermine the intellectual edifice of contemporary physics, and by the dawn of the twentieth century a serious crisis had developed in the physical sciences.

In a series of experiments beginning in 1887, American physicist Albert A. Michelson (1852–1931) failed to detect the motion of the earth relative to any ether. Theory predicted that the speed of light ought to change ever so slightly as measured from the earth moving relative to a stationary ether; that is, over a six-month period the orbiting earth moves in opposing directions around the sun, and so, from the frame of reference of a fixed ether, over that period of time the same beam of light ought to move first with the earth and then contrary to the earth's motion, and that difference in motion ought to be detectable. But through experiments of exquisite exactness Michelson and his colleague E. W. Morley consistently obtained a negative result. In retrospect, this "failure" can easily be portrayed as paving the way for Einstein's theory of relativity, in which the null result of the Michelson-Morley experiment is expected and easily explained by the constancy of the speed of light independent of relative motion. In point of fact, however, Einstein came to relativity through a different route, and contemporary physicists, rather than abandon classical principles on the basis of the Michelson-Morley experiment, characteristically rushed to "save the phenomena" and to patch up the existing system by improvising ad hoc explanations that preserved the ether.

Several other developments added to a mounting sense of crisis in contemporary physics. In the fall of 1895 the German experimentalist Wilhelm Roentgen (1845–1923), working with a standard piece of laboratory apparatus, uncovered X-rays, a new type of radiation. While not wholly outside the bounds of classical physics, this discovery extended the range of electromagnetic radiation well beyond its conventional limits and called into question traditional assumptions about the spectrum and about accepted laboratory procedures.

The discovery of the electron constituted a related and more serious problem. Already in the 1870s scientists recognized that an electric current passing through an evacuated receiver produced "rays" of some sort called cathode rays. In 1897 British physicist J. J. Thomson (1856–1940) demonstrated that cathode rays were particulate in nature; that is, they were distinct particles nearly 2,000 times smaller than an atom of hydrogen, the smallest atom. Thus, the traditional, indivisible atom was not the smallest unit of matter after all. How was that to square with received notions?

A year earlier, French physicist Antoine-Henri Becquerel (1852–1908) accidentally noticed that uranium ore clouded unexposed photographic plates, and he thus uncovered still another unexpected natural phenomenon. In 1898 the acclaimed Franco-Polish scientist Marie Curie (1867–1934) coined the term *radioactivity,* and thereafter it became clear that certain heavy elements spontaneously emitted several different sorts of radiation, including electrons, superenergetic electro-magnetic waves (gamma rays), and subatomic particles called alpha particles. By 1901 the phenomenon of radioactive decay became apparent, wherein one element, say uranium, transformed itself through radioactive emissions into another element, say lead. Such transformations clearly violated the fundamental principle of the immutability of atoms. The fixity of atoms broke down like the fixity of species and provided yet another great riddle for scientific explanation.

Two other highly technical puzzles compounded this crisis: the so-called photoelectric effect and the mathematics of "black body radiation." The photoelectric effect, discovered in 1887 by Heinrich Hertz, concerned the apparent paradox that light shining on certain materials would induce an electric current, but only above a certain wavelength. Even intense amounts of light below the requisite threshold would not trigger a current. The black body problem concerned research that implied that an ideal system might emit more radiant energy than it received as input; that is, if the electromagnetic spectrum were truly continuous, then an initial wave of light or radiant heat might be redistributed into an infinite number of smaller waves and hence produce an infinite amount of energy. This was an obviously absurd finding that contradicted empirical experiments and the received laws of thermodynamics. The interpretations that began to emerge, notably those enunciated after 1901 by the German physicist Max Planck (1858–1947), involved suggestions that light (or radiation in general) came in discrete energy packets or quantum units and did not exist in the infinitely graded energy continuum required by classical physics.

Albert Einstein (1879–1955) achieved intellectual maturity in the foment of physics at the turn of the twentieth century. The son of an unsuccessful businessman, Einstein displayed no precocious talents as a child. At 16 he dropped out of the *gymnasium* (or high school) in Munich to join his family, who had moved to Italy. After some difficulties getting in, Einstein attended and in 1900 graduated from the Federal Polytechnic School in Zurich. Excluded because of his Jewish ancestry from becoming a schoolteacher, Einstein and his first wife moved to Bern, Switzerland, where he took a minor post in the Swiss patent office. He received a doctorate in physics from the University of Zurich in 1905, and he remained in Bern until 1909 pursuing physics in his spare time. Einstein was thus perfectly positioned to effect a revolution in contemporary physics: he was well educated technically in the central dogmas of the field, yet he was young enough and professionally marginal enough as an outsider not to be locked into established beliefs.

Einstein published an extraordinary series of papers in 1905 which redirected modern physics. The most dramatic paper dealt with special relativity or the physics of bodies moving uniformly relative to one another. While highly technical, their conceptual novelties are easy to summarize. In essence, by positing that nothing can move faster than the speed of light, Einstein reformulated Newtonian mechanics, which contains no such restriction. The upshot was special relativity, an interpretation of motion that dispensed with the reference frames of absolute space and absolute time fundamental to Newtonian physics and the Classical World View. In Einstein's interpretation the cosmos contains no privileged frames of reference, no master clock. All observations (such as when an event takes place, how long a ruler is, or how heavy an object) become relative and depend on the position and speed of the observer. As a feature of his new physics, Einstein posited his celebrated formula, $E = mc^2$, which equated mass *(m)* and energy *(E),* elements held scrupulously distinct in Classical physics, with the speed of light *(c)* a constant in the equation.

Sometimes, particularly in physics classes, Newtonian physics is portrayed as simply a special case of Einsteinian physics—slow-moving bodies supposedly obey Newton's laws while bodies approaching the speed of light follow Einstein's. Such a view, while it facilitates science teaching, distorts the historical record and obscures the revolutionary transformation effected by Einstein's 1905 papers. For Newton and Classical physics, space and time are absolute: an "Archimedean" point exists somewhere out there against which all motions may be measured; somewhere a standard pendulum beats out universal time; mass and energy are not interconvertible; objects can go faster than the speed of light. Einstein reached categorically different conclusions. Thus, simply because the same letter *m* (standing for mass) appears in the Newtonian formula, $F = ma$, and in the Einsteinian one, $E = mc^2$, one should not confuse the two categorically different concepts of mass or the different physics behind them.

Einstein's special relativity of 1905 concerned uniform motions. In 1915 he published on general relativity, or the physics of accelerated motions, wherein he equated gravity to acceleration. In a highly imaginative thought experiment Einstein supposed that inside an elevator one could not distinguish gravitational forces induced by an imaginary planet approaching from below from forces produced by the upward acceleration of the elevator; both events would produce the same nervous flutter in a passenger. As a consequence of this equivalence, the nature of space underwent a profound transformation. Uniform, three-dimensional Euclidean space—yet another absolute of Newtonian physics and the Classical World View—became obsolete, replaced by the four-dimensional continuum of Einsteinian space-time. As a consequence of this reinterpretation, it came to be understood that bodies distort the shape of space. Gravity—a "force" in Newtonian mechanics—becomes only an apparent force in Einstein's general relativity, the result of the curvature of space warped by heavy bodies within it. Planets orbit the sun not because a gravitational force attracts them, but because they must follow the shortest path through curved space. Observations of a total solar eclipse in 1919 seemed to confirm Einstein's prediction that the mass of the sun should bend starlight, and similar, very precise calculations of Mercury's orbit around the sun brought like agreement with general relativity. By the 1920s Classical physics, with its absolutes and its ether, had become a thing of the past. Physicists led by Einstein had created a conceptually new world.

Another set of scientific developments—having to do with atomic theory and the physics of the very small—complemented relativity and likewise proved highly consequential for twentieth-century natural philosophy. Einstein was a major contributor to this line of work, too, especially with another paper that appeared in 1905 on the photoelectric effect and that supported the notion that light comes in discrete bundles and not in continuous waves. Yet, when the implications of mature quantum theory came to the fore—implications concerning the impossibility of visualizing phenomena, an inherent uncertainty in knowing nature, and probabilistic limits to the behavior of particles—Einstein recoiled, objecting in a celebrated phrase that "God does not play dice with the universe." Nevertheless, quantum mechanics and other, more recent work in particle physics flourished without the master's blessing.

Once the discovery of the electron and radioactivity undermined the indivisible and immutable nature of the classical atom, atomic theory became a prime focus of experimental and theoretical research. J.J. Thomson, coincident with his discovery of the electron in 1897, proposed a model of the atom having negatively charged electrons sprinkled about like raisins in a cake. Using radioactive emissions as a tool to investigate the interior structure of the atom, in 1911 Ernest Rutherford (1871–1937) announced that atoms were composed mostly of empty space. Along with the Danish physicist Nils Bohr (1885–1962), Rutherford proposed a model for the atom that had electrons orbiting a solid nucleus, much like planets orbit the sun. In the 1920s problems with this solar-system model (e.g., why electrons maintained stable orbits or why, when excited, atoms radiated energy in a discontinuous manner) led to the formulation of the so-called new theory of quantum mechanics and yet another radical transformation of our understanding of nature.

The paradoxical principles of quantum mechanics were difficult to accept, yet empirical studies supported the theory, as did the social network that arose around Bohr and the Institute for Theoretical Physics that he headed in Copenhagen from 1918 until his death in 1962. While of great mathematical and technical sophistication, the basic ideas behind the "Copenhagen interpretation" of quantum mechanics are not difficult to comprehend. Essentially, quantum theory replaces the deterministic mechanical model of the atom with one that sees atoms and, indeed, all material objects not as sharply delineated entities in the world but, rather, as having a dual wave-particle nature, the existence of which can be understood as a "probability wave." That is, quantum mechanical "waves" predict the likelihood of finding an object—an electron or an automobile—at a particular place within specified limits. Everything that is, is a probability wave.

The power of this counterintuitive analysis was greatly strengthened in 1926 when two distinct mathematical means of expressing these ideas—matrix mechanics developed by Werner Heisenberg (1901–76) and wave equations developed by Erwin Shrödinger (1887–1961)—were shown to be formally equivalent. In 1927 Heisenberg proposed his famous Uncertainty Principle that extended the conceptual grounds of quantum theory in surprising ways. In short, Heisenberg stated that in principle one cannot simultaneously determine the position of a body and its speed (or momentum) with equal accuracy. In other words, in contradistinction to the determinism of Classical physics, where in theory, given initial conditions, we should be able to predict all future behavior of all particles, quantum mechanics reveals an inherent indeterminacy built into nature and our understanding of it. With quantum mechanics, chance and randomness were shown to be an intrinsic part of nature. We can say nothing with certainty, but can only make probabilistic predictions. The Heisenberg Uncertainty Principle made plain, furthermore, that the act of observing disturbs the subjects of observation. The key implications were that nondisturbing or "objective" observation is impossible, that the observed and observer form part of one system, and that probability waves "collapse" into observed reality with the act of observation. In other words, when we look, within limits we find, but when we are not looking nothing exists but clouds of possibility.

By the 1930s, therefore, with such bizarre quantum possibilities gaining intellectual respectability, the Classical World View of the nineteenth century was a relic of the past, and knowledge of subatomic particles developed rapidly thereafter. In 1930 Wolfgang Pauli postulated the existence of a virtually massless uncharged particle he christened the neutrino. Despite the inordinate difficulties of observing the neutrino—detected only in 1954—it, too, soon entered the pantheon of new elementary particles. In 1932 the discovery of the neutron—a neutral body similar to the proton—complemented the electron and the proton. In the same year the detection of the positron—a positively charged electron—revealed the existence of antimatter, a special kind of matter that annihilates regular matter in bursts of pure energy. Based on the work of the flamboyant American physicist Richard Feynman (1918–88), quantum theory is now known as quantum electrodynamics or quantum field theory and, hand in hand with experimental high energy physics, it has revealed an unexpectedly complex world of elementary particles. With the use of particle accelerators of ever greater energy as research probes, nuclear physicists have created and identified over 200 different types of subatomic particles, most very short-lived. The Tevatron accelerator at the Fermi National Accelerator Laboratory in Batavia, Illinois, for example, currently the world's most powerful, attains energy levels of 1 trillion electron volts.

Today, physicists classify elementary particles into three main groups, each with its matter and antimatter components. Heavy particles such as neutrons and protons, for example, are now understood to be composed of tripartite combinations of six even smaller units called quarks. In a remarkable confirmation of the theory (called quantum chromodynamics) behind quarks, in 1995 physicists at Fermi Lab demonstrated the existence of the elusive "top quark." The electron is an example of the second class of particles known as leptons, which are generally lighter than neutrons and protons but which still have mass. Particles called bosons form the third group of elementary particles, which include photons or light quanta of electromagnetism which lack rest mass. Bosons are thought to be vectors for carrying the four known forces of nature: electromagnetism, gravity, and the strong and weak forces—the latter two governing radioactive decay and the binding of particles in the nuclei of atoms.

The quest for understanding the material stuff of the world—a quest that began with the Milesian natural philosophers in the fifth century B.C.—has achieved an unprecedented level of sophistication in our own time, and active research continues within the general theoretical frame established by contemporary quantum field theories. For example, elusive particles known as the Higgs boson and the graviton (posited to mediate the force of gravity) are now prime quarry. As part of this effort, theorists have worked to understand the forces of nature on a deeper, more unified level. They have achieved some success in this area, notably in the 1970s in conceptually unifying electromagnetism and the weak nuclear force into the so-called electro-weak force. An as-yet-unrealized Grand Unified Theory will perhaps add the strong nuclear force, but the Holy Grail of an ultimate fundamental theory of nature uniting all the forces of the universe, including a quantum theory of gravity, remains a distant goal. Such theorizing, while of great philosophical import, remains utterly useless for practical purposes.

Cosmology

Cosmology represents another area where theorists forged profound conceptual novelties in the twentieth century. In the eighteenth century, astronomers had broached the idea that the Milky Way was an "island universe," and observations of nebulous bodies in the nineteenth century suggested that many galaxies may populate the cosmos outside the Milky Way. Also in the nineteenth century the spectroscope, an optical instrument to analyze the light spectrum of the sun and stars, indicated

Fig. 17.1. Natural philosophy and big science. Today, discoveries in such fields as high-energy physics require large and expensive equipment and teams of scientists working together. Pictured is a portion of the four-mile main accelerator at the Fermi National Accelerator Laboratory in Illinois. The tunnel is 20 feet underground. The Tevatron accelerator (the lower ring) uses superconducting magnets to accelerate protons and antiprotons to very high energies. Sophisticated detectors display the results of particle collisions which are then "read" by technical experts.

a single chemistry common to the entire universe. Of even greater intellectual consequence, that lines in a spectrograph seemed to shift as a function of the motion of stars opened the possibility after 1870 that the universe may be expanding. (The application of dry-plate photography to astronomical practice in the later 1870s greatly facilitated spectrographic and related work.) Only in the 1920s, however, due particularly to the work of the American astronomer Edwin Hubble (1889–1953), did the extragalactic nature of "nebulous" bodies, the immense distances involved, and the apparent expansion of the universe become established among cosmologists.

Beginning in the 1930s, relativity and particle physics greatly impacted on cosmology. Einstein's equation of matter and energy, along with evolving understandings of nuclear processes, not only led to practical applications in the atomic and hydrogen bombs, but also to theoretical understandings of thermonuclear fusion as the energy source powering the sun and the stars. Such theorizing ultimately led to the recognition that chemical elements more complex than hydrogen and helium, including those necessary for life, arose in stellar furnaces. In addition, the discovery of thermonuclear processes, by replacing ordinary combustion models, greatly extended the age of the sun and the solar system and clinched the argument against Lord Kelvin and his critique of Darwinism premised on a comparatively short lifetime for the sun.

How to account for the apparent expansion of the universe remained a great debate among cosmologists through the 1950s. Essentially, two mutually exclusive views divided the allegiance of theorists. One, the "steady state" model, held that new matter appeared as space expanded, with the density of the universe a resulting constant. The alternative theory, first articulated by the Belgian abbé Georges Lemaître in 1931 and developed by the Russian emigré physicist George Gamow and colleagues in the 1940s and early 1950s, proposed that the universe originated in an incredibly hot and dense "Big Bang" and in a universe that continues to expand.

Both camps had their supporters and arguments, although most cosmologists seemed predisposed to steady-state models throughout the first half of the twentieth century. The debate was ultimately decided only after 1965 with the almost accidental discovery by two Bell Laboratory scientists, Arno Penzias and Robert Wilson, of the so-called 3° background radiation. The idea that explains their discovery is elegant. If the universe began as a hot fireball with a big bang, then it should have "cooled" over time, and calculations could predict what should be the residual temperature of the universe today. And it was this relic heat, measured by an omnipresent background radiation at roughly three degrees above absolute zero (2.73° K), that Penzias and Wilson stumbled upon. Their discovery won them the Nobel Prize in 1979 and sounded the death knell for steady-state cosmology.

The watershed discovery of the 3° background radiation showed the theoretical power forthcoming from the unification of particle physics and cosmology. In tandem, the science of the very small and the study of the largest astronomical systems proved extraordinarily fruitful in elaborating the coherent "Standard Model" of the universe. In the "inflationary" variant of that model generally accepted today, the universe began 10 to 15 billion years ago in a violent explosion out of what theorists extrapolate to have been a virtually unimaginable "singularity" when the universe was infinitely small. The known laws of nature do not apply at the moment of creation, but particle physics provides a reasoned account of the evolution of the cosmos from the merest fraction of a microsecond after the big bang: an initial union of energy, matter, and the forces of nature; a period of stupendous inflation as the universe ballooned seventy-five orders of magnitude in size coincident with a "phase shift" akin to a change of state, like boiling; the subsequent decoupling of energy and matter after a still comparatively brief period of 100,000 years as the primordial universe continued to expand and cool; and the ensuing evolution of galaxies and stars over the following billions of years, ultimately leading to the formation of our own nondescript solar system 5 billion or so years ago.

Many uncertainties affect this picture, and research continues in a number of key areas. Precisely how old is the universe? Experts seek a more exact value for the so-called Hubble constant to help settle the question. What will be the fate of the universe? Will it reach a limit and collapse back into itself? Or will it keep on expanding, as recent evidence suggests, to peter out in a "big chill"? Answers hinge on the amount of mass in the universe, and investigations are currently under way to discover the "missing mass" required to prevent the universe expanding forever. While the cosmos is extraordinarily homogeneous on a large scale, how can we explain the obvious nonuniformities evident in galaxies and, indeed, ourselves, on a smaller scale? How did the universe begin out of nothing? Work in quantum cosmology concerning "imaginary time," "virtual particles," quantum "tunneling," and the chance coming-into-being of matter out of the quantum vacuum may prove useful in addressing these issues. Research into "black holes"—those extraordinarily dense celestial objects, whose gravity prevents even the escape of light, but that "evaporate" energy—may also illuminate these ultimate questions. Today, theorists entertain such other esoteric entities as massive high-energy vacua called cosmic strings, "superstrings" collapsed from ten spatial dimensions of an original creation, multiple universes, and equally provocative, but perhaps intellectually productive, concepts. Such exciting research indicates the continuing vitality of pure science today. But to repeat a point made throughout this volume, the value of inquiries such as these derives principally from the intellectual and spiritual satisfaction they provide and not from any potential utility. They represent the continuation of the venerable Hellenic tradition, alongside the increasingly practical bent of scientific research in the twentieth century.

Life after DNA

Physics and cosmology have always been of prime significance in providing the basic reference frames for comprehending the world around us. But explaining life has been of no less importance, and in this domain biologists in the twentieth century have reformulated our view of living organisms. The uncovering of the double helical structure of DNA in 1953 by Watson and Crick, discussed in the previous chapter, provided a conceptional breakthrough for understanding reproduction, the nature of heredity, and the molecular basis of evolution. Also in 1953 a noted set of experiments that produced amino acids—the chemical building blocks of life—by electrifying a broth of more simple compounds supported the idea that life itself originated in the primeval conditions of the early earth. Alternative views include the idea that life arose from the catalytic action of clays, while some scientists, notably Crick himself, propose an extraterrestrial origin for life on Earth through spores evolved elsewhere and deposited from space.

Once again, in examining contemporary science we encounter not final answers, but a process of inquiry.

Along with molecular biology have come refined accounts of the history of life on Earth. As research on the details of plant and animal evolution have mounted over the last decades, the fundamental perspective has remained the Darwinian one of evolution through natural selection. Understanding the molecular basis of inheritance has provided important new ways to analyze evolutionary history and to classify living plants and animals. New approaches, including "cladistics," "molecular clocks," and the study of mitochondrial DNA, permit scientists to measure the rate of evolutionary change and evolutionary distances between species. Does everyone agree? Hardly, and biological scientists today pursue vigorous debates over a range of topics, including, for example, whether or not birds are the direct descendants of dinosaurs and whether evolution proceeds at a constant, slow pace or whether periods of evolutionary stability are "punctuated" by periods of rapid change. Fundamentalist Christian groups have seized on the latter debates as evidence of the failure of Darwinian evolutionary theory, but to do so misses the point that such debates constitute the norm in science.

Fundamental discoveries in paleontology and the history of human evolution figure among related developments in twentieth-century life science with strong natural philosophical implications. Although Darwin had postulated the evolution of humans through various stages from simian ancestors, a coherent story of human evolution remains decidedly the product of twentieth-century scientific thought. As noted earlier, the initial discoveries of the Neanderthal variety of *Homo sapiens* and of Paleolithic cave art (produced by anatomically modern humans) were not generally accepted until the turn of the twentieth century. The first *Homo erectus* fossil was unearthed in 1895, and the first *Australopithecus* surfaced only in 1925. The "discovery" of the notorious "Piltdown man" in 1908, with his big brain and robust, ape-like jaw, argued strongly against human evolution from more primitive, small-brained ancestors, and only in 1950 was the Piltdown artifact conclusively shown to be a fraud. Since that time, an increasingly clearer picture has emerged of the stages of human evolution, from *A. afarensis* ("Lucy" and her kin first uncovered by Donald Johanson in 1974), through *H. habilis, H. erectus,* and then varieties of *H. sapiens.* Exciting research is ongoing in this area, too, and we have had to introduce subtle reinterpretations because of new findings reported even while writing this book.

A categorical distinction between humans and the rest of the living world formed an essential principle of previous worldviews which has now been undermined by the life sciences in the twentieth century. In one area, however, resistance remains to extending to humans and human social behavior scientific findings taken from the animal world.

We refer to sociobiology and evolutionary psychology, fields which postulate that patterns of culture may be explained on Darwinian evolutionary principles. Thus, for example, evolutionary accounts have been proposed for altruism, aggression, cooperation, incest taboos, alcoholism, gender differences, homosexuality, and attitudes toward children and strangers. In the 1960s and 1970s such ideas were attacked as offensive to liberal democratic ideals, for they seemed to limit the range of social reform, but increasingly the research is gaining acceptance from scientists and the informed public alike.

As the volume of scientific activity has mushroomed in the twentieth century, so, too, has the range of theoretical innovation expanded. In geology, for example, the idea that the continents may actually be "plates" riding on the earth's mantle is of considerable theoretical import. This idea, generally rejected after it was first put forward in 1915 by the German geologist Alfred Wegener (1880–1930), gained virtually universal acceptance in the 1960s for a variety of technical and social reasons. Understanding plate tectonics and continental drift has allowed scientists to "run the film backwards" and thus to recapitulate the geological history of the earth, with all the implications that holds for both geology and biology.

Another noteworthy example concerns evidence uncovered in 1980 suggesting that a catastrophic meteorite or cometary impact caused a mass extinction, including the dinosaurs, at the end of the Cretaceous period 65 million years ago. The significance of this finding lies in what it implies about the role of accident in the history of life, including the ultimate fate of humans. Similarly, planetary space science along with pictures sent back from space—including the glorious shot of the "pale blue dot" of the earth taken from the Voyager spacecraft 4 billion miles in space—have transformed how we visualize the earth and its planetary near neighbors.

Elements from these and still other disciplines combine to forge today's scientific worldview. Is what we think today in some final sense true? Clearly not. Still, science represents the best means we have for saying things about the world, and today's story is demonstrably better than any other account given to date. Of course, that realization does not mean that the story will not change again in the future, and undoubtedly it will.

CHAPTER 18

Applied Science and
Technology Today

How, then, did science and technology achieve the merger and integration that we observe today? That was the question with which we began this book, and in the course of this historical survey we have collected more than a few clues to its answer. In the beginning there was only technology. Then, 6,000 years ago in the first civilizations, science originated in the form of written traditions of mathematical and astronomical knowledge. But this development occurred in the social context of state-level societies where the central government saw to it that science was applied to the needs of a complex agricultural economy. This pattern of limited state support for useful knowledge repeated itself wherever and to whatever extent strong central states appeared. In classical Greece, where the state was weak, science in the form of natural philosophy and technology as craft remained estranged. Later centralized societies influenced by Greek philosophy merged pure and applied science traditions, but, again, without the great body of contemporary technology being affected by either. Only in the nineteenth and twentieth centuries, and only slowly and grudgingly, have governments and an increasing number of industries fully recognized the possibilities of applying theoretical research to the problems of technology and industry. The result is a dramatic expansion in the applications of science to technology, this time under the rubric of R&D—research and development.

Technical Careers

An elaborate system of social practice surrounds the activities of scientists today. In the early nineteenth century scientific careers were haphazardly arrived at and concerned only a handful of people. Today, the social role of the scientist and what it takes to become a scientist are fairly rigidly defined, and the pursuit of science is a full-time occupational concern of almost 1 million people in the United States alone who engage in R&D. Indeed, although wide latitude exists in scientific employment, the social pathways available for scientific training are quite narrow. In many respects, whether someone will be a scientist is already determined in his or her middle school years. Almost universally, one has to graduate from high school with the requisite science courses, attend college and complete an undergraduate major in a field of science, pursue graduate work and, on the basis of original research, obtain a Ph.D., and then, normally, continue training for a period of years after the Ph.D. in temporary positions known as post-docs.

The Ph.D. constitutes the scientist's license, and from there career paths diverge. Traditionally, the norm has been to pursue academic careers of research and teaching in universities. But increasingly, young men and women of science find productive lives for themselves in government and private industry. Also, what scientists do today spans a broad range of practice: from disinterested pure science research in universities or specialized research facilities to more mundane applied science work in industry. Generally speaking, whether in academe, industry, or government, younger scientists tend to be the more active producers of scientific knowledge. The more one moves up the career ladder in science today, the more one tends to leave active research and turn toward scientific administration and directing the research efforts of others.

The status of women in science changed dramatically in the twentieth century. Women have never been completely absent from the social or intellectual histories of science, and as they gained admission to universities in the nineteenth century, gradually increasing numbers of women have become scientists. Heroic accounts often single out Marie Curie, who became the first female professor at the Sorbonne and the only woman ever to win two Nobel Prizes—the first in physics in 1903 shared with her husband, Pierre Curie, for their work on radioactivity, and the second in 1911 on her own for her discovery of radium. Like Einstein, the Austrian physicist Lise Meitner (1878–1968) held a professorship at the University of Berlin before leaving Nazi Germany; she played a seminal role in articulating the theory behind atomic fission and the atomic bomb, and her case exemplifies the increasing presence and professionalism of women in modern science. With her mastery of X-ray diffraction techniques, physical chemist Rosalind Franklin (1920–58) was instrumental in the discovery of the double helix of DNA, but her story presents cautionary lessons as well, because the social system of British science in the 1950s did little to welcome her or give her credit for her work. It remains difficult for women to pursue research and applied-science careers in academe and industry, but mirroring other shifts in Western societies over the last several decades, opportunities have improved for women, and the idea of a female scientist is no longer exceptional.

Mainstream scientists are expected to publish research results in scientific journals, and the old adage of "publish or perish" applies to life in contemporary science. Participation in scientific societies and at scientific meetings comprises another requisite part of scientific life today. Normally scientists belong to several societies that represent their specialized interests and their general professional status. For example, in addition to joining smaller organizations concerned with specialized research areas, a physicist will probably belong to the American Physical Society (1899), a chemist to the American Chemical Society (1876), an astronomer to the American Astronomical Society (1899), and they will all belong to the American Association for the Advancement of Science (AAAS, 1848). Such organizations usually hold annual conventions where scientists gather to present their research, to participate in the organizational life of the society, and to party.

Most scientific research today is an expensive enterprise; getting grants forms another essential aspect of normal scientific practice. Scientists often spend a great deal of time on a grant treadmill that calls for them to write proposals to get funds to do research to support more proposal writing and more research. Grants typically include funds for principal investigators' salaries, graduate and post-doc support, and equipment. In what amounts to government and foundation subsidies of institutions, a significant element of all such budgets are so-called overhead or indirect costs, on the order of one-third of a total budget, paid not to support research but to underwrite the grantee's institution. Private and public agencies, such as the National Science Foundation (NSF), solicit proposals and judge them through intricate procedures that involve outside referees, internal study groups and panels, and financial decisions by program officers and governing boards. Today, the overall success rate of applications to the NSF stands at about 30 percent, although only 15 percent of submissions to other more competitive programs are funded.

Honorary organizations such as the National Academy of Science (1863) in the United States and like institutions in other countries cap the elaborate social organization of contemporary science. A number of international organizations and commissions, such as the International Council of Scientific Unions (1931), govern science on a world level. In the public mind at least, at the very top of the social and institutional pyramid of contemporary science stand the Nobel Prizes, given annually from 1901 in several fields of science—physics, chemistry, and medicine/physiology.

Today, engineering and technology are also fully professional occupations that share similarities with the world of science. Both technology and science are highly competitive enterprises involving research, and it usually takes sophisticated technical education and training to enter the world of technology today. Technology and engineering are, like science, organized into a host of specialized and professional societies, such as the American Society of Mechanical Engineers (ASME,

1880) and the National Academy of Engineering (1964), while any number of specialized periodicals also serve the needs and interests of practicing engineers.

These similarities notwithstanding, the contrasts between the world of science and the training and practice of engineers and technologists remain profound and revealing. In the matter of education, for example, university training of engineers arose only in the nineteenth century, and today an undergraduate degree in engineering is a terminal degree satisfactory for entering professional ranks. A master's degree in a field of engineering or technology is the only advanced degree that practitioners occasionally seek to round out their educations, while the Ph.D. is usually reserved for engineers who wish to go into university research and teaching.

Both engineer/technologists and scientists pursue research, but here, too, fundamental differences distinguish the enterprises of science and technology even today. Scientific research, for example, most often focuses on very narrowly defined problems, such as the structure of a particular protein or the intensity of magnetic fields at the southern pole of the sun. Such research generally takes into account only a limited number of other recently published scientific papers, and it is most often directed at a restricted "invisible college" of fellow research practitioners. Engineering and technical research, by contrast, ordinarily encompasses a broader problem set (e.g., creating practical technologies for videoconferencing or electric cars). Solutions very often involve a more diverse range of elements—from scientific knowledge to choosing materials, the aesthetics of design, manufacturing considerations, financing, and marketing. And the consumers of engineering solutions are not usually other engineers or scientists but, more often, governments, corporations, and the general public.

Science and technology have drawn closer together in the twentieth century through increased understanding and exploitation of science for applied ends. That said, however, it may be analytically useful today to distinguish pure science on the one hand from a merged applied science and technology on the other. That is, a social and institutional separation today exists not, as in the past, between science and technology but between theoretical scientific inquiries and scientifico-technological applied science. The product of scientific research, for example, is new knowledge; applied scientists, technologists, and engineers, on the other hand, strive to produce useful material things or processes. The results of scientific research are generally not of immediate economic value. Scientists get professional credit when other people cite and use their work, and therefore scientists are inclined to "give away" the results of their research in the hopes of such nonpecuniary social rewards. In contrast, the products of engineers and applied scientists possess real economic value. Thus engineers and the companies that employ them tend to keep work secret until they receive a patent that secures their economic rights. This difference in work product—papers versus patents—is telling of differences between the world of pure science and the worlds of applied science and technology. By the same token, a certain blurring of those distinctions becomes evident in cutting-edge industries like biotechnology and computers where, these days, new scientific discoveries are regularly patented rather than published and made publicly available to other researchers.

Explosive Growth

In assessing science in the twentieth century and its relation to contemporary technology, we need to recognize that more is involved than simply the linear evolution of scientific ideas or successive stages in the social and professional development of science. The exponential growth of science represents another characteristic feature of the history of modern science. By every indication, science has grown at a geometric rate since the seventeenth century, outpacing other social indicators such as population. In figure 18.1, for example, one sees that the scale of the scientific enterprise has increased a millionfold since the seventeenth century, doubling in size about every 15 years.

Several paradoxical consequences follow from the exponential growth of science in the modern era. For example, a high proportion of all the scientists who ever lived—some 80–90 percent—are alive today. And, given exponential growth, the scientific enterprise clearly cannot continue to expand indefinitely, for such growth would sooner or later consume all resources, human, financial, and otherwise. Indeed, as predicted, an exponential growth rate for science has not been maintained since the 1960s and 1970s. In any particular area of science, however—especially a hot new one such as superconductivity or AIDS research—exponential growth and an ultimate plateau remain the typical pattern.

Other metrics, notably citation studies, add to these basic understandings of the character of the contemporary sciences. Scientists back up their results by referencing other papers, and much can be learned from studying citation patterns. Such studies reveal, for example, that a significant percentage of published work is never cited or actively used. In essence, much of the production of contemporary science disappears into "black holes" in the literature. (But that fraction is considerably less than in the humanities.) Citation studies also show a drastically skewed inequality of scientific productivity among researchers. That is, for any selected group of scientists, a handful will be big producers, while the vast majority will produce little work, if any. Typically, for 100 scientific authors, two will write 25 percent of the papers. In all, the top 10 will be responsible for 50 percent of the papers, while 90 will produce the remaining 50 percent. Most scientists produce only one or two papers in a career. A corollary of this differential productivity affects institutions, too, in that a handful of elite research universities attract top producers and dominate the production of science in the United States.

Citation studies also show an unusual "half-life" of scientific information. That is, new work in science tends to evolve out of other recent work, and therefore the citation and presumably the utility of scientific knowledge fall off with time. Older scientific results prove less useful to practicing scientists than more recent science, and so, in contrast with the traditional humanities, the scientific enterprise displays a characteristic present-mindedness. The works of Shakespeare or of Homer, for example, still speak to the literary critic or to a writer active today, whereas Newton's works, much less Aristotle's, are of no value to the practicing scientist. As a result, old science is consistently ignored in the teaching of science, except insofar as it is filtered through textbooks, which serve to convey the content—and not the history—of science to students. Science teachers also often associate some historical figure with a scientific law, such as the analysis of the action of a spring that goes under the name of Hooke's law. Taken together, the half-life of scientific information revealed in citation studies, the fact of scientific revolution, the demographic preponderance of living scientists, and pedagogical practices, all make plain the distinctive fact that science repudiates its past.

Big Science and the Bomb

The development and use of the atomic bomb by the United States during World War II marks a watershed in the history of modern science and technology. The reasons are twofold. One, the case dramatically revealed the practical potential of science or what could emerge from the turning of theory to useful ends. Second, it demonstrated what might be forthcoming when government supported large-scale scientific R&D with abundant resources. Neither of these considerations was entirely new in the 1940s. The application of theory to practice had already occurred in the nineteenth century, as evidenced in the dye and electrical industries, among others. Government support for potentially useful research was evident in World War I, just as it had been in ancient Egypt. The novelty of the atomic bomb case—and what it portended for the future of science and technology in the second half of the twentieth century—stemmed from the combination of these two factors: large-scale government initiatives to exploit scientific theory for practical ends. It set a new pattern that changed not only traditional relations between science and government, but how we think about applied science in general.

The story of the atomic bomb is well known and can be quickly told. The scientific theory that made the bomb possible emerged only in 1938 and 1939. In 1938 German physicist Otto Hahn (1879–1968) demonstrated that certain heavy elements (such as uranium) could fis-

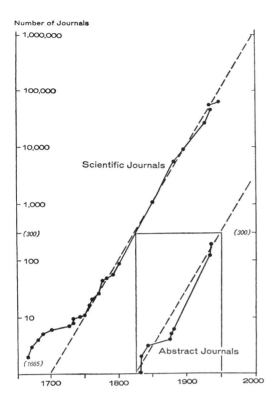

Number of Journals

Fig. 18.1. Exponential growth. The scientific enterprise has grown exponentially over the three centuries following the Scientific Revolution, as indicated in this logarithmic graph of the growth of scientific journals since the seventeenth century. The quantitative study of parameters affecting science can shed useful light on the nature of the scientific enterprise.

sion or be split into more simple components, and then in 1939 Lise Meitner, an Austrian physicist who had emigrated to Sweden from Nazi Germany, proposed a theoretical explanation for fission and calculated the immense amounts of energy that in principle could be released from an explosive nuclear chain reaction. With war under way in Europe, Allied physicists recognized the destructive potential of a nuclear bomb and the threat that Germany might develop one. In a historic letter dated August 2, 1939, Albert Einstein wrote to President Franklin Roosevelt about the matter, and, as a result, Roosevelt authorized a small exploratory project. In the fall of 1941, on the eve of the U.S. entry into World War II, Roosevelt gave the go-ahead for a major initiative to create atomic weapons. The result was the largest science-based R&D venture in history. The resulting Manhattan Project, under the overall command of U.S. general Leslie Groves, involved 43,000 people working in thirty-seven installations across the country, and it ended up costing 2.2 billion contemporary dollars. In December 1942 the Italian emigré scientist Enrico Fermi succeeded in creating the first controlled nuclear chain reaction beneath a football stadium at the University of Chicago. And, on July 16, 1945, the team directed by American physicist J. Robert Oppenheimer set off the world's first atomic explosion at the Trinity site near the Los Alamos labs in New Mexico. On August 6 the *Enola Gay* dropped a uranium-235 bomb on Hiroshima, Japan, killing 70,000 people, and on August 9 a plutonium-239 bomb fell on Nagasaki. Japan surrendered five days later.

The atomic bomb brought World War II to a dramatic end. At the same time the bomb launched the Cold War that followed. The political, scientific, and military-industrial establishments of the United States and the Soviet Union raced to develop larger atomic weapons and then, from 1952, even more powerful hydrogen or thermonuclear bombs that derived their energy not from the fission of heavy elements but from the fusion of hydrogen into helium. World War II gave a push to a number of other government-funded, applied-science projects, such as radar, penicillin production, jet engines, and the earliest electronic computers. The war established a new paradigm for science and government relations that has endured to this day, that of comparatively large-scale government investment in pure and applied science in the hope of large-scale payoffs in industry, medicine, and military technologies. The success of the Manhattan Project—in so closely linking novelties in theory to an immediately useful application—likewise forged a new image of connections between science and technology. These endeavors, which in so many ways—historically, institutionally, sociologically—had for so long been largely separate enterprises, became fused in the public mind. After World War II it became hard to think of technology other than as applied science.

In many respects, too, the Manhattan Project typified a new way of doing science, that of the industrialization of scientific production or what has been called Big Science. In the nineteenth century, the individual scientist working alone or with a few coworkers in a small laboratory represented the dominant mode for the production of scientific knowledge. With the development of nuclear physics in the twentieth century, however, that age-old pattern changed. Research began to necessitate large installations and expensive equipment, increasingly beyond the resources of individual experimenters or even universities or private research facilities. Teams of scientific researchers began to replace the labors of individual scientists. Each team member became a specialized science worker in charge of one aspect of a complicated research endeavor. Scientific papers issuing from such team-based science came to be authored sometimes by hundreds of individuals. For example, the discovery of the "top quark" in 1995 was produced at the Fermi National Accelerator Laboratory by two separate teams, each numbering 450 scientists and technicians staffing two detectors costing $100 million apiece. For another example, the National Institutes of Health, a federal agency, employ over 16,000 people. Individual or small-group research continues in many fields, such as botany, mathematics, and paleontology, but in other areas, such as particle physics, biomedicine, or space exploration, Big Science represents an important novelty of the twentieth century.

Science as a Mode of Production

The promise of socially useful knowledge is an old theme that harks back to the origins of civilization. Throughout this study we have seen that governments, rulers, states, and political entities always and everywhere valued useful knowledge and supported experts who possessed (or claimed to possess) such knowledge. Utility in the service of the state was the goal, and, to a degree, the state got its money's worth in cadres of specialists who could keep lists, tally the harvest, calculate the days and seasons, follow the sun and moon, collect taxes, construct altars, forecast the future, heal the sick, tell time, direct public works, maintain inventories, make maps, and indicate the direction of Mecca, to list just a few activities undertaken by scientific specialists over the millennia. As noteworthy as such activities were for the development of science and society, up until the nineteenth century the practical effect of science and medicine on the organization and functioning of states and societies remained comparatively small. That is, expert activities such as reckoning, healing, and forecasting were important and perhaps even necessary for civilization, but civilizations were nonetheless overwhelmingly traditional societies where agriculture and customary crafts provided the mainsprings of civilized life. Although "science" may have played a part in the history of civilizations, until recently that part remained a small one, and state support for scientific expertise thus remained at a comparatively and correspondingly low level.

Consider, as an example, France during the era of its Sun King, Louis XIV, whose personal reign extended from 1661 to 1715. At the time, France was technologically and culturally the most advanced civilization in Europe, if not in the world, and no other contemporary state supported the natural sciences to the degree that France did. The Royal Academy of Sciences in Paris stood as the premier scientific institution in the world. It was an institution devoted explicitly to the sciences; its list of members reads like a Who's Who in the world of late seventeenth- and eighteenth-century science; its expeditions and scientific work remained unparalleled, its publications unrivaled. Nevertheless, the Academy was seriously underfunded in the last decades of the seventeenth century, its members paid late, if at all. Fine arts and literary academies received much higher subventions than did the science academy, and scientific academicians held a comparatively lower social status than their literary and fine arts counterparts. Indeed, a smaller and ephemeral academy of technology received relatively more government support, and the Academy of Sciences was constantly pressured to turn its attentions to useful ends, since it otherwise had essentially no impact on the tax base or the growing of grain in France. Thus, as much as we need to locate the modern origins of government support for science in seventeenth- and eighteenth-century France, we need also recognize the comparatively limited impact of science on society until more recent times.

The promise of state patronage of applied science fully materialized only in our own century. Here again the Manhattan Project proved pivotal, in demonstrating that support for theoretical science on a large scale can pay off in big ways for patrons and for society. That pattern and the fully modern model of the "union" of science and technology continued in the post–World War II world, the second half of the twentieth century. For example, in 1930 the United States supported science to the tune of a mere $160 million and 0.2 percent of its Gross National Product. By 1945 those figures stood at $1.52 billion and 0.7 percent of GNP. By 1965 they had risen to over $20 billion ($15 billion public, $5 billion private) and 3 percent of GNP. In 1995 $73 billion in federal dollars went to scientific R&D, a figure that alone represents 2.6 percent of GNP. The nation's total public and private investment in basic research now hovers at $160 billion.

Military and perceived defense needs overwhelmingly drive government patronage for science in the United States. Indeed, the science budget for the U.S. Department of Defense is over two and one-half times that of the next largest consumer of federal science dollars, the Department of Health and Human Services (99 percent of whose research funds go to the National Institutes of Health [NIH]). Defense-related science expenditures continue to receive nearly half of all federal science spending. Similarly, federal dollars overwhelmingly go to support applied science and technology, on the order of an 80/20 applied and pure split overall, with applied science in defense-related spending approaching 100 percent (96.5 percent for FY1997). The descending order of funding from the Defense Department, to DHHS, to the Department of Energy, to NASA, and to the National Science Foundation (NSF) is also revealing in this regard in that funding for defense, health, energy, and NASA space projects (which have always entailed significant political and employment goals) come well before support of the agency nominally charged with the promotion of pure scientific research, the NSF (founded in 1950). And even there only $2.479 billion of the NSF's overall FY 1995 budget of $3.027 billion goes to research; indeed, the budgetary philosophy at the NSF increasingly directs research and dollars into so-called strategic areas linked to national needs and to harness science to economic growth.

The divergent fates of two government-sponsored Big Science projects highlight the forces shaping science today. The first concerns the cancellation in 1994 of the Superconducting Supercollider, a gigantic, publicly funded particle accelerator then under construction in Texas designed to pack 20 times the energy of the Tevatron at Fermi Laboratory in Illinois. Although scientists made vague claims as to its potential utility (to "cure cancer," for example), most recognized that such a fantastic machine would be of primary importance for natural phi-

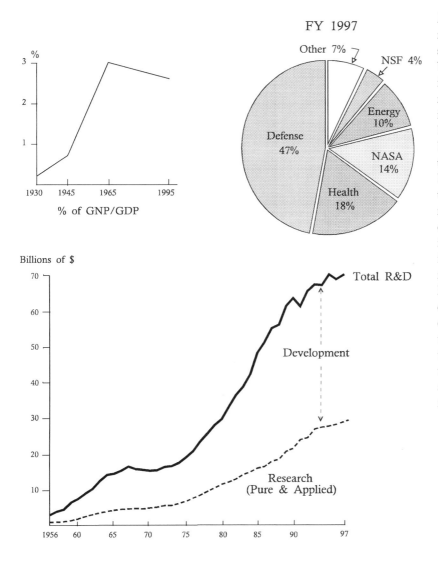

Fig. 18.2. Federal research and development spending. The level of U.S. government support for science and scientific research has risen dramatically since the 1940s, both in terms of the number of dollars and relative to the overall federal budget. Since the 1970s the proportion of federal monies allocated for development has increased compared to funds earmarked for research. In a pattern that harks back to the earliest civilizations federal expenditures on science today are heavily skewed toward useful applications, notably in defense and health-related matters.

losophy—that it would help produce exciting new knowledge of the fundamental constituents of the universe. When the costs of the Supercollider escalated to $11 billion in public dollars, politicians could no longer justify its expense, even as a public works project, and they shut it down. In contrast, the human genome project, the effort to map the panoply of human DNA, promises to be of great practical utility for society at large, in discovering and possibly curing genetic diseases such as Huntington's disease, cystic fibrosis, or muscular dystrophy. Although projected federal funding for the genome project for FY 1998 stands at a modest $202 million, one can be confident that its funding will never be in jeopardy.

These same patterns and pressures are also apparent in other countries that support scientific research through government subventions. Although science is currently funded at unprecedentedly high levels by government and industry, the rationale for such support of scientific knowledge essentially has not changed since ancient Babylonia, the hoped-for practical benefits that may be derived from higher learning.

A new and fruitful paradigm of applied science has thus emerged to alter connections between science and technology in the second half of the twentieth century, and it will doubtless continue to shape the course of history in the twenty-first. The vision of a scientific medicine, for example—a vision articulated by René Descartes in the seventeenth century—today seems well on its way toward fulfillment. The twentieth century has seen stunning developments in medical science which represent the application of fundamental research in biology, chemistry, and physics to the identification, understanding, and treatment of disease. Although penicillin was discovered by accident in 1928, today many new antibiotics and drugs are developed through systematic experimentation and "designer" chemistry. The medical application of X-rays followed almost immediately upon their discovery in 1895, and today such powerful diagnostic tools as MRIs (magnetic resonance imaging) and PET scans (positron emission tomography) spring from the field of nuclear physics. To repeat a point, the discovery of the double helical structure of DNA by Watson and Crick in 1953 marked a milestone in the scientific understanding of the nature of life and inheritance; today, the burgeoning fields of DNA testing and applications in medicine, agriculture, and forensics evidence the practical and economic significance of their discovery.

As remarkable as have been the accomplishments of scientific medicine in the twentieth century, its legacy remains mixed. The problem of resistant bacteria, for example, results directly from the widespread use of antibiotics and pesticides. Improved treatments for infertility have raised a host of unprecedented ethical and legal issues concerning the rights of children and biological and nonbiological parents. The growing expense of sophisticated medical procedures seems to have divided industrial societies into subgroups that can and cannot afford access to advanced medicine. Genetic testing threatens discrimination in medical and insurance coverage. New genetic databases available to law enforcement agencies promise more reliable identification of felons but raise troubling questions for civil liberties. The cloning, genetic engineering, and the patenting of life forms now pose other issues.

Computers and computer engineering represent another area where applied science has transformed society. Scientists working at Bell Laboratories created the first solid-state transistor in 1947, an achievement for which they won the Nobel Prize in physics in 1956. Such solid-state devices eventually supplanted vacuum tubes (which regularly failed), and they permitted the development of the first practical large computers in the 1950s and 1960s. Since that time computer science—a hybrid field encompassing logic, electronics, and raw experimentation—has developed across a broad range of practice and theory that includes artificial intelligence as well as new computing devices and games. Today, computers are a common consumer item, and large and small they have revolutionized entire areas of our social existence, from banking, to the stock market, to airline and concert reservations.

One overriding conclusion stands out: the promise of applied science that first manifested itself with government support of science and scientific experts in the pristine civilizations beginning in the fourth millennium B.C. became substantially realized only in the twentieth century. The result has been that support for science and applied knowledge from government and industry has grown to match the long-standing rhetoric of the utility of science. As characteristic and significant as the intellectual and social rapprochement of technology with science has been in the twentieth century, the precise bonds between contemporary science and technology need to be pursued. In many areas today, as indicated above, technology represents the direct application of scientific advance and theory. In many other areas, however, even today it is misleading to envision technology as simply or directly "applied science." For example, the education and on-the-job training of technologists and engineers entail significant doses of science, yet rarely does that component consist of advanced theoretical science from the research front. Rather, "boiled down" science usually suffices for the practicing engineer or science-based technologist. NASA scientists and engineers had no need of and did not apply cutting-edge research in relativity or quantum mechanics in launching the Apollo moon missions or any of the other outstanding successes of the U.S. space program. The old celestial mechanics proffered by Newton in the seventeenth century and perfected by Laplace at the turn of the nineteenth century provided an entirely sufficient basis for calculating trajectories in the space program.

Then, even where scientific knowledge forms the basis of a device or a technology, so much else is ordinarily involved that it is often still misleading to characterize a new technology as simply applied science. For example, in 1938 Chester A. Carlson, an American inventor using his "boiled down" knowledge of optics and photochemistry, produced the first xerox, a dry-process photocopy method. Necessity was hardly the mother of invention in this case, as Carlson spent fruitless decades trying to persuade various backers, including a skeptical International Business Machines Corporation (IBM), of the utility of his substitute for carbon paper. In the end, it took engineering and design improvements that had little to do with science to perfect a practical photocopying machine, and, in fact, a marketing decision to lease rather than sell the machines proved the key to the stunning success of the first Xerox copiers in the 1960s. Only at that point did a perceived need for photocopies multiply and the photocopy machine become a technological commonplace. In this instance invention was once again the mother of necessity.

The fact that many different styles and brands of copy machine exist on the market today reveals another difference between contemporary science and technology. That is, the scientific community generally recognizes only one solution to a given scientific puzzle or problem. A particular protein, for example, will have only one agreed-upon chemical makeup and structure. In technology, by contrast, even in science-based technologies, multiple design and engineering solutions are commonplace. Sometimes different designs are intended for different uses, such as personal copiers versus large office copiers. Other times, in the variety of VCRs or personal computers presently available, one gets the sense that multiple designs exist merely for marketing purposes.

Science and technology have indeed forged historically significant new types of interaction in the twentieth century. These range from what has been labeled a "strong" direct application of scientific theory to an applied-science product or process, such as the atomic bomb, to "weak" or "boiled down" interactions of science and technology of the sort illustrated by Chester Carlson and his copy machine. By the same token, the traditional independence of technology from science and natural philosophy also continues today. Technical innovations continue to crop up which have nothing to do with the world of science or theory and which represent continuations of independent traditions in technology, engineering, and the crafts characteristic of technology since its roots in prehistory. In 1994, for example, Navy veteran Theodore M. Kiebke received patent 5,361,719 from the U.S. patent office for a new variety of cat litter made out of wheat. His invention has potentially significant consequences for a $700 million-a-year industry, yet it seems unreasonable to claim that Mr. Kiebke's new cat litter represents any kind of applied science.

Nonetheless, science and technology together have become indisputably significant forces in shaping the modern world. Marxist thinkers in the former Soviet Union recognized the novelty of this development, labeling it the twentieth century's "scientific-technological revolution." The term has not been taken up generally, but it does capture much of what is at issue here: the effective merger of science and technology in the twentieth century and the increasing importance of science-based technologies and technology-based sciences for the economic and social well-being of humanity today.

In the period following World War II science enjoyed an unquestioned moral, intellectual, and technical authority. Through the operations of its apparently unique "scientific method," theoretical science seemed to offer an infallible path to knowledge, while applied science promised to transform human existence. Paradoxically, or perhaps not, just as the effects of a fused science-technology culture began to refashion advanced societies—with the bomb, television, interstate highways, computers, the contraceptive pill—so, too, beginning in the 1960s did a wave of antiscience and antitechnological social reactions manifest themselves. The back-to-nature movement of the hippie counterculture of the 1960s and 1970s represents one example, and historical parallels are strong between such movements and romantic antiscience reactions in the early nineteenth century. More recently, technological failures such as the explosion of the nuclear reactor at Chernobyl, perceived threats from depletion of the ozone layer, and the spread of new diseases such as AIDS or the Ebola virus, all make many people suspicious of the material benefits of science and technology. Increasing societal concerns with ecology, recycling, "appropriate technologies," and "green" politics derive from such suspicions.

Intellectual critiques of science that emerged in the 1960s likewise have to be understood in the antiscience context of those times. Detailed work then and since in the philosophy of science and the sociology of science and scientific knowledge have challenged any uniquely privileged position for science as a means of knowing or staking claims about any ultimate truth. Most thinkers now recognize that scientific knowledge claims are relative, fallible, human creations, and not final statements about an objective nature. Although some would seize on this conclusion to promote intellectual anarchy or the primacy of theology, paradoxically no better mechanism exists for understanding the natural world around us than the human enterprise of science and natural philosophy.

From most historical perspectives, then, the social and intellectual circumstances of science and technology today would seem unique. Although the patronage and institutionalization of science as useful knowledge originated at the dawn of civilization itself, governments and industries today funnel unprecedented resources into the support of pure and applied science at a categorically higher level. Science as natural philosophy had its origins with the Greeks, and, although that enterprise also continues today, the content of today's science is radically different not only from what the ancient Greeks thought but, more to the point, from fundamental scientific conceptions universally held as recently as the beginning of the twentieth century. As we have seen, technology has even deeper roots in our biological heritage and in prehistory, yet the Industrial Revolution of the eighteenth and nineteenth centuries created an utterly new mode of human existence in the form of Industrial Civilization, and the full merger of science with the engines of economic production constitutes yet another novelty of the twentieth century.

The Medium of History

Our study has shown that technology has functioned as a fundamental driving force in human history. Technology proved decisive in fashioning and maintaining human societies in the Paleolithic and Neolithic eras and in every civilization since. Undoubtedly, then, as long as humankind exists and inhabits planet Earth, humans will continue to shape their world using their technologies.

We likewise saw that science-based technologies operated only since the first civilizations as governments recruited experts and expert knowledge in the service of running the state—in mathematics, astronomy/astrology, engineering, alchemy, medicine, and, later, cartography. The strong link between science and *industry* that we are familiar with is a comparatively new connection forged since the Industrial Revolution. As each of us experiences on virtually a daily basis, this science-industry connection represents a potent historical novelty with both positive and negative effects. The application of knowledge from the biomedical sciences, for example, has undoubtedly improved the human condition; the long-term effects of the chemical or science-based armament industries are more problematic. In the short term, the pace of "progress" in these domains is likely to continue. The long-term future of applied science in industry and in the military is cloudy, and to some extent worrisome.

What of the future of science itself? Oddly reminiscent of discussions that arose at the end of the nineteenth century concerning closure of the Classical World View, renewed talk is heard these days of the "end of science." Usually what is meant is that science will shortly figure everything out, that it will reach a conceptual conclusion of some sort. The "end of science" scenario derives from asking established scientists about problems remaining for their disciplines to solve. But that approach does not reveal the intellectual future of science. We do not now know the problem areas in biology or physics or cosmology or other yet-undreamed-of disciplines that may lead to radically new formulations. If science is the reasoned story humans tell about nature, then, as long as natural philosophy exists as a social and intellectual enterprise, that story would seem bound to continue to change. Indeed, one secure lesson of the history of science is that virtually every scientific formulation to date has failed—to be replaced by a better one.

There is one way in which we do need to consider the possibility of science's demise: the scientific enterprise viewed as a social endeavor may well come to an end, even as human societies continue. Science is a historical phenomenon. It came into being, and it may well pass away, just as ancient Greek science and medieval Islamic science did in their turn. There would seem to be nothing inevitable about the continuation of secular natural philosophy: indeed, on a global plane, the vision of the world articulated by the natural sciences today is probably a minority view and a fragile one. As the popular mentality turns away from the traditions of science and the human enterprise of trying to decode the world around us, science could well lose its central position in high culture.

We have largely left out of this account any explicit consideration of the discipline of history and the research enterprise that leads us to tell this story and not another. We have made this omission reluctantly but deliberately, in order to simplify the presentation for a particular audience, but the serious reader who has come this far needs to recognize that what we have said regarding science and technology in history comes to us, not through some objective or final appeal to unquestioned realities of the past, but through the living interpretative medium of history as practiced by historians. The reader who consults works in the accompanying guide to further reading will discover not truth and unanimity but a vibrant craft practice, lively debates concerning explanations of historical change, and sophisticated research undertaken to support or undermine one argument or another. We cannot therefore offer timeless conclusions for the edification of readers. Instead, we need to acknowledge the historiographical bases and biases of this account and the limitations inherent in all narrative.

As the study of the past, the discipline of history provides an unreliable guide to thinking about the future. The often-quoted aphorism of the philosopher George Santayana (1863–1952), "Those who cannot remember the past are condemned to repeat it," has its appeal both in suggesting that historical study can be useful and in warding off the notion that history is pursued by historians only for the pleasure of pure understanding. In the final analysis, however, Santayana's commandment rings hollow because it posits that what happened in the past is applicable to changed circumstances in the future. But there is no predictable direction to history. History does not cyclically repeat itself—the present differs from the past, the future will differ from the present, and, therefore, what we may learn of the past can be of only limited utility in understanding, much less affecting, the future.

Enlightenment thinkers of the eighteenth century first popularized the idea that there was a secular, progressive direction to history. The Enlightenment's view of human progress and social betterment predominated well into the second half of the twentieth century, and histories of science and technology were called on to exemplify that progress. From today's critical perspectives, however, and from the long view encompassed in this study, it seems clear that progress is neither inevitable nor necessarily sustainable. In particular, the Industrial Revolution and its consequences over the last two centuries have transformed historical circumstances so rapidly and in such profound ways that the current modes of intensified industrial existence are not likely to continue.

Guide to Resources

General Orientations

Given the nature of this book as a textbook and as an introduction to the histories of science and technology for the general reader, it was thought best not to clutter the pages with citations and references. Instead, we are providing this supplementary guide for those readers who may wish to delve further into the subject or who may seek clarification of some episode of the story. An excellent resource, available in many libraries, that covers the history of science (and in some measure technology) through the 1970s is Charles C. Gillispie, ed. *Dictionary of Scientific Biography*, 16 vols. (New York: Scribner's, 1970–80). Other useful general sources include Sheila Jasanoff, Gerald E. Markle, James C. Petersen, and Trevor Pinch, eds., *Handbook of Science and Technology Studies* (Thousand Oaks, Calif.: Sage Publications, 1995); Helge Kragh, *An Introduction to the Historiography of Science* (Cambridge: Cambridge University Press, 1987); R. C. Olby, G. N. Cantor, J.R.R. Christie, and A.M.S. Hodge, eds., *Companion to the History of Science* (London: Routledge, 1990); William F. Bynum, ed. *Dictionary of the History of Science* (Princeton: Princeton University Press, 1985). Helaine Selin, *Science across Cultures: An Annotated Bibliography of Books on Non-Western Science, Technology, and Medicine* (New York: Garland Publishing, 1992), is also valuable for its global perspective. Regarding technology, see Donald Cardwell, *The Norton History of Technology* (New York: Norton, 1995). Although it appeared after the present work was essentially complete, another book, edited by Selin, *Encyclopedia of the History of Science, Technology, and Medicine in Non-Western Cultures* (Dordrecht: Kluwer Academic Publishers, 1997), is highly recommended.

Readers are encouraged to encounter the writings of Thomas Kuhn whose seminal work redirected the understanding of science and its history; an acquaintance with his approach may be made through Thomas S. Kuhn, "The History of Science," in *The Essential Tension: Selected Studies in Scientific Tradition and Change* (Chicago: University of Chicago Press, 1977). A full development of the geographical thesis that recurs throughout this book will be found in Harold Dorn, *The Geography of Science* (Baltimore: Johns Hopkins University Press, 1991). See also *ISIS*, Official Journal of the History of Science

Society (USA) and its annual *Critical Bibliography* and *Technology and Culture: The International Quarterly of the Society for the History of Technology.*

Cool Web Sites

The World Wide Web Virtual Library: History of Science, Technology and Medicine:
   http://www.asap.unimelb.edu.au/hstm/hstm_ove.htm
History of Technology Resources Available on the Internet:
   http://www.englib.cornell.edu/ice/lists/historytechnology/
   historytechnology.html#topics
STS Links:
   http://www2.ncsu.edu/ncsu/chass/mds/stslinks.html
   http://nextwave.sunyit.edu/~sts/stslinks.htm
   http://www.dcu.ie/staff/hsheehan/sts/stslinks.htm
Virtual STS:
   http://helix.ucsd.edu/~bssimon/index.html
Society for Social Studies of Science (4S) Homepage:
   http://www.lsu.edu80/guests/ssss/public_html/
History of Science Society Homepage:
   http://weber.u.washington.edu/~hssexec/
4000 Years of Women in Science:
   http://www.astr.ua.edu/4000WS/4000WS.html

CHAPTER 1. Humankind Emerges

The evolutionary origin of human beings is inherently fascinating and is now the subject of several well-written, semipopular books. The field of study has made great strides over the past forty years based on both a series of remarkable fossil discoveries and developments in molecular biology. Ian Tattersall's *The Human Odyssey: Four Million Years of Human Evolution* (New York: Prentice Hall, 1993) and Jean Guilaine, ed., *Prehistory: The World of Early Man* (New York: Facts on File, 1991), provide solid introductions. An account of one of the most sensational discoveries is Donald Johanson and Maitland Edey, *Lucy: The Beginnings of Humankind* (New York: Warner Books, 1981). Prehistory—the Old and New Stone Ages—receives classic treatment in V. Gordon Childe, *Man Makes Himself* (New York: NAL/Dutton, 1983). Alexander Marshack, *The Roots of Civilization* (Wakefield: Moyer Bell, 1992), is a pioneering work on the earliest records of astronomical interest among Old Stone Age populations. And a reliable survey of the entire field across the world is Robert J. Wenke, *Patterns in Prehistory: Mankind's First Three Million Years,* 3rd ed. (New York: Oxford University Press, 1990). Readers may find the journal *Archaeoastronomy* worth consulting.

Cool Web Sites

ArchNet: WWW Archaeology:
   http://www.lib.uconn.edu:80/ArchNet/Topical/Topical.html
Paleo-Psychology:
   http://watarts.uwaterloo.ca/~acheyne/paleoimg.html
National Center for Prehistory (France):
   http://dufy.aquarel.fr:8001/html/index.html

Paleolithic Art Gallery:
   http://watarts.uwaterloo.ca/~cpshelle/neo-gallery.html
Origins of Humankind:
   http://www.dealsonline.com/origins/

CHAPTER 2. The Reign of the Farmer

The Neolithic era is mainly the province of the archeologist. An authoritative overview is Colin Renfrew and Paul Bahn, *Archaeology: Theories, Methods, and Practice* (New York: Thames and Hudson, 1991). Astronomy in the preliterate Neolithic is accessible only through its architectural monuments. Anthony Aveni, *World Archaeoastronomy* (Cambridge: Cambridge University Press, 1989), covers the subject extensively. On Stonehenge, see Jean-Pierre Mohen, *The World Megaliths* (New York: Facts on File, 1990), and Christopher Chippindale, *Stonehenge Complete* (New York: Thames and Hudson, 1983); Gerald S. Hawkins and John B. White, *Stonehenge Decoded* (New York: Delta, 1965), remains worthwhile consulting. Jo Anne Van Tilburg, *Easter Island: Archaeology, Ecology, and Culture* (Washington, D.C.: Smithsonian Institution Press, 1994), provides an authoritative introduction.

Cool Web Sites

The Iceman:
   http://dm2.uibk.ai.at/c/c5/c504/iceman_en.html
How the Shaman Stole the Moon by William H. Calvin:
   http://weber.u.washington.edu/~wcalvin/bk6.html
Stone Pages: Megalithic Sites of Europe:
   http://joshua.micronet.it/utenti/dmeozzi
Stones of Wonder: Megalithic Sites and Astronomy:
   http://www.geocities.com/SoHo/2621/stones1.htm
Virtual Stongehenge:
   http://avebury.arch.soton.ac.uk/LocalStuff/Stonehenge/index.html
Easter Island Homepage:
   http://www.netaxs.com/~trance/rapanui.html

CHAPTER 3. Pharaohs and Engineers

The advent of civilization saw the origin of the centralized state, large cities, and higher learning. It was the beginning of the "historical" era—the period that, for the first time, allows the historian to examine documents. An informative survey of early civilizations across the world is C. C. Lamberg-Karlovsky and Jeremy A. Sabloff, *Ancient Civilizations: The Near East and Mesoamerica* (Prospect Heights, Ill.: Waveland Press, 1987). A presentation of various theories that attempt to account for the origin of the earliest states is Ronald Cohen and Elman R. Service, eds., *Origins of the State* (Philadelphia: Institute for the Study of Human Issues, 1978). Karl Wittfogel, *Oriental Despotism* (New Haven: Yale University Press, 1957), is a learned, albeit idiosyncratic, approach to the subject. Monumental building in the centralized state was, in complexity, an order of magnitude beyond Neolithic structures: on the Egyptian case, see J.-P. Lepre, *The Egyptian Pyramids: A Comprehensive,*

Illustrated Reference (Jefferson, N.C.: McFarland and Co., 1990); Zahi A. Hawass, *The Pyramids of Ancient Egypt* (Pittsburgh: Carnegie Museum of Natural History, 1990); and for an attempt by a physicist to explain the purposes of the ancient pharaohs in undertaking their great building projects: Kurt Mendelssohn, *Riddle of the Pyramids* (New York: Thames and Hudson, 1986). Otto Neugebauer, *The Exact Sciences in Antiquity,* 2nd ed. (New York: Dover Publications, 1969), provides an expert standard treatment of the subject. For a more detailed view of the Egyptian case, see Marshall Clagget, *Ancient Egyptian Science,* 2 vols. (Philadelphia: American Philosophical Society, 1989–95). And the reader who wishes to see original ancient documents colorfully reproduced should consult Gay Robins and Charles Shute, *The Rhind Mathematical Papyrus: An Ancient Egyptian Text* (New York: Dover Publications, 1987).

Cool Web Sites

Ancient World Web: Meta Index:
   http://atlantic.evsc.virginia.edu/julia/AW/meta.html
Ancient Near East Web Sites:
   http://eawc.evansville.edu/www/nepage.html
Ancient Egyptian Web Sites:
   http://eawc.evansville.edu/www/egpage/htm
The Egyptian Pyramids (NOVA):
   http://www.pbs.org/wgbh/pages/nova/pyramid/
The Oriental Institute:
   http://www-oi.uchicago.edu/OI/default.html
Egyptology Page:
   http://www.newton.cam.ac.uk/egypt/
Near and Middle East Archaeology:
   http://www.global.org/bfreed/archeol/ne-arch.html
Indus Civilization:
   http:/www.harappa.com/welcome.htm

CHAPTER 4. Greeks Bearing Gifts

Greek science is the subject of many scholarly studies. G.E.R. Lloyd, *Early Greek Science: Thales to Aristotle* (New York: Norton, 1970), and his *Greek Science after Aristotle* (New York: Norton, 1973) provide a standard popular introduction. Still valuable is Marshall Clagett, *Greek Science in Antiquity,* rev. ed. (New York: Barnes and Noble, 1994). The reader who needs a more detailed survey will find it in George Sarton, *Ancient Science through the Golden Age of Greece* (New York: Dover, 1993) and *Hellenistic Science and Culture in the Last Three Centuries B.C.* (New York: Dover, 1993). Joseph Ben-David, *The Scientist's Role in Society: A Comparative Study,* with a new introduction (Chicago: University of Chicago Press, 1984), leans toward a sociological approach to the subject. Lois N. Magner, *A History of the Life Sciences* (New York: Marcel Dekker, 1994), provides a summary of ancient views of matters biological. On technology in antiquity, see J. G. Landels, *Engineering in the Ancient World* (Berkeley: University of California Press, 1978), and L. Sprague de Camp, *The Ancient Engineers* (New York: Ballantine Books, 1988).

Cool Web Sites

Ancient Greece Web Sites:
   http://eawc.evansville.edu/www/grpage.htm
The Perseus Project:
   http://www.perseus.tufts.edu
The Internet Classics Archive:
   http://classics.mit.edu
Diotima: Women and Gender in the Ancient World:
   http://www.uky.edu/ArtsSciences/Classics/gender.html
History of Mathematics: Greeks
   http://alepho.clarku.edu/~djoyce/mathhist/greece.html
The Philosophers Page:
   http://www.bookstore.uidaho.edu/Philosophy/
Hypatia of Alexandria:
   http://cosmopolis.com/people/hypatia.html
The Catapult Museum Online:
   http://www.nzp.com/o2contents.html
Classics and Mediterranean Archaeology Page:
   http://rome.classics.lsa.umich.edu/welcome.html
Science in Antiquity:
   http://www.mala.bc.ca/~mcneil/s1.htm

CHAPTER 5. The Enduring East

The reader who wishes to dip more deeply into Islamic science and civilization will find the following helpful: Seyyed Hossein Nasr, *Science and Civilization in Islam* (New York: Barnes and Noble Books, 1992); George Saliba, *A History of Arabic Astronomy* (New York: New York University Press, 1994); Aydin Sayili, *The Observatory in Islam* (New York: Arno Press, 1981); Michael Adas, ed., *Islamic and European Expansion: The Forging of a Global Order* (Philadelphia: Temple University Press, 1993); and, for technology, Ahmad Y. al-Hassan and Donald R. Hill, *Islamic Technology: An Illustrated History* (Lanham: UNIPUB, 1992), and Donald R. Hill, *Islamic Science and Engineering* (Chicago: Kazi Publications, 1996). Of particular value in placing Islamic science in a comparative context is Toby E. Huff, *The Rise of Early Modern Science: Islam, China, and the West* (New York: Cambridge University Press, 1993).

Cool Web Sites

Early Islam Web Sites:
   http://eawc.evansville.edu/www.ispage.htm
Islamic Texts and Resources Meta Page:
   http://wings.buffalo.edu/sa/muslim/isl/isl.html
Muslim Scholars Page:
   http://www.oman_net.com/msp/main.htm
Medieval Science:
   http://www.mala.bc.ca/~mcneil/s2.htm
Byzantine Studies on the Internet:
   http://www.bway.net/~halsall/byzantium.html

CHAPTER 6. The Middle Kingdom

No student of the history of science should neglect Joseph Needham's many-splendored volumes, *Science and Civilization in China,* 16 vols. to date (Cambridge: Cambridge University Press, 1954–95). Needham's magisterial work is accessible in the abridgement by Colin A. Ronan, *The Shorter Science and Civilization in China: An Abridgement of Joseph Needham's Original Text,* 5 vols. (Cambridge: Cambridge University Press, 1978–95). Less formidable but no less valuable is Derk Bodde, *Chinese Thought, Society, and Science: The Intellectual and Social Background of Science and Technology in Pre-Modern China* (Honolulu: University of Hawaii Press, 1991). And for insightful comparative approaches see Huff, *Rise of Early Modern Science* (see chap. 5 above), and G.E.R. Lloyd, *Adversaries and Authorities: Investigations into Ancient Greek and Chinese Science* (Cambridge: Cambridge University Press, 1996).

Cool Web Sites

The Joseph Needham Home Page:
  http://www.soas.ac.uk/Needham/Home.html
Ancient China Web Sites:
  http://eawc.evansville.edu/www/chpage.htm
China 5000 Pro Timeline:
  http://www.china5000pro.com/timeline.html
China Room:
  http://www.chinapage.com/china-rm.html
History of China:
  http://www-chaos.umd.edu/history/toc.html
  http://www.chinacommercial.com/chinahist.htm
Condensed China: Chinese History for Beginners:
  http://www.hk.super.net/~paulf/china.html
Chinese Philosophy Page:
  http://www-personal.monash.edu.au/~sab/index.html

CHAPTER 7. Indus, Ganges, and Beyond

Readings in the histories of science and technology in precolonial India are limited. Among the few surveys of the precolonial period are S. Balaachandra Rao, *Indian Mathematics and Astronomy* (Bangalore: Jnana Deep Publications, 1994); Debiprasad Chattopadhyaya, ed., *Studies in the History of Science in India,* 2 vols. (New Delhi: Editorial Enterprises, 1982), and, in various editions with various titles and subtitles, O. P. Jaggi, *History of Science, Technology, and Medicine in India,* 15 vols (Delhi: Atma Ram and Sons, 1969–86). David Pingree, "History of Mathematical Astronomy in India" in *Dictionary of Scientific Biography,* ed. C. C. Gillispie (New York: Scribner's, 1978), 15:533–633, is specialized, but still essential reading. The *Indian Journal of History of Science* is another source. On the Khmer empire, see Eleanor Mannikka, *Angkor Wat: Time, Space, and Kingship* (Honolulu: University of Hawaii Press, 1996).

Cool Web Sites

WWW Sites Relating to Ancient India:
   http://eawc.evansville.edu/www/inpage.htm
History of India:
   http://www.ib_net.com/links/history.htm
   http://www.incore.com/india/history.html
   http://www.indiagov.org/culture/history/intro.html
Information on Ayurveda:
   http://www.boii.com/ayurfrme.htm

CHAPTER 8. New Worlds

Ancient American civilizations have become an especially active field of research and that activity is reflected in several recent publications. See especially Jeremy A. Sabloff, *The Cities of Ancient Mexico* (London: Thames and Hudson, 1989), and his *The New Archaeology and the Ancient Maya* (New York: W. H. Freeman, 1990); Michael D. Coe, *The Maya,* 5th ed., fully revised and expanded (New York: Thames and Hudson, 1993); Brian M. Fagan, *Kingdoms of Gold, Kingdoms of Jade: The Americas Before Columbus* (London: Thames and Hudson, 1991); Craig Morris and Adriana von Hagen, *The Inka Empire and Its Andean Origins* (New York: American Museum of Natural History/Abbeville Press: 1993); Michael E. Moseley, *The Incas and Their Ancestors* (London: Thames and Hudson, 1993); David Muench and Donald G. Pike, *Anasazi: Ancient People of the Rock* (New York: Harmony Books, 1974). Linda Schele and Mary Ellen Miller, *The Blood of Kings* (New York: George Braziller, 1986), has contributed to the revision that now sees aggression and violence in those civilizations where they were formerly thought to be pacific.

On science in pre-Columbian America: Anthony F. Aveni, *Empires of Time: Calendars, Clocks, and Cultures* (New York: Basic Books, 1989); Bernard R. Ortiz de Montellano, *Aztec Medicine, Health, and Nutrition* (New Brunswick, N.J.: Rutgers University Press, 1990), and William Fash, *Scribes, Warriors, and Kings: The City of Copán and the Ancient Maya* (London: Thames and Hudson, 1991).

Cool Web Sites

Maya/Aztec/Inca Center:
   http://www.realtime.net/maya/
Maya Links:
   http://www.ruf.rice.edu/~jchance/link.html
A Mesoamerican Archaeology WWW Page:
   http://copan.bioz.unibas.ch/meso.html
The Maya Astronomy Page:
   http://www.astro.uva.nl/michielb/maya/astro.html
Archaeology of Teotihuacán:
   http://archaeology.la.asu.edu/vm/mesoam/teo/index.htm
The Inca Trail and Machu Picchu:
   http://www.tardis.ed.ac.uk/~angus/Gallery/Photos/SouthAmerica/Peru/Inca
   Trail.html

NOVA Online: Ice Mummies of the Inca:
   http://www.pbs.org/wgbh/pages/nova/peru
Chaco Canyon National Monument:
   http://www.chaco.com/park/index.html
Cahokia Mounds Historic Site:
   http://medicine.wustl.edu/~kellerk/cahokia.html.

CHAPTER 9. Plows, Stirrups, Guns, and Plagues

Over the past fifty years the history of medieval and early modern European technology has been a flourishing field of research with the result that the reader who wishes to explore the subject further may choose from a wealth of publications. A classic work, Lynn White Jr., *Medieval Technology and Social Change* (Oxford: Oxford University Press, 1966), interprets the rise of feudalism in terms of technological innovation; Arnold Pacey, *The Maze of Ingenuity: Ideas and Idealism in the Development of Technology,* 2nd ed. (Cambridge: MIT Press, 1992), focuses on the early modern period; Jean Gimpel, *The Medieval Machine: The Industrial Revolution of the Middle Ages* (New York: Penguin Books, 1983), remains valuable. And a readable survey is Frances and Joseph Gies, *Cathedral, Forge, and Waterwheel: Technology and Invention in the Middle Ages* (New York: HarperCollins, 1994).

Michael Roberts, *The Military Revolution, 1560–1660* (Belfast: Queen's University Press, 1956), opened a field of research on the unique military technology and techniques of early modern Europe and their social consequences. He has been followed by other historians including Geoffrey Parker, *The Military Revolution: Military Innovation and the Rise of the West, 1500–1800* (New York: Cambridge University Press, 1988); Carlo M. Cipolla, *Guns, Sails, and Empires: The Technological Innovation and the Early Phases of European Expansion, 1400–1700* (New York: Pantheon Books, 1966); and William H. McNeill, *The Pursuit of Power: Technology, Armed Force, and Society since A.D. 1000* (Chicago: University of Chicago Press, 1984).

On medieval science itself, see David C. Lindberg, *The Beginnings of Western Science* (Chicago: University of Chicago Press, 1992); David C. Lindberg, ed., *Science in the Middle Ages* (Chicago: University of Chicago Press, 1978); Edward Grant, *The Foundations of Modern Science in the Middle Ages* (New York: Cambridge University Press, 1996) and his earlier *Physical Science in the Middle Ages* (New York: Cambridge University Press, 1977).

Cool Web Sites

Medieval Europe Web Sites:
   http://eawc.evansville.edu/www/mepage.htm
Labyrinth: A WWW Server for Medieval Studies:
   http://www.georgetown.edu/labyrinth/labyrinth-home.html
Medieval Studies Resources:
   http://info.ox.ac.uk/departments/humanities/med.html
Orb: On-Line Text Materials for Medieval Studies:
   http://orb.rhodes.edu/
The Medieval Science Page:
   http://members.aol.com/mcnelis/medsci_index.html

Medieval Technology and Everyday Life: Web Resources:
http://scholar.chem.nyu.edu/~medtech/medweb.html

CHAPTERS 10–12. The Scientific Revolution—General

The story of the Scientific Revolution is one of the centerpieces of the history of science. H. Floris Cohen, *The Scientific Revolution: A Historiographical Inquiry* (Chicago: University of Chicago Press, 1994), provides a sweeping entrée into the well-developed literature surrounding the Scientific Revolution. I. B. Cohen, *Puritanism and the Rise of Modern Science: The Merton Thesis* (New Brunswick, N.J.: Rutgers University Press, 1990) is another essential piece of historiography.

Among the more recent retellings and interpretations, see Steven Shapin, *The Scientific Revolution* (Chicago: University of Chicago Press, 1996); David C. Lindberg and Robert S. Westman, eds., *Reappraisals of the Scientific Revolution* (Cambridge: Cambridge University Press, 1990); Roy Porter and Mikuláš Teich, eds., *The Scientific Revolution in National Context* (Cambridge: Cambridge University Press, 1992). Also useful is Norriss S. Hetherington, *Cosmology: Historical, Literary, Philosophical, Religious, and Scientific Perspectives* (New York: Garland Publishing, 1993). Older, but still excellent, accounts are Arthur Koestler, *The Sleepwalkers: A History of Man's Changing Vision of the Universe* (New York: Viking Penguin, 1990); Herbert Butterfield, *The Origins of Modern Science, 1300–1800,* rev. ed. (New York: Free Press, 1965); I. Bernard Cohen, *The Birth of a New Physics,* revised and updated (New York: Norton, 1985); and Richard S. Westfall, *The Construction of Modern Science: Mechanism and Mechanics,* 2nd ed. (Cambridge: Cambridge University Press, 1977).

More recent works that have recast the interpretative context for the Scientific Revolution include William Eamon, *Science and the Secrets of Nature: Books of Secrets in Medieval and Early Modern Culture* (Princeton, N.J.: Princeton University Press, 1994); Pamela Smith, *The Business of Alchemy: Science and Culture in the Holy Roman Empire* (Princeton, N.J.: Princeton University Press, 1994); David C. Goodman, *Power and Penury: Government, Technology, and Science in Philip II's Spain* (Cambridge: Cambridge University Press, 1988); Frank J. Swetz, *Capitalism and Arithmetic: The New Math of the 15th Century* (LaSalle, Ill.: Open Court, 1987); Carolyn Merchant, *The Death of Nature: Women, Ecology, and the Scientific Revolution* (San Francisco: HarperSanFrancisco, 1990).

CHAPTER 10. Copernicus Incites a Revolution

On Copernicus, Tycho, and Kepler see Bruce Stephenson, *Kepler's Physical Astronomy* (Princeton, N.J.: Princeton University Press, 1994), and his companion *The Music of the Heavens: Kepler's Harmonic Astronomy* (Princeton, N.J.: Princeton University Press, 1994); Max Caspar, *Kepler,* trans. and ed. C. Doris Hellman with a new introduction and references by Owen Gingerich (New York: Dover Publications, 1993); Victor E. Thoren, *The Lord of Uraniborg: A Biography of Tycho Brahe* (Cambridge: Cambridge University Press, 1990). Thomas S. Kuhn, *The Copernican Revolution* (Cambridge: Harvard University Press, 1957), remains insightful.

Renaissance Science:
   http://www.mala.bc.ca/~mcneil/s3.htm
Copernicus, *Revolution of the Heavenly Spheres* (Dedication):
   http://pluto.clinch.edu/history/wciv1/civ1ref/revolu.htm
Tycho Brahe Page in English:
   http://inet.uni-c.dk/~nefts/tycho.htm
Johannes Kepler and More Kepler:
   http://www.physics.virginia.edu/classes/109N/1995/lectures/kepler.html
   http://www.physics.virginia.edu/classes/109N/1995/lectures/morekepl.html
Kepler's Laws:
   http://scruffy.phast.umass.edu/a114/lectures/lec03/lec03.html
   http://www.itsnet.com/home/bmager/public.html/pluto/kepler.html
History of Mathematics:
   http://www-groups.dcs.st-and.ac.uk/~history
Museum of the History of Science (Oxford)
   http://info.ox.ac.uk/departments/hooke
The Leonardo da Vinci Museum:
   http://cellini.leonardo.net/museum/gallery.html#start

CHAPTER 11. The Crime and Punishment of Galileo Galilei

Galileo's scientific works are among the few in the history of science that can be read and enjoyed by the general reader. Stillman Drake, *Discoveries and Opinions of Galileo* (Garden City, N.Y.: Doubleday Anchor, 1990), provides a sampler. Also available are Galileo's major publications: *Galileo on the World Systems: A New Abridged Translation and Guide,* trans. and ed. Maurice A. Finocchiaro (Berkeley: University of California Press, 1997) and *Two New Sciences,* trans. S. Drake (Madison: University of Wisconsin Press, 1992); the former is also available as *Dialogue Concerning the Two Chief World Systems,* trans. S. Drake (Berkeley: University of California Press, 1967). For a discussion of Galileo's scientific work, Stillman Drake, *Galileo at Work: His Scientific Biography* (New York: Dover Publications: 1995), is the standard source. And for his persecution by the Inquisition for his Copernicanism, Maurice Finocchiaro, *The Galileo Affair: A Documentary History* (Berkeley: University of California Press, 1989), provides an essential starting point; Georgio de Santillana, *The Crime of Galileo* (Alexandria, Va.: Time-Life Books, 1981), is still a valuable account. Also to be consulted are Mario Biagioli, *Galileo Courtier: The Practice of Science in the Culture of Absolutism* (Chicago: University of Chicago Press, 1993); Pietro Redondi, *Galileo Heretic* (Princeton, N.J.: Princeton University Press, 1987); Michael Segré, *In the Wake of Galileo* (New Brunswick, N.J.: Rutgers University Press, 1991).

Cool Web Sites

The Galileo Project:
   http://es.rice.edu/ES/humsoc/Galileo
Museum of the History of Science, Florence:
   http://galileo.imss.firenze.it/museo/4/index.html

Jesuits and the Sciences, 1600–1800:
  http://www.luc.edu/libraries/science/jesuits/index.html
The Art of Renaissance Science: Galileo and Perspective:
  http://www.cuny.edu:80/multimedia/arsnew/arstoc.html
Descartes, *Discourse on Method:*
  http://www.wsu.edu:8080/~wldciv/world_civ_reader/world_civ_reader2/de
  scartes.html

CHAPTER 12: "God said, 'Let Newton be!'"

The science and career of Isaac Newton are the subjects of a major scholarly industry. Richard S. Westfall, *Never at Rest: A Biography of Isaac Newton* (Cambridge: Cambridge University Press, 1983), a masterpiece of scientific biography, is the essential starting point; see also its abridgement, *The Life of Isaac Newton* (Cambridge: Cambridge University Press, 1993). A. Rupert Hall, *Isaac Newton, Adventurer in Thought* (Oxford: Blackwell, 1992), also gives a standard account. Betty Jo Teeter Dobbs and Margaret C. Jacob, *Newton and the Culture of Newtonianism* (Atlantic Highlands, N.J.: Humanities Press, 1995), provides a ready introduction to a broad range of themes. I. Bernard Cohen and Richard S. Westfall eds., *Newton: Texts, Backgrounds, Commentaries,* A Norton Critical Edition (New York: W. W. Norton, 1995), is a similarly important introductory source. John Fauvel et al., *Let Newton Be!* (Oxford: Oxford University Press, 1988), contributes accessible and up-to-date perspectives on the great man, his work, and its social impact. For the general historical context, Margaret C. Jacob, *The Newtonians and the English Revolution* (New York: Gordon and Breach, 1990). B.J.T. Dobbs, *The Janus Faces of Genius: The Role of Alchemy in Newton's Thought* (Cambridge: Cambridge University Press, 1991), provides a detailed scholarly study of Newton's private science.

Cool Web Sites

Newton Biographies:
  http://www.maths.tcd.ie/pub/HistMath/People/Newton/RouseBall/
  RB_Newton.html
  http://euler.ciens.ucv.ve/English/mathematics/newton.html
  http://www.newton.cam.ac.uk/newtlife.html
Science During the Age of Newton:
  http://www.mala.bc.ca/~mcneil/s4.htm
Newtonia:
  http://home.cern.ch/~mcnab/N/index.html
The Sir Isaac Newton Home-Page:
  http://newton.gws.uky.edu/cover.html
"The Official Isaac Newton Homepage:"
  http://www2.andrews.edu/~ganos/newton.html
The Alchemy Virtual Library:
  http://www.levity.com/alchemy/home.html
Early Modern English and the Scientific Revolution:
  http://marie.mit.edu/~bruen/EME.html

CHAPTER 13: The Industrial Revolution

Studies of the Industrial Revolution and its global impact have become more sophisticated in recent years. See George Basalla, *The Evolution of Technology* (Cambridge: Cambridge University Press, 1989); Arnold Pacey, *Technology in World Civilization* (Cambridge: MIT Press, 1992); Peter N. Stearns, *The Industrial Revolution in World History* (Boulder, Colo.: Westview Press, 1993); Vaclav Smil, *Energy in World History* (Boulder, Colo.: Westview Press, 1994). Alfred Crosby, *Ecological Imperialism: The Biological Expansion of Europe, 900–1900* (New York: Cambridge University Press, 1993); Daniel Headrick, *Tools of Empire: Technology and European Imperialism in the Nineteenth Century* (New York: Oxford University Press, 1981). A traditionalist account of the European Industrial Revolution is T. S. Ashton, *The Industrial Revolution, 1760–1830* (Westport, Conn.: Greenwood Press, 1986). On the cultural impact of science on the processes of industrialization, see Margaret C. Jacob, *Scientific Culture and the Making of the Industrial West* (New York: Oxford University Press, 1997).

Cool Web Sites

The Industrial Revolution:
   http://www.anglia.co.uk/angmulti/indrev/contents.html
   http://www.stedwards.edu/cfpages/stoll/iw/industrl.htm
Internet Resources for Economic Historians:
   http://cs.muohio.edu/Other/other-services.shtml
Canadian Economic History:
   http://www.upei.ca/~rneill/course-outline.htm
Railroad-Related Internet Resources:
   http://www-cse.ucsd.edu/users/bowdidge/railroad/rail-home.html
Railroad History:
   http://www.rrhistorical.com/index.html
National Railway Museum:
   http://www.nmsi.ac.uk/nrm
Ford Historical Library:
   http://www.fmcc.com/archive/

CHAPTER 14: The Road to Modern Science: Pure and Applied

The interpretation put forth in this chapter takes off from Thomas S. Kuhn, "Mathematical versus Experimental Traditions in the Development of Physical Science," in *The Essential Tension: Selected Studies in Scientific Tradition and Change* (Chicago: University of Chicago Press, 1977). A related view of post-Newtonian science is presented in I. Bernard Cohen, *Revolution in Science* (Cambridge: Belknap Press of Harvard University Press, 1985); see also his earlier *Franklin and Newton: An Inquiry into Speculative Newtonian Experimental Science* (Philadelphia: American Philosophical Society, 1956). For an entrée into the intricacies of nineteenth-century science, see Christa Jungnickel and Russell McCormmach, *Intellectual Mastery of Nature,* 2 vols. (Chicago: University of Chicago Press, 1990); still valuable is Edmund Whittaker, *A History of the Theories of Aether and Electricity* (New York: Dover

Publications, 1989). And for technology, particularly as the United States took center stage, Ruth Schwartz Cowan, *A Social History of American Technology* (New York: Oxford University Press, 1997), and Thomas Parke Hughes, *American Genesis: A Century of Invention and Technological Enthusiasm* (New York: Viking, 1989), are recommended.

Cool Web Sites

Victorian Science and Technology: An Overview:
  http://www.stg.brown.edu/projects/hypertext/landow/victorian/science/sciov.html
Classical Science, 1750–1820:
  http://www.mala.bc.ca/~mcneil/s5/htm
Romantic Science, 1820–1900:
  http://www.mala.bc.ca/~mcneil/s6.htm
The Development of Mechanics:
  http://www.chembio.uoguelph.ca/educmat/chm386/rudiment/tourclas/tourclas.htm
Noted Figures in Physics, Engineering, and Astronomy:
  http://144.26.13.41/phyhist/homepage.htm
A Few Famous Physicists:
  http://www.physics.gla.ac.uk/introPhy/Famous/
Glass Bead Game:
  http://userwww.sfsu.edu/~rsauzier/Biography.html
Time Line of the History of Radio Technology:
  http://www.antique-radio.org/timeline/time.html

CHAPTER 15. Life Itself

As in the case of Galileo some of Darwin's writings can and should be read, especially *The Origin of Species,* 1st ed., which is widely available in various imprints. Other of Darwin's writings are accessible in Philip Appleman, ed. *Darwin,* A Norton Critical Edition, 2nd ed. (New York: Norton, 1979) and Thomas F. Glick and David Kohn, eds., *Darwin on Evolution* (Indianapolis, Ind.: Hackett Publishing, 1996). E. J. Browne's multivolume *Charles Darwin: A Biography* (Princeton, N.J.: Princeton University Press, 1996–) promises to become the standard modern study. Peter J. Bowler's work, especially his *Charles Darwin: The Man and His Influence* (Cambridge: Cambridge University Press, 1996) and *Evolution: The History of an Idea,* rev. ed. (Berkeley: University of California Press, 1989), provides solid historical overviews to Darwin and the development of evolutionary theory as well as an entrée into the prodigious Darwin scholarship. The reader who requires assistance with the biological principles and their development in the nineteenth and twentieth centuries will find it in Maitland A. Edey and Donald C. Johanson, *Blueprints: Solving the Mystery of Evolution* (New York: Penguin Books, 1989). David Kohn, ed., *The Darwinian Heritage* (Princeton, N.J.: Princeton University Press, 1985), is a major collection of essays on the subject. The principle of natural selection receives a spirited presentation in Richard Dawkins, *The Blind Watchmaker: Why the Evidence of Evolution Reveals a Universe Without Design* (New York: Norton, 1996); similar but more philosophical is

Daniel C. Dennet's *Darwin's Dangerous Idea* (New York: Simon and Schuster, 1995). Jonathan Weiner, *The Beak of the Finch: A Story of Evolution in Our Time* (New York: Knopf, 1994), provides a gracefully written account of evolution in action. Carl N. Degler, *In Search of Human Nature: The Decline and Revival of Darwinism in American Social Thought* (New York: Oxford University Press, 1991), presents a thorough narrative of Darwinian controversies in the social sciences. And Edward O. Wilson, *On Human Nature* (Cambridge: Harvard University Press, 1978), is a popular discussion of sociobiology, one of the controversial results of evolutionary theory.

Cool Web Sites

Enter Evolution: Theory and History:
  http://www.ucmp.berkeley.edu/history/evolution.html
Darwin: *On the Origin of Species* and *Voyage of the Beagle*:
  http://www.literature.org/works/Charles-Darwin/
Darwin: *The Descent of Man*:
  gopher://gopher.vt.edu:10010/02/69/1
Darwin: *The Expression of Emotions in Man and Animals* (Excerpts):
  http://paradigm.soci.brocku.ca/!lward/SUP/DARWIN00.html
Down House:
  http://www.nhm.ac.uk/museum/Downhse/downhse.html
Artificial Life Online:
  http://alife.santafe.edu/
The Tree of Life:
  http://phylogeny.arizona.edu/tree/phylogeny.html.
Mendel Web:
  http://www.netspace.org/MendelWeb
The WWW Virtual Library: Evolution:
  http://golgi.harvard.edu/biopages/evolution.html.

CHAPTER 16. Toolmakers Take Command

See sources listed for chapter 13, above.

CHAPTER 17. The New Aristotelians

The physical sciences in the twentieth century are presented in many popular and semipopular books. The following is a small sample: Russell McCormmach, *Night Thoughts of a Classical Physicist* (Cambridge: Harvard University Press, 1982); Stephen W. Hawking, *The Illustrated A Brief History of Time,* updated and expanded (New York: Bantam Books, 1996); Steven Weinberg, *The First Three Minutes: A Modern View of the Origin of the Universe,* updated ed. (New York: Basic Books, 1993); George Gamow, *Thirty Years That Shook Physics: The Story of Quantum Theory* (New York: Dover, 1985); David Lindley, *The End of Physics: The Myth of a Unified Theory* (New York: Basic Books, 1993); John Horgan, *The End of Science: Facing the Limits of Knowledge in the Twilight of the Scientific Age* (Reading, Mass.: Helix Books, 1996).

On the substantial literature dealing with Albert Einstein, see David C. Cassidy, *Einstein and Our World* (Atlantic Highlands, N.J.: Humanities Press, 1995); Ronald William Clark, *Einstein: The Life and Times* (New York: Wings Books, 1995); Abraham Païs, *"Subtle is the Lord . . . ": The Science and Life of Albert Einstein* (Oxford: Oxford University Press, 1982); Jeremy Bernstein, *Albert Einstein and the Frontiers of Physics* (New York: Oxford University Press, 1996), and his *Einstein* (New York: Penguin, 1976).

James D. Watson's story of the discovery of the structure of DNA details not only that landmark of twentieth-century science but the realities of contemporary scientific practice; it is best accessible in Gunther S. Stent, ed. *The Double Helix,* A Norton Critical Edition (New York: Norton, 1980).

Cool Web Sites

Albert Einstein Online:
  http://www.sas.upenn.edu/~smfriedm/einstein.html
Modern Science:
  http://www.mala.bc.ca/~mcneil/s7.htm
Nobel Laureates in Physics, 1901–1996:
  http://www.slac.stanford.edu/library/nobel.html
Hubble's Constant and the Age of the Universe:
  http://www.mathsoft.com/astronomy/hubble.html
Marie Curie:
  http://myhero.com/science/curie.asp
History of the Lawrence Livermore National Lab:
  http://www.llnl.gov/llnl/history/history.html

CHAPTER 18. Applied Science and Technology Today

The twentieth-century merger of science and technology has been approached from many directions. Sociological analyses are presented in Ben-David, *The Scientist's Role in Society* (see chapter 4), and Derek J. da Solla Price, *Little Science, Big Science . . . and Beyond* (New York: Columbia University Press, 1986). The paradigmatic event, the invention and building of the atomic bomb, is described in Richard Rhodes, *The Making of the Atomic Bomb* (New York: Simon and Schuster, 1986). See also David Dickson, *The New Politics of Science,* with a new preface (Chicago: University of Chicago Press, 1988); John Ziman, *The Force of Knowledge* (Cambridge: Cambridge University Press, 1976), is still of value as an introduction. For innovative approaches to the study of contemporary science and technology, see Bruno Latour and Steve Woolgar, *Laboratory Life: The Construction of Scientific Facts* (Princeton, N.J.: Princeton University Press, 1986); Bruno Latour, *Science in Action* (Cambridge: Harvard University Press, 1987); Wiebe E. Bijker, Thomas P. Hughes, and Trevor Pinch, *The Social Construction of Technological Systems* (Cambridge: MIT Press, 1989); and Weibe E. Bijker and John Law, *Shaping Technology/Building Society: Studies in Sociotechnical Change* (Cambridge: MIT Press, 1992).

Cool Web Sites

Atomic Bomb Information:
  http://astro.uchicago.edu/home/web/jeffb/abomb.html

Trinity Atomic Web Site:
   http://www.envirolink.org/issues/nuketesting
City of Hiroshima:
   http://www.city.hiroshima.jp/City/2-1.html
Nagasaki Atomic Bomb Museum:
   http://www.us1.nagasaki-noc.or.jp/~nacity/nabomb/museum02e.html
A Career Planning Center for Beginning Scientists and Engineers:
   http://www2.nas.edu/cpc/index.html
American Physical Society Careers/Employment Information:
   http://www.aps.org/jobs/index.html
Employment and the U.S. Mathematics Doctorate:
   http://www.ams.org/committee/profession/etfreport-text.html
National Science Foundation, Science Resources Studies Home Page:
   http://www.nsf.gov/sbe/srs/stats.html

Illustration Credits

Fig. 1.2. © 1982 Jay H. Matternes. Reprinted with permission.

Fig. 1.4a. Gerald S. Hawkins, *Beyond Stonehenge* (1973). Reprinted by permission of the author.

Fig. 2.2. Kathleen Mary Kenyon, *Digging Up Jericho: The Results of the Jericho Excavations, 1952–1956.* © 1957 by Frederick A. Praeger, Inc. Reprinted by permission of Henry Holt and Co., Inc.

Fig. 3.1. Archive Photos.

Fig. 4.9. © Estate of Mary E. and Dan Todd, reprinted with permission.

Fig. 5.1. Arabic astrolab, 10th century. © Museum of the History of Science, Oxford, U.K. Reprinted with permission.

Fig. 6.2. Joseph Needham, *Science and Civilization in China,* vol. 2 (Cambridge: Cambridge University Press, 1956), facing p. 363. Reprinted with permission from Cambridge University Press.

Fig. 6.3. Arnold Pacey, *Technology in World Civilization* (Cambridge, Mass.: MIT Press, 1992), 65. Reprinted with permission.

Fig. 6.5. Joseph Needham, *Science and Civilization in China,* vol. 4, pt. 2 (Cambridge: Cambridge University Press, 1965), 449. Reprinted with permission from Cambridge University Press.

Fig. 6.6. Joseph Needham, *Science and Civilization in China,* vol. 3 (Cambridge: Cambridge University Press, 1959), 629. Reprinted with permission from Cambridge University Press.

Fig. 8.1. Reprinted with permission from J. Justeson and T. Kaufman, "A Decipherment of Epi-Olmec Hieroglyphic Writing," *Science* 259 (19 March 1993). © 1993 American Association for the Advancement of Science.

Fig. 8.5. Archive Photos.

Fig. 10.1. Vesalius, *De humani corporis fabrica* (Basel, 1555). Reprinted with permission from the Berg Collection, the New York Public Library, Astor, Lenox, and Tilden Foundations.

Fig. 10.4. Jean Bleau, *Geographia* (Amsterdam, 1662). Reprinted with permission from the Spencer Collection, the New York Public Library, Astor, Lenox, and Tilden Foundations.

Fig. 12.4. Robert Boyle, *Nova experimenta physico-mecanica* (Rotterdam, 1669). Reprinted with permission from the Wheeler Collection, Rare Book

Division, the New York Public Library, Astor, Lenox, and Tilden Foundations.

Fig. 13.3. Archive Photos.

Fig. 14.1, a and b. Abbé Nollet, *Leçons de physique expérimentale* (Paris, 1745–48). Reprinted with permission from the Wheeler Collection, Rare Book Division, the New York Public Library, Astor, Lenox, and Tilden Foundations.

Fig. 14.3. Archive Photos.

Fig. 15.3a. Archive Photos.

Fig. 16.1. Archive Photos.

Fig. 16.2. Archive Photos.

Fig. 17.1. Fermilab Visual Media Services.

Fig. 18.1. D. Price, *Science Since Babylon* (New Haven: Yale University Press, 1961), 166. Reprinted by permission from Yale University Press.

Index

Heliocentrism, 82–83, 115, 137, 203, 209–214, 216–218, 220–221, 229, 231, 243, 259

Hellenic period, 55–80, 93, 103, 107, 122, 129, 145, 155, 183, 245, 289, 309, 343, 352

Hellenistic period, 52, 55, 79–91, 93, 99

Henry VIII, 194

Henry the Navigator, 200

Henslow, John, 319

Heraclides of Pontus, 82

Heraclitus, 64

Herding, 13, 15, 17–20

Hermes Trismegistus, 85

Hermeticism, 206, 245

Hernández, Francisco, 201

Hero of Alexandria, 86

Herodotus, 43, 49, 60

Herschel, William, 294

Hertz, Heinrich, 304, 310, 345

Hieroglyphs, 38, 48–50, 155–157, 163

Hildegard of Bingen, 189

Hinduism, 141, 143, 145, 147–149, 151

Hippocrates, 65, 91, 99, 101, 106, 111, 184

Hippopede, 69–71

Hiroshima, atomic bombing of, 361

Hittites, 42

Holy Roman Empire, 182, 198, 200, 226

Homo erectus, 6–9, 353

Homo habilus, 6–8, 353

Homo neanderthalensis, 6–11, 326, 353

Homo sapiens, 6–7, 9–10, 12, 330, 353

Hooke, Robert, 244, 256–257, 262, 271, 273, 288, 305

Hooker, Joseph, 323

Hopewell culture, 167

Horses, 19–20, 42, 101, 147, 178–180

Horticulture, 13, 15–19, 22

Hospitals, 99, 101, 111, 151

House of Trade, 201

House of Wisdom, 106, 109

Hsia dynasty, 36, 131

Hubble, Edwin, 350, 352

Humanism, 204, 226–227, 267

Humors, 65, 74, 147

Hundred Years' War, 192, 194–195

Hunting, 9–13, 16, 18–19, 22

Hutton, James, 317

Huxley, Julian, 329

Huxley, Thomas Henry, 327

Huygens, Christiaan, 243, 264, 301

Hwang-Ho, 32, 36–37, 117–118, 123, 127

Hydrodynamics, 200, 258

Hypatia, 94

Ibn Sīnā, 109, 111–112

Ice Age, 9, 11, 15, 17

I Ching, 138

Idealism, 62, 67–68, 72–73

Imperial Academy, 129–130

Imperial Medical College, 134

Incas, 40–42, 164–167

Inclined plane, 235, 239–240

Index (prohibited books), 229, 233, 242

India, 42, 51, 78, 102, 106, 111–112, 114, 131, 133, 141–153, 188, 194, 197, 199, 334–336, 341

Indian Ocean, 126, 128, 134, 148, 151, 199

Induction, 271

Indus River, 32, 34–36, 42, 49, 78, 142–144, 148

Industrial Revolution, 21, 31, 94, 124–125, 148, 277–292, 303, 309, 333–334, 342, 369, 371, 373

Inertia, 190, 240–241, 257

Infantry, 195–196

Inquisition, 223, 227–229, 231–234, 242

Institute for Theoretical Physics, 347

Institution of Civil Engineers, 290

Interchangeable parts, 285–286, 338

International Council of Scientific Unions, 357

Ionia, 57, 59–62, 78

Ireland, 182

Iron, 125, 148, 278–281, 283, 285, 287, 290–291

Iron Age, 21, 26, 86

Irrigation, 18, 32–33, 35–41, 102–103, 113–114, 117–119, 123, 141, 143–144, 148–149, 151–152, 162–165, 169–170

Isidore of Seville, 95

Islam, 71, 78, 94, 101–115, 124, 127, 131–134, 137–138, 141, 143, 146, 148–149, 180–185, 188, 193–194, 197, 199, 255, 372

Italy, 62, 95, 183–184, 205, 208,

Norfolk system, 279
North America, early cultures in, 167–172
North Star, 172, 211, 267

Observatories, 110–111, 133–134, 159, 161, 256, 262, 268
Obsidian, 39, 41
Oersted, Hans Christian, 300–301
Oil, 285, 340–342
Oldenburg, Henry, 253
Olmecs, 29, 38, 155
Oppenheimer, J. Robert, 361
Optics, 85, 112, 188, 223–224, 228, 243–245, 249, 252, 263–264, 269, 295, 301–302
Oresme, Nicole, 187–188, 190–191, 238
Orreries, 133
Osiander, Andreas, 214
Ottoman empire, 99, 102, 194
Oxen, 42, 178
Oxford University, 183, 187, 189, 238, 245, 292, 307, 327
Oxygen, 298
Ozone layer, 369

Padua, 223–225
Paleolithic era, 5, 9–18, 20–23, 29–30, 33, 38, 167–168, 172–173, 353, 371
Paleontology, 353
Paley, William, 314–315
Papacy, 187, 192, 198, 208, 229–230, 233
Paper, 109, 124
Papyrus, 49
Parallax, stellar, 83, 213, 215–216
Paranthropus, 6–8
Paris, 183, 187, 189–190, 253, 255–256, 296, 309
Parmenides, 64, 77
Particle physics, 347–350
Particle theory of light, 264, 301
Pascal, Blaise, 242, 270
Pasteur, Louis, 306, 310–311
Pastoral nomadism, 17–20, 103, 105
Patronage, 55–56, 79–81, 99, 102–103, 105, 109–110, 112, 128, 131, 135, 145–147, 149, 151, 164, 223–227, 229–230, 233, 242, 262, 364
Pauli, Wolfgang, 348
Peasantry, 33, 37, 286
Pendulums, 236, 242, 258

Penzias, Arno, 351
Peregrinus, Petrus, 189
Perkin, William, 311
Persia, 62, 78, 93, 100–103, 106, 111, 114, 133, 143
Petty, William, 268
Pharaohs, 31, 35, 42–45, 54
Philip II (of Macedon), 71, 78, 87–88
Philip II (of Spain), 195, 200–201, 245–246
Philoponus, John, 77, 101, 190
Philosophical Transactions, 254, 289
Phlogiston, 298–299
Phoenecia, 49
Photocopy machines, 367–368
Photoelectric effect, 345, 347
Physics, 72–77, 82, 101, 115, 189–191, 249, 257–260, 294, 305, 310, 326–327, 343–349
π, 50, 130, 146
Piaget, Jean, 331
Pico della Mirandola, 245
"Piltdown man," 353
Pisa, 223, 236–239
Pizzaro, Francisco, 167
Plains Indians, 168
Planck, Max, 345
Planets, 52, 67–71, 81–85, 132, 146, 159, 161–162, 165, 209–212, 215–221, 224, 228, 231, 256, 259–260, 294, 346–347, 354
Plato, 56, 59, 65–73, 78, 80–82, 84, 86, 94–95, 99, 141, 189, 212, 220, 235
Playfair, John, 292
Pliny the Elder, 89–90
Plows, 18, 31, 37, 178
Pollution, 338, 341–342
Polo, Marco, 131–132, 193
Polyhedrons, 66, 68, 72, 218–219
Polynesians, 29–30
Pope, Alexander, 265
Population, 12, 15–16, 18, 31, 33, 39, 57, 103, 117, 119, 151, 162, 169, 177, 179, 192, 278, 281, 287, 321, 324, 326, 337, 341
Portugal, 126, 198–201, 267
Potatoes, 17, 19, 40
Pottery, 21, 23, 88, 124, 289
Priestley, Joseph, 289, 298–299
Prime Mover, 78
Printing, 109, 124–125, 204–205
Probability waves, 348

Library of Congress Cataloging-in-Publication Data

McClellan, James E. (James Edward), 1946–
    Science and technology in world history : an introduction / James E.
McClellan III and Harold Dorn.
        p.   cm.
    Includes bibliographical references and index.
    ISBN 0-8018-5868-2 (alk. paper). — ISBN 0-8018-5869-0 (pbk. : alk. paper)
    1. Science—History.   2. Technology—History.   3. Tool and die makers—
History.   I. Dorn, Harold, 1928– .   II. Title.
Q125.M414   1999
509—DC21   98-28898
                                                                          CIP